# MINERALS OF SCOTLAND

# MINERALS OF SCOTLAND
## PAST AND PRESENT

**Alec Livingstone**

National Museums of Scotland

First published in Great Britain in 2002

This edition published in 2006 by
NMSE Publishing
a division of NMS Enterprises Limited
National Museums of Scotland
Chambers Street
Edinburgh EH1 IJF

Format © Trustees of the National Museums of Scotland 2006
(original publication NMS Publishing Limited 2002)

Text © Alec Livingstone

**British Library Cataloguing in Publication Data**
A catalogue record of this book
is available from the British Library

ISBN (10 digit) 1 901663 46 9
ISBN (13 digit) 978 1 901663 46 4

Original typographical design by Elizabeth Robertson.

Layout by Jim Farley.

Printed and bound by
Craft Print International Limited, Singapore.

## ERRATA from Original Edition

page 14: 'NMS Library' (not archives);

p.29 (col. I, para. 4): should read '*Edinburgh New Philosophical Journal*';

p.59 (col. I, para. 5): should read '... the Patrick Dudgeon Memorial Hall was named, after his grandson Patrick Wellwood Dudgeon.';

p.95 (col. I, para. 3): should read 'Elsewhere found within the Unst serpentinite ...';

p.109 (col. III, para. I): '(NMS G1994.20.40)';

p.135 (col. II, para. 2): 'Morvern'; see also p.177 (col. I, under 'Okenite'); and p.203 (col. II, under 'Walker, G. P. L.');

p.151: photograph 'Romanechite': 'specimen size 15 x 11 x 4cm';

p.167 (col. II): under 'Geikielite', should read 'qandilite'; see also p.180 (col. I); p.190 (appendix 3); and p.210 (index);

p.174 (col. II): 'Magnesiotaramite' should be 'Magnesiohastingsite $NaCa_2(Mg,Fe^{+2})_4Fe^{+3}Si_6Al_2O_{22}(OH)_2$'; see also p.209 (index);

p.192 (col. II): under 'Cameron, W. E.', should read 'pp.497-513';

p.203 (col. I): under 'Temple, A. K. (1956)', should read 'pp.85-113';

p.203 (col. II): under 'Walker, G. P. L.', should read '*Commemorative*'.

# CONTENTS

The Glencrieff Mine, Wanlockhead,
Dumfriesshire, which produced
many fine mineral specimens, is
now a scene of dereliction which
contrasts markedly against the
Southern Uplands' rolling topography.
A pipeline excavation ran down the
valley prior to landscaping during
spring 1991. (Painting by the author,
acrylics.)

# FOREWORD

*Minerals of Scotland, past and present* is a modest title for a majestic work that includes a definitive account of the pioneering role of Scottish scientists and collectors in the development of mineralogy. This book is about minerals and the men who discovered, collected and examined them. Driven by the energy of seventeenth and eighteenth-century natural historians, mineralogy developed into a science; and in John Walker, Robert Ferguson, Thomas Thomson and Robert Jameson, Scotland boasts gentlemen scholars and early mineralogists of international stature. No account of Scottish mineralogy would be complete without a tribute to Professor Matthew Heddle, Professor of Chemistry at St Andrews, whose 1901 treatise on *The Mineralogy of Scotland* is a fitting memorial to that most famous of Scotland's mineralogists. Through the work of Heddle, Jameson, Brewster and Thomas Thomson, the foundation of the inextricable link between mineralogy and the physical sciences is evident.

Minerals are, of course, the focal point of the book and a list of the first documented reports of the 552 minerals native to Scotland embodies the meticulous research that makes this such a valuable compilation. Sixty-one of these minerals are selected for more detailed analysis and the description of the human dimension to their discovery gives life to these accounts. For instance strontianite, discovered in 1791 at Strontian, Argyll, is one of the 29 mineral species first discovered in Scotland; classic specimens of dendritic silver hail from the famous mines at Alva, whereas those at Hilderston were appropriated by King James VI when he heard of their wealth; while 'mullite' was discovered in 1924 by Bowen, one of the fathers of experimental mineralogy.

Modern mineralogists are heavily indebted to museum collections and curators for providing classic specimens for detailed investigation, and in turn the museums of Scotland have benefited from donations of private collections over the years. Information on these benefactors (not the least of whom was William Hunter) and the provenance of their collections is documented here, as are the details of historical collections which are now, sadly, lost. Many of Scotland's present-day mineral collections are privately owned and a selection of these is also recorded and their important exhibits highlighted. It is no surprise that there are so many references to the most fertile of Scotland's mineral locations: the lead mines at Leadhills-Wanlockhead.

The author has accessed a wide range of sources, and the combination of information derived from learned journals, museum archives, estate papers, and quotations from contemporary correspondents, as well as a large number of illustrations, all contribute to the richness of this book. One hundred years after Heddle's masterpiece, we have another work which ensures that the history of the science of mineralogy in Scotland is preserved.

*Dr Richard Pattrick*
Reader in Mineralogy, University of Manchester

A group of Wanlockhead miners
with their manager Mr Mitchell,
c.1900. Courtesy of the Trustees
of the Museum of Lead Mining,
Wanlockhead.

# ACKNOWLEDGEMENTS

As an X-ray mineralogist I have been privileged throughout my 35-year career to operate sophisticated equipment capable of generating highly accurate results. Within this relatively short time-span instrument development and capabilities have been wide ranging. Imagine travelling back in time to the pre-X-ray days when optics and chemistry were standard validatory methods, and beyond, when only a few people in the whole of Europe could perform chemical analysis of minerals. My historical researches opened new horizons and have taken me back to the very beginnings of mineralogy in Scotland and to the personalities who helped to lay down the foundations of the new science.

I am grateful to many people in numerous organisations within the United Kingdom and abroad who so freely gave of their time in various ways. In particular I wish to thank Mr Michael Jewkes and Dr John Faithfull of the Hunterian Museum, Glasgow; Professor Brian Bluck, Department of Earth Sciences, University of Glasgow; Mr Peder Aspen, Museum of the Department of Geology and Geophysics, University of Edinburgh; Mr Alistair Gunning, Glasgow Museums, Kelvingrove, Glasgow; Mrs Anne Abernethy, formerly of Perth Museum, Perth; Mr Charles J. Woodward, Dick Institute, Kilmarnock; Mr Stephen Moran, Inverness Museum, Inverness; Dr C. J. Stanley, Department of Mineralogy, Natural History Museum, London; Dr Alison Sheridan, Department of Archaeology, National Museums of Scotland (NMS), Edinburgh; Dr William Burch, Department of Mineralogy, Museum of Victoria, Melbourne, Australia; Herr H. Hofmann and Frau Heidrun Fuchs of the Bergakademie, Freiberg, Germany; Mr J. Williams, Dumfries; Mr S. C. Brownlee of Cumnock and Doon Valley District Council; Mr R. Hunter and Mr Kevin Wilbraham of the City Council Archives, Edinburgh; Dr Henri-Jean Schubnel, Muséum National d'Histoire Naturelle, Paris; Mr Robert Munro Ferguson, Kirkcaldy; Dr Clive M. Rice, Department of Geology and Petroleum Geology, University of Aberdeen; Dr D. I. Green of the Manchester Museum, University of Manchester; and Mr Stuart Allison, Department of Geography and Geosciences, University of St Andrews. To many other individuals, too numerous to mention, I express grateful thanks for their assistance.

For permission to reproduce diagrams, documents, photographs or portraits I am grateful to many people who kindly searched archives. I would like to thank both the organisations and individuals involved, especially Mrs Morag Williams of the Crichton Royal Museum, Dumfries; Mrs Margaret De Motte, Central Library, Manchester; Miss Julie Coleman, Department of Special Collections, Glasgow University Library, Glasgow; Mrs S. Mackenzie and Miss Yeo, National Library of Scotland, Edinburgh; Dr Jean Alexander, British Geological Survey, Nottingham; Mr J. Alexander Speer, Mineralogical Society of America, Washington; Dr Andrew M. Clark, Mineralogical Society of Great Britain and Ireland; Dr Peter G. Hill and the Editorial Board, Scottish Journal of Geology, Edinburgh; Sarah Robinson, Image Bank, Edinburgh; Mrs Susanna Kerr, Scottish National Portrait Gallery, Edinburgh; Mrs Pringle and Mr Iain Milne of the Royal College of Physicians, Edinburgh; Mr David Scarrat, Huntly House Museum, Edinburgh; Miss Helen Nicoll, National Galleries of Scotland, Edinburgh; Mr James Mann, Philips Maps, London; Mr Santiago Jagot of the Ordnance Survey, Southampton; Mr Peter Black and Denise Pulford, Hunterian Museum and Art Gallery, Glasgow; Alison Morrison-Low, Department of History of Science (NMS), Edinburgh; Mr Andrew Bethune, City Libraries, Edinburgh; Lord and Lady Moray, Moray; Mr William Whyte, Nairn Literary Institute, Nairn; Miss Jean Archibald, Special Collections, Edinburgh University Library, Edinburgh; Sandra Cumming, the Royal Society, London; R. W. Tschernich of Snohomish, Washington, U.S.A.; and Dr Iain Allison and the Geological Society of Glasgow.

Special thanks are due to Michael P. Cooper, Robert J. Reekie, Dr Suzanne Miller and especially to Mrs Suzie Stevenson for their photographic contributions. This work would not have been made possible without the typing proficiencies of Mrs Stella Doddie, Mrs Elizabeth Rennie, Mrs Dot Hartley and Mrs Vicky Blair who expertly deciphered the author's 'notorious writing'.

The assistance of Helen Kemp, Liz Robertson, Cara Shanley and Lesley Taylor of NMS Publishing Limited is gratefully acknowledged. Museum colleagues Dr Mark Shaw, Brian Jackson, William J. Baird, Peter J. Davidson, David Herd, Dr Suzanne Miller, Dr Roberta L. Paton and Mrs Suzie Stevenson cheerfully assisted me upon my numerous perturbations into their daily working orbits. To them, and to Emma Robinson in the NMS Library, I am very grateful indeed.

Thanks are due to Dr Charles D. Waterston and the late Dr Harry G. Macpherson of the Department of Geology (NMS) for their organisational skills and sterling work prior to my arrival in the department during 1973, from which I instantly directly benefited. Their foundation work enabled me immediately to commence X-ray investigations on superbly housed specimens, in well-curated mineral collections, and without the results this work would be considerably impoverished. To them, and for my museum career, I owe a great debt of gratitude.

Financial support was gratefully received from the Edinburgh Geological Society and also the British Micromount Society, British Zeolite Association, British Zeolite Corporation, the Russell Society and the Museum Trust of the Museum of Lead Mining, Wanlockhead, to help defray publication costs.

*Alec Livingstone*

Glencrieff Mine, Wanlockhead, with the Daisy Bell engine house in the foreground, c.1920s. Courtesy of the Trustees of the Museum of Lead Mining, Wanlockhead.

# INTRODUCTION

Following the Middle Ages an awakening of interest in the natural environment had occurred for a variety of reasons; economic, social and scholastic, or simply personal or social interest. People began to collect and classify natural history specimens, including minerals. Scotland's long history of mineral collectors extends back to the early seventeenth century.

From that time onwards, and more especially the last quarter of the eighteenth century, people from all walks of life created collections, dealt in minerals for financial gain, instituted popularist mineral and natural history museums, lectured, authored works on mineralogy, and formed societies to promote mineralogy and geology. A coterie of Scots, many living and working in and around Edinburgh, contributed immensely to the early beginnings of mineralogy. Rapid expansion of the new science relied heavily upon good mineral specimens and collections. Edinburgh became a European capital for mineralogy and geology.

Collections attained a zenith in the late eighteenth and early nineteenth centuries, then declined. With the passage of time, many collections inevitably disappeared, although others remained in private hands. Some outstanding collections fortunately found their way into national museums (the Allan-Greg Collection, Natural History Museum, London; the Dudgeon and Heddle collections, National Museums of Scotland; and Thomas Thomson's collection, National Museum of Victoria, Australia) and became national treasures, numerous aspects of which reflect a bygone metalliferous mining era in Scotland.

The first part of this book details briefly Scotland's physiography: the Highlands, Lowlands and Islands, each with distinctive features and enthralling scenery. Then follows a simplified account of Scotland's geological evolution from its beginnings in the southern hemisphere. Scottish geology is extremely complex, with many internationally renowned aspects. Plate tectonic models enabled an unravelling, and facilitated an understanding, of Scotland's geological evolution from the formation of rocks in the north-west Highlands 3300 million years (m.y.) ago to post-Ice Age events less than 10,000 years from present time. Scotland's voyage through time to its present-day position in the northern hemisphere highlights geological processes caused by the coming together of numerous disparate crustal fragments. For collection purposes, mineral specimens are obtained from highly localised areas, although they may well represent concentrated end products resulting from seemingly unrelated widescale geological events.

The introduction to the first major section of the book, *Collectors and Collections*, describes the main reasons for collections, whether for scientific investigation, educational or public purposes, financial gain or aesthetic pleasure. The fruitful relationships between amateur and professional mineralogists are also explored, highlighting the people involved in formulating mineral collections in Scotland. Selection criteria, wherever possible, focused upon mineral rather than rock/agate collections housed in Scotland, although included in the work are several photographs of Scottish specimens held in collections outwith Scotland. Sixty collectors are detailed – 21 are arranged chronologically by year of birth from 1600, 27 alphabetically (largely from the eighteenth century) and twelve are contemporary collectors. Collectors' biographical details, including date of death (where unsourced predominantly deriving from the *Dictionary of National Biography* 1885-1900, 63 volumes, Smith Elder and Co., London), generally precede collection information and scientific achievements. Eight collectors achieved mineralogical immortality by having minerals named after them. The current mineral collecting scene is explored by highlighting a group of more recent Scottish collectors. Whether by donation to the national collection, or publications, their contribution to Scottish topographical mineralogy is significant. Current repositories of mineral collections, whether universities, museums, institutions, or popularist displays, are also noted.

The second major section, *The Minerals*, covers descriptions,

a brief history for some species, and occurrence. For the minerals, emulation of Heddle's two-volume classic work *The Mineralogy of Scotland* (1901) was never intended. Sixty-one minerals are detailed, including many for which Scotland is mineralogically renowned. The text discusses not just minerals exclusively occurring in mining areas, but those from the Precambrian to recent formations, *ie* from Scotland *sensu lato*. Complete literature and locality coverage for each species never figured in the compilation strategy. (The Trustees of the National Museums of Scotland wish to encourage public use of the Scottish Mineral Collection database. In a few cases specific locality information remains confidential at the donor's request.) Twelve plates follow *The Minerals*; each highlights fine Scottish mineral specimens not featured in the text. Although Scotland is rich in agates of unsurpassed beauty and variety, these are not dealt with in the work. The superbly illustrated publication *Agates* (Macpherson, 1989) deals comprehensively with

agates found throughout Scotland, whereas the relatively recent discovery of large agates (up to 60 x 30cm) at Ardownie Quarry, Monifieth, Tayside, is reported by Ingram (1994).

Three appendices conclude the work:

(1) An up-to-date glossary of 552 species known to occur in Scotland for which ideal formulae, associated minerals and brief locality details are stated.

(2) Minerals (29) new to science first described from Scotland (such as caledonite, lanarkite, leadhillite, susannite, greenockite, brewsterite and strontianite), for which Scotland is mineralogically famous, together with their ideal formula, how named, and type locality details.

(3) A listing of 32 species, with references, included in Appendix 1 that have come to light from January 1991 to December 1996.

The loneliness and beauty of a great wilderness, Loch Lurgainn, 15km north of Ullapool, Ross and Cromarty. Image Bank, Edinburgh.

Where quotations are given, the original has been adhered to as closely as possible, including apparent misspellings, old place names, or mineral names (*cf* modern usage). Mineral localities and places are usually followed in the text by the old county name, old regional or present county names, as appropriate, depending on the original author's usage. Additionally, within the National Museums of Scotland (NMS) Scottish Mineral Collection localities use predominantly old county names to an extent.

## SCOTLAND, THE COUNTRY, PHYSIOGRAPHY

Scotland, a land of rugged mountains, glens, rolling hills, lochs and precipitous coastal cliffs, covers some 77,205 square kilometres. Enthralling scenery has captivated and inspired poets, artists, musicians, authors and scientists to create masterpieces or classic works in their own right: it was the sound of the sea echoing through the depths of Fingal's Cave (a large cavern in the island of Staffa, eroded in basalt columns rising from the sea like huge organ pipes and first described by Sir Joseph Banks in 1772) which inspired Mendelssohn, on his visit to Scotland in 1829, to compose *The Hebrides* overture.

The second half of the eighteenth century saw an awakening awareness of natural phenomena within Scotland that contributed to the beginnings of earth sciences, mineralogy being no exception. Inquisitiveness supplemented by controversy transformed initial observations into scientific facts to lay down the foundations of Scottish geology and mineralogy. An interest in the scenery may have triggered many active minds. Scottish rocks, some of great antiquity, formed under wide-ranging temperature-pressure regimes,

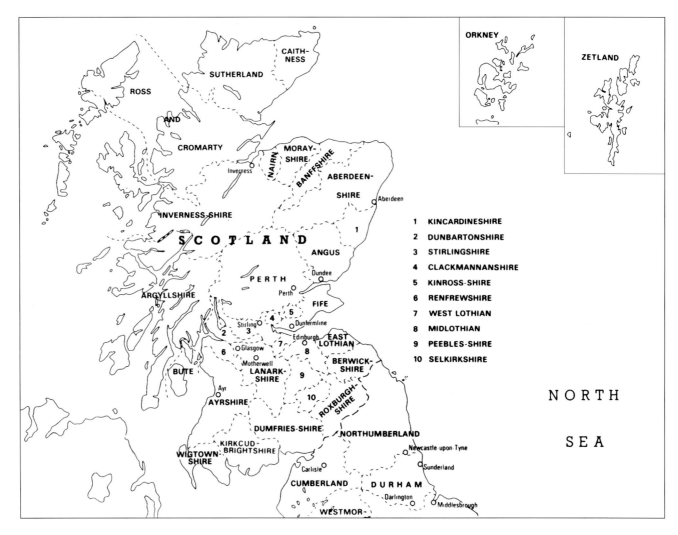

County boundaries pre-1974.
Ordnance Survey © Crown
Copyright.

The following labels appear on the map:

ORKNEY

ZETLAND

CAITHNESS

SUTHERLAND

ROSS

AND

CROMARTY

MORAYSHIRE

NAIRN

BANFFSHIRE

ABERDEEN-SHIRE

Inverness

INVERNESS-SHIRE

SCOTLAND

Aberdeen

ANGUS

Dundee

PERTH

Perth

FIFE

ARGYLLSHIRE

Stirling

Dunfermline

Edinburgh

EAST LOTHIAN

Glasgow

BERWICK-SHIRE

Motherwell

LANARK-SHIRE

BUTE

Ayr

AYRSHIRE

ROXBURGH-SHIRE

DUMFRIES-SHIRE

NORTHUMBERLAND

KIRKCUD-BRIGHTSHIRE

WIGTOWN SHIRE

Newcastle-upon-Tyne

Carlisle

Sunderland

CUMBERLAND

DURHAM

Darlington

Middlesbrough

WESTMOR-

NORTH SEA

1 KINCARDINESHIRE
2 DUNBARTONSHIRE
3 STIRLINGSHIRE
4 CLACKMANNANSHIRE
5 KINROSS-SHIRE
6 RENFREWSHIRE
7 WEST LOTHIAN
8 MIDLOTHIAN
9 PEEBLES-SHIRE
10 SELKIRKSHIRE

encompassing complex igneous and metamorphic processes to surface formation. Great topographical and mineralogical diversity resulted as a consequence of the complicated geological evolution of Scotland.

Long before railways or Ordnance Survey maps, mineral collections had been accumulated for personal gratification, scientific pursuits, commercial or educational purposes. Many specimens were extracted under arduous conditions, or transported great distances in the days of sail. An understanding of Scotland's terrain is imperative in order to appreciate the sheer physical effort inherent in amassing mineral collections whether past or present. Mineral quests extended from mountain tops, of which there are 284 over 914.4m (3000ft)

known as Munros (McNeish, 1996), to almost inaccessible sea cliffs and the deepest mines.

Physiographically, Scotland may be subdivided into three distinct units: the Highlands, the Lowlands and the Islands.

The Highlands

Precise demarcation between the Highlands and Lowlands is unclear. Although, in geological terms, the area north of the Highland Boundary Fault, *ie* north of a line from Dumbarton to Stonehaven, is the Highlands, appreciable expanses of this terrain are low-lying. The term 'Highlands' is self-explanatory, although altitudes in absolute terms are not great compared

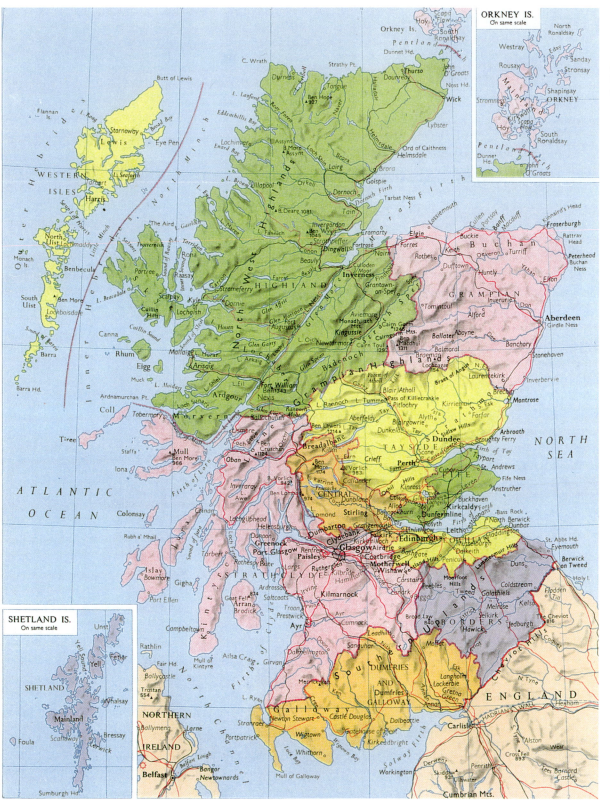

Administrative regions, post-1974.
© George Philip and Son Ltd. 1977,
cartography by Philip's.

ORKNEY IS.
On same scale

SHETLAND IS.
On same scale

XV

with mountain chains throughout the world. While high and low land closely interdigitate, creating the appearance of mountainous landscapes, only 6% of the land rises above 610m (Haynes, 1983). The highest point, Ben Nevis (1347m), towers majestically above Loch Linnhe near Fort William. Running down the western side of the Scottish mainland from Durness to Loch Lomond is a deeply dissected, mountainous backbone – the north-west Highlands. Greatest relief occurs in the western part, where the high land of Lochaber is close to the sea. To the east, extending from south-west to north-east, lie the Grampian mountains.

The Highlands thus offer a range of spectacular scenery: mountains, glens, lochs, rugged coastal scenery and wild, desolate Rannoch Moor and Glen Coe. An impressive line of deep lochs running south-west to north-east from Fort William to Inverness marks the site of the Great Glen, a geological wrench of 105km displacement. Loch Maree, one of Scotland's most beautiful lochs, is set in wild, mountainous terrain. Many areas of the Highlands possess a forbidding appearance, and not surprisingly they contain the lowest population density within the British Isles, at eight per square kilometre.

The Grampians, at the south-eastern extremity of the Highlands, range in altitude from 610m to 1214m at Ben Lawers, to Britain's second-highest mountain Ben Macdhui (1309m) in the north-east Cairngorm massif. Character-istically, the Cairngorms are rounded, large, flat-summitted granite mountains with precipitous corrie faces. Geikie (1887) comments:

> It would hardly be an exaggeration to say that there is more level ground on the tops of these mountains than in areas of corresponding size in the valleys below.

Collectively these two mountainous areas occupy 44% of Scotland (Haynes, 1983).

## The Lowlands

This term is generally applied to all of Scotland south of the Highlands, though more appropriately the name is sometimes given just to that mainland portion lying between the Highlands and Southern Uplands. A salient feature of the Lowlands is the fault-bounded Midland Valley, with Glasgow and Edinburgh in the west and east respectively. In a broad sense, the Midland Valley is a swathe of lower ground abundantly dotted with hills between the southern Highlands flank and the northern Southern Uplands.

The Lowlands are scenically rich, though less rugged than the Highlands. Pastoral land from west to east gives way in a southerly direction across the Clyde and Forth estuaries to the Southern Uplands. South of the Midland Valley, virtually to the English border, high rounded hills (to 792m) and thickly wooded river valleys offer a softer scenery than the Highlands. The Southern Uplands, with their south-west-north-east Caledonian fold belt lineation, form approximately 16% of the country (Haynes, 1983).

## The Islands

Some 780 islands, comprising 13.5% of the land surface, are dotted around the mainland; nearly 130 are inhabited. The islands fall into four distinct geographical groups, with the Inner and Outer Hebrides to the west and the Orkney and Shetland islands off the northern tip. Two smaller clusters are located in the firths of Clyde and Forth.

The Outer Hebridean chain of islands extends for more than 200km and includes five principal islands: Lewis with Harris, North Uist, Benbecula, South Uist and Barra. Rounded and conical hills extending to over 610m rise from undulating intensely glaciated rocky platforms – the Scottish analogue to parts of the Baltic and Canadian shields. Of the Inner Hebrides, the enchanting Isle of Skye is renowned for its mythical past and history which is encompassed in the famous *Skye Boat Song*. Rhum, Eigg, Canna, Coll and Tiree islands, distinctive in their own right, enhance Scotland's scenic beauty. Iona (where St Columba landed in AD563), the cradle of Scottish Christianity, nestles off Mull. Closer to the mainland lies Bute, and Arran with its northern rugged parts rising to 873m at Goatfell. Across the notorious Pentland Firth off the far northern tip of Scotland are the Orkney and Shetland islands, constellations of 67 and over 100 islands respectively.

# SCOTLAND'S GEOLOGICAL EVOLUTION
## THE VOYAGE OF SCOTLAND THROUGH TIME

In order to appreciate Scotland's mineralogical diversity, it is necessary to gain an understanding of how Scotland evolved geologically. Mineral specimens tend to be collected from highly localised areas, although they may represent concentrated end products resulting from major, seemingly unrelated, widespread geological events. Disseminated lead sweated out during regional metamorphism could be sufficiently concentrated to form economic deposits. Isotopic ages of Scottish galena range from 910 ± 80 million years (m.y.) to 45 m.y. (Moorbath, 1962; Gallagher, 1964), thus spanning numerous geological periods. For such a small area, Scotland possesses a geology more complex than any other country in the world.

With the acceptance of plate tectonics, and of continents drifting over the globe, it becomes appropriate that we should ask the question: where and at what period in geological history did Scotland form? Plate tectonic models are based only on the geology of the last 200 m.y., *ie* the age of the oldest extant ocean floor. When plates collide head-on, one plate may be subducted deep into the mantle and partially melt. The molten products rise to form volcanoes and large granite and granodiorite masses. However, if overriding/underriding plates slide obliquely against each other, large fault-bounded basins, source blocks and accretionary prisms may become displaced from their original position of development. These large displaced blocks are termed *terranes*. Projecting plate tectonics back in geological time is highly interpretative. Palaeomagnetic studies, together with studies of fossils, enable palaeolatitudes of rocks to be ascertained; hence continental wandering can be unravelled and aspects of plate tectonics supported.

Scotland represents an assemblage of geologically diverse fragments of the earth's crust accreted at various times over nearly 3300 m.y. (Amphibolites which occur in the Gruinard Bay area of northwest Ross and Cromarty and dated at 3300 m.y. by Burton *et al.* [1994] are the oldest known rocks in Scotland.) From its origin in the southern hemisphere, the Scottish crust drifted into equatorial regions around 300 m.y. ago, then

moved to its present northerly position some 50 m.y. ago. James Hutton (1726-97), the initiator of geological thought in Scotland, in 1788 postulated the rise and fall of mountains and development and disappearance of oceans. Although a holistic view, in essence, he simplistically eclipsed later discoveries.

The following account of Scotland's geological evolution is taken mainly from Craig (1991), Bluck (in Lawson and Weedon, 1992, and *pers. comm.*, 1996) and Dunning *et al.* (1978). From its beginnings over 3300 m.y. ago, the ancient Scottish Precambrian crust lay some 30°S of the Equator. Together with Labrador and Greenland, the Outer Hebrides and the north-west Highlands west of the Moine Thrust formed part of an Archaean craton which underwent a major supracrustal-plutonic event 3100-2900 m.y. ago.

The oldest sedimentary rocks (Torridonian) of north-west Scotland rest unconformably on highly deformed crystalline rocks – the Lewisian gneisses. Several metamorphisms have obscured the latter's origin. Chemical studies of the dominant grey, quartzo-feldspathic gneisses indicate a calc-alkaline (granodiorite, tonalite, andesite) parentage. Humphries and Cliff (1982) showed from Sm-Nd whole-rock measurements that the calc-alkaline protolith of the Lewisian granulites separated from a previously undifferentiated chondritic mantle 2920 ± 50 m.y. ago. Throughout, the gneisses are tectonically thinned disrupted lenses and layers of metamorphosed layered gabbroic and ultrabasic bodies. Minor metasediments are represented by marbles, quartzites, pelites and semi-pelitic assemblages. Interlayered with, and cross-cutting, the grey gneisses are granitic rocks, migmatites and acid pegmatites. Extending over the entire mainland Lewisian, and throughout the Outer Hebrides, a suite of basic dykes intruded 2400-2000 m.y. ago acts as a chronological marker. Pre-dyke formations are Scourian and Badcallian in age, whereas post-dyke formations are Laxfordian.

The Lewisian rocks possess multistage histories: metamorphic episodes, tectonic intercalations, crustal thickenings and later retrogressions and uplift. Granulite facies mineral

1

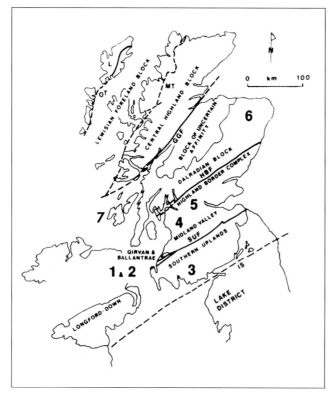

A map of the Palaeozoic tectono-stratigraphic blocks within Scotland, after Bluck (In Lawson and Weedon, 1992). Reproduced by kind permission of the Glasgow Geological Society.

1. Ballantrae ophiolite
2. Girvan cover sequence
3. Southern Uplands
4. Midland Valley
5. Highland Border Complex
6. Dalradian Block
7. Islay-Colonsay Terrane

HBF = Highland Boundary Fault
SUF = Southern Uplands Fault
IS = Iapetus Suture
L = Isle of Lewis
OT = Outer Isles Thrust
MT = Moine Thrust
GGF = Great Glen Fault

assemblages demonstrate that reactions occurred at up to 1000-1100°C and 10-15kbar, corresponding to burial depths of 25-40km.

Crustal stability was achieved and erosion of vast areas occurred as the Scourian landscape became finally eroded to a peneplain, or to its present surface around 1000 m.y. ago. Torridonian rocks were deposited on this surface to start another cycle of Precambrian accumulation. The Torridonian was part of a large continent called Rodinia.

Torridonian rocks crop out west of the Moine Thrust from Cape Wrath in the north to the island of Rhum. Around Stoer in the northern section, breccia fans overlie the ancient hilly landscape of Lewisian gneiss with clasts derived exclusively from the Lewisian; rounding characteristics suggest short transport distances. Fanglomerates pass into pebbly red sandstones and red shales. Torridonian sediments may have accumulated to 7km thickness, with the upper 5km responsible for the spectacular scenery of north-west Scotland, especially the mountains around Torridon. The sediments – arkoses, thin sandstones and shales – formed on a vast desert-like plain threaded by great rivers fed from a landmass of which the Outer Hebrides and Greenland formed a part. Shallow, invading seas periodically flooded the barren plain.

Moine rocks occur from the Moine Thrust in the north-west Highlands south-eastwards to where they dip under the Dalradian block. This large swathe of Scotland consists of polymetamorphosed sediments, silts, sandstones and thin limestones, together with basalts and some granites. Shallow-water depositional features are found in widely separated Moine rocks: cross-bedding, ripple marks and slump structures are common. The sediments were probably deposited on the edge of a continental mass and have undergone polyorogenic episodes. Lewisian basement underlies both Torridonian and Moine rocks and was actively involved in the orogenic deformation of the latter. Crustal shortening generated polyphase folding and metamorphism around 1000 m.y. ago: the early phase of Caledonian orogenesis and intrusion of early granitic plutons in the Highlands, *eg* Ardgour granite (870 m.y. old). Pegmatitic sheets (800-750 m.y. old) were also intruded into the metamorphic rocks and at 590 m.y. old the Carn Chuinneag granite also. Late Caledonian orogenesis continued into the Lower Palaeozoic and generated a suite of granites which generally young in age southwards to the Galloway granites, around 400-393 m.y. old. The Caledonian orogeny gave rise to extensive regional metamorphism. Classic work by Barrow (1893, 1912) revealed that regionally metamorphosed pelitic rocks changed in response to increasing temperature and pressure regimes. A series of zones from low to high temperatures and pressures were recognised, each zone demarcated by the incoming of a critical index mineral. Barrow recognised the following six zones: chlorite, biotite, garnet, staurolite, kyanite and sillimanite. Temperature-pressure parameters range from 300-700°C, with pressures corresponding to burial depths of 35-40km. During the climactic period of regional metamorphism, the Younger Gabbros of north-east Scotland were intruded.

Moine rocks underwent a succession of fold movements spanning the 500 m.y. between Precambrian and Silurian. Radiometric dating places the major tectonic feature, the Moine Thrust, at 430-425 m.y. ago. The dislocation is the youngest of a system of Caledonian thrusts which carried Moine rocks with their pressure slides over the Foreland. (The Foreland sequence of Lewisian, Torridonian and Cambrian rocks formed a passive margin to the Laurentian continent after it split from Rodinia.) The Moine Thrust is a detachment surface traceable for 500km from Shetland to beyond Mull, with a displacement of at least 70km.

Dalradian rocks comprise mostly metamorphosed marine sediments of late Precambrian and Lower Palaeozoic age, which formed on a continental shelf. Structural extension of the shelf may be represented by lavas and volcanogenic sediments. Tillite deposits (eg Port Askaig tillite) suggest that a late Precambrian glaciation was unusual in low latitudes, for some Dalradian sediments are warm-water dolomites. (A considerable volume of evidence is accumulating to support the view that during the Neoproterozoic era the world was largely sheathed in ice (Walker, 1999). A turbidite sequence forms the southern extremity of the block. The Dalradian block may have traversed an ocean, for it is unrelated to North American geology; at the present time its source is unknown.

Cambrian and Lower Ordovician rocks occur in the north-west Highlands and form a narrow 200 x 20km belt running south-south-west from Durness in the north to the Isle of Skye. Cambrian fossils from Scotland belong to a western and central American assemblage, in contrast to English and Welsh fossils which are related to an eastern American fauna. The 'Proto-Atlantic' or 'Iapetus' Ocean separated the two faunas.

The Highland Border Complex, mostly of Ordovician rocks, including ophiolite, occurs south of the Dalradian tract as a series of discontinuous small, faulted exposures associated with the Highland Boundary Fault. The outcrop of the Highland Border Complex is nowhere greater than 1km wide, yet black shales, cherts and pillow lavas are traceable from Arran to Stonehaven. Within this suite, the oldest known rocks, close to 490 m.y. old, contain clasts of serpentinite, gabbro, dolerite and spilite, ie classic ophiolite assemblages. The ophiolite would need to have been obducted, exposed and eroded to provide material for the 490 m.y.-old rocks (Bluck, 1985). The ophiolite represents an oceanic fragment probably obducted onto a continental landmass at c. 530 m.y. ago, whereas the Ordovician sediments and volcanic rocks (pillow lavas, tuffs and breccias) are interpreted as a portion of a separate terrane. Obducted ophiolite may have been responsible for the burial and subsequent regional metamorphism of the Dalradian rocks.

A second ophiolite complex, just north of and situated at the south-western extremity of the Southern Uplands Fault, covers some 80 square kilometres and includes black shales, cherts and serpentinites. Beneath part of the serpentinite lies a metamorphic aureole which comprises garnet metapyroxenites, amphibolites and greenschists. Other rocks in the complex include pyroxene granulites and amphibolites. Within the complex are rocks of differing ages and origins. Blueschist blocks imply a metamorphic style unrelated to that of the garnet pyroxenites. Igneous rocks in the complex show little evidence of contamination with continental crust, and sedimentary rocks show virtually no signs of derivation from continental basement. The ophiolite is essentially oceanic in aspect, with components from at least 60km depth, yet there is abundant evidence for shallow-water accumulation and terrestrial lava extrusion (Bluck, 1985). Above this ophiolite lies a sequence of Ordovician conglomerates and turbidites derived from the Midland Valley, which is interpreted as a volcanic-plutonic arc in Ordovician times. Rocks of 576 m.y. age have been recorded from the area, yet the complex was thought to have been finally obducted around 480 m.y. ago onto the Midland Valley. The complex may well have arisen from structurally assembled different blocks.

Throughout the Southern Uplands, rock formations are greywacke, silty rocks, black graptolitic shales, turbidites and volcanic rocks. Ordovician vulcanicity was extensive and contributed much to the sediments. Towards the close of the period, up to 8km of greywacke sands and muds had accumulated.

At this stage, Scotland lay on the south-eastern edge of a continent separated from a landmass, from which England and Wales arose, by the Iapetus Ocean, c. 1000km wide (Piper, 1978). As the two continents converged, the ocean became consumed and the highly folded Ordovician and Silurian rocks of the Southern Uplands represent the off-scraping of the descending plate. Ultimately the continents fused in sub-tropical latitudes around 410 m.y. ago, although continental merging had generated repercussions long before actual contact.

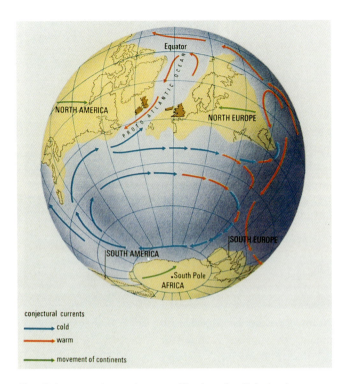

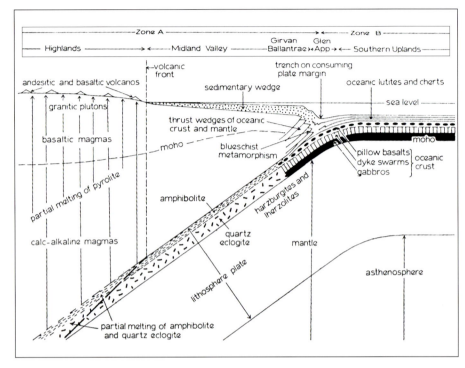

During the Caledonian orogeny, the Midland Valley was probably a magmatic-plutonic arc, with basins to the north and south, founded on an old metamorphic basement. A large elongated basin, into which torrential Devonian rivers carried vast quantities of sand and pebbles, existed between the Highlands and Southern Uplands. Gentle folding and development of a fault system created a large tectonic trough, the Midland Valley, although present fault boundaries are Late Caledonoid thrust features developed just prior to the Carboniferous.

Southern Uplands rocks may have formed as an accretionary prism and, together with the Midland Valley plus the Girvan-Ballantrae area, they resemble tectonic units of a destructive plate margin. Bordering the destructive margin was a landmass called Laurentia. During the Ordovician period, the Highland Border Complex, Midland Valley and Southern Uplands were separate displaced destructive terranes formed on the southern edge of Laurentia. By Devonian times they had moved together, accreted and, towards the close of the period, docked with the Dalradian. Prior to this event, the Highlands were far to the west of the Midland Valley terrane.

On the Devonian globe three super continents, including Euramerica, existed some 400–350 m.y. ago. At the commencement of the Devonian, Britain lay between 20°S and 30°S, slowly drifting northwards to between 10°S and 20°S by the end of the period. The topography of the Scottish area was controlled in Lower Devonian times by the final closure of the Iapetus Ocean. At low latitudes, palaeoclimatic indicators suggest a warm to hot, semi-arid to arid environment with torrential downpours. Devonian landscape consisted of a northerly mountainous region with a vast southerly plain covered in alluvial deposits, and basins which passed into the Devonian sea. Old Red Sandstone (ORS) molasse deposits, derived from the emerging Caledonian mountains of Alpine to Himalayan proportions, accumulated in alluvial fans on floodplains. (Old Red Sandstone is a term used to denote terrestrial sediments and volcanic rocks comparable in age to the marine Devonian deposits of south-west England and Europe.) Lavas and ashes poured out from volcanic chains ultimately to form the Ochil and Pentland Hills, and also around the Cheviot Hills. The Late Caledonian granite plutons were intruded in the Grampians and around the Cheviot Hills and Galloway.

Following the granitic intrusive phase, north-east Scotland submerged under a great land-locked body of water – Lake

Orcadia – in which fish-life abounded. Middle ORS deposits are extensive in Caithness, where they attain 4km in thickness, and in Orkney (2km). Essentially formed from sands, silts and limey muds, they ultimately gave rise to the flagstone rocks of this period. In the Midland Valley, 4-5km of sediments accumulated with abundant conglomerates, especially on the northern flanks. The conglomerates originated from aprons of coalescing separate fans along active fault escarpments close to the Highland Boundary and Southern Uplands faults.

During the Upper Palaeozoic Hercynian compression from the south, large Caledonoid (ie north-east-south-west) tear faults resulted with sinistral movement. An outstanding example is the Great Glen Fault, which possesses a displacement probably of several tens to a few hundreds of kilometres. It has remained at much the same brittle level in the crust since Middle Devonian times. Thus the Great Glen Fault is not thought to be a terrane junction, for the previously separate Highland Border Complex, the Midland Valley and Southern Uplands terranes behaved as a single entity from Siluro-Devonian times. Earth movements associated with the Iapetus Ocean closure ceased in ORS times, for Upper ORS deposits were laid down over the three conjoined terranes during a period of crustal stability.

At this stage, around 340 m.y. ago, early Carboniferous Scotland lay just south of the Equator and formed part of a southern margin of a new North America-North Europe craton. With the convergence of Gondwanaland and Euramerica, which ultimately formed one great super continent, Pangaea, Gondwanaland continued to rotate clockwise and move northwards, whereas Euramerica moved in a north-easterly direction.

The Devonian continent had subsided and a shallow, warm sea advanced northwards as far as the Highlands. Large islands remained to the south, especially the Southern Uplands which formed a barrier between the Midland Valley and northern England. To the north, the Caledonian mountains provided sediments for alluvial plains, deltas and shallow seas. With vast outpourings of lava and ashes, and the reactivation of older faults and basin formation, crustal instability returned.

Scottish Carboniferous rocks are restricted to the Midland Valley and Sanquhar areas; in parts of the former area, Upper ORS deposits pass conformably into Lower Carboniferous sandstones, shales and carbonate rocks. Sedimentalogical

studies suggest that sand-bar, lagoonal and beach environments existed during the opening of the Carboniferous period in Scotland. The littoral-lagoonal phase rapidly drew to a close with vast outpourings of the Clyde Plateau lavas (up to 1km thick), which now form, in the west, the Campsie, Kilpatrick and Renfrew Hills. Volcanic activity in the eastern Midland Valley gave rise to the Garleton and Bathgate hills, and to Edinburgh's volcano: Arthur's Seat. Carboniferous-Permian vulcanism, which generated a range of lava compositions from ankaramite to phonolite, was structurally controlled within the Midland Valley graben. Volcanic centres were confined to east-west faults and crests or hinges of upfolds within the graben.

Isolated basins surrounded by volcanic rocks formed with lagoon-like expanses rich in algal and plant remains, which ultimately gave rise to the famous oil-shales of West Lothian and the Edinburgh environs. Limestones developed with abundant faunal assemblages of sea-lillies (crinoids), brachiopod shells, corals and sea mats. Tropical forests grew around large deltas, which had replaced the Devonian alluvial cover, and on coastal plains. Cycles of shales, mudstones, sandstones and coals repeat many times as sedimentation rates generally matched subsidence. Up to 3.5km of coal-bearing sediments accumulated in Fife and the Lothians. The Midland Valley sediments became mildly folded, with intrusion of large quartz-dolerite sills into east-west fractures.

Around 280 m.y. ago, by the end of the Carboniferous, the two supercontinents, Gondwanaland and Euramerica, had fused to form Pangaea, and Scotland lay too far north to suffer major folding, unlike southern England and Europe. Hot, arid conditions returned to Scotland, which crossed the Equator and lay just to the north around 250 m.y. ago, in Permo-Triassic times.

Within the vast Pangaea supercontinent, Scotland formed part of a new red desert. The Late Palaeozoic sea had retreated southwards and hot, arid desert persisted for 30 m.y. Early Permian deserts became increasingly arid, erosion reduced the hills and dune seas covered large plains. Aeolian sandstone studies indicate that winds blew from the east and north-east. Upland areas, which supplied coarse-grained water-laid sediments, flanked the dune terrain. Scottish Permo-Triassic rocks (the New Red Sandstone) occur in a series of basins, particularly in south-western areas around Mauchline,

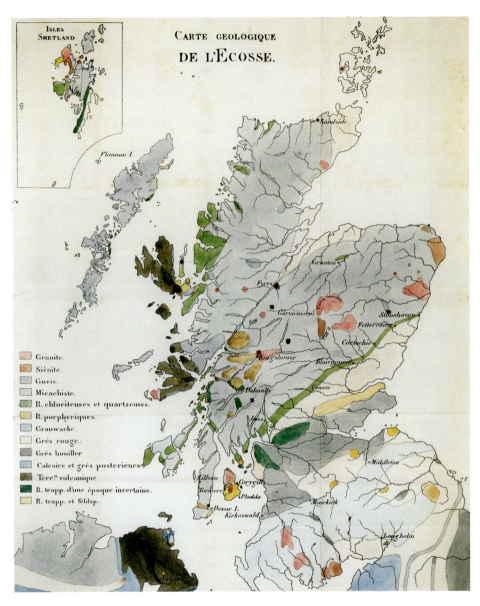

CARTE GEOLOGIQUE
DE L'ECOSSE.

ISLES
SHETLAND

Flannan I.

Granite.
Siènite.
Gneis.
Micachiste.
R. chloriteuses et quartzeuses.
R. porphyriques.
Grauwacke.
Grés rouge.
Grés houiller
Calcaire et grés posterieurs.
Terr.ᵉ volcanique.
R. trapp. d'une époque incertaine.
R. trapp. et feldsp.

The earliest published geological map showing the whole of Scotland, from *Essai Géologique Sur L'Écosse* (Boué, 1820). Glasgow University Library, Special Collections. In 1808 L. A. Necker de Saussure had prepared a manuscript sketch-map of the geology of Scotland but this work was not published (Flett, 1936).

warm, shallow sea, the Zechstein Sea, invaded the land, extending from the north of England to Poland. Scotland remained a stable upland area never to be fully submerged again. From Scotland's position just north of the Equator in the Permian, by Late Triassic times it had drifted to 30°N.

Early rifting of Pangaea occurred in the Permian, with seas invading widening rifts until individual continents developed in the Triassic. Mega-fragmentation of Pangaea and drifting of the continents to their present locations on the globe occupied the last 220 m.y. of geological time. During the Jurassic and Cretaceous periods, the main continental separation took place, for in the Cretaceous the new North Atlantic Ocean separated America from Greenland. Tensional rifting around Britain came to a halt. Concomitant with sharp uplift, a warm, seasonally wet climate prevailed, for Britain had drifted to about 40°N. At this time dinosaurs roamed the land, including parts of Scotland.

Scottish Jurassic and Cretaceous deposits are sparse, being limited mainly to the north-west and north-east regions. Important restricted outcrops occur in the Inner Hebrides, particularly on Skye and Mull, and on the easterly northern mainland coast between Golspie and Helmsdale. Sands, clays and chalk deposits indicate marine transgressions into these areas by the Tethys Ocean, which separated continental masses to the north and south. Deposits also include thin limestones, shales and the Jura oolitic ironstone. In the North Sea Basin, to which Scotland forms an edge, thick accumulations of Jurassic sandstones with interbedded volcanics became major oil reservoirs. Plants flourished in the Jurassic to the extent that thin coal seams formed. On land, the Scottish Jurassic and Cretaceous sediments represent remnants of a once-extensive series now removed by Tertiary and Quaternary erosion.

In Jurassic times, the development of an extensive submarine fault scarp and fault system (typical of the North Sea) off Helmsdale, together with a major volcanic province in the eastern Moray Firth Basin, ushered in changes. The break-up of Pangaea commenced in the Late Triassic-Early Jurassic 180 m.y. ago (Windley, 1977), and the last continent separated in Early Tertiary times. Fracturing of Pangaea, combined with rising lava plumes, caused the crust to buckle upwards. As Pangaea fragmented, rifts opened up which ultimately formed the Atlantic Ocean. During the Tertiary era Britain lay in the

Thornhill and Lochmaben. A sparse fossil flora has been discovered at the base of the Mauchline Basin, while reptilian footprints occur in the Cornockle sandstone of the Lochmaben Basin.

In Late Triassic times, a marine transgression in the Hebrides occurred. Alluvial fan-facies and floodplain deposits inter-finger in the Stornoway rocks. Elsewhere, in floodplain sequences, streams became highly sinuous indicating a lowering of relief with a change to less arid conditions. A

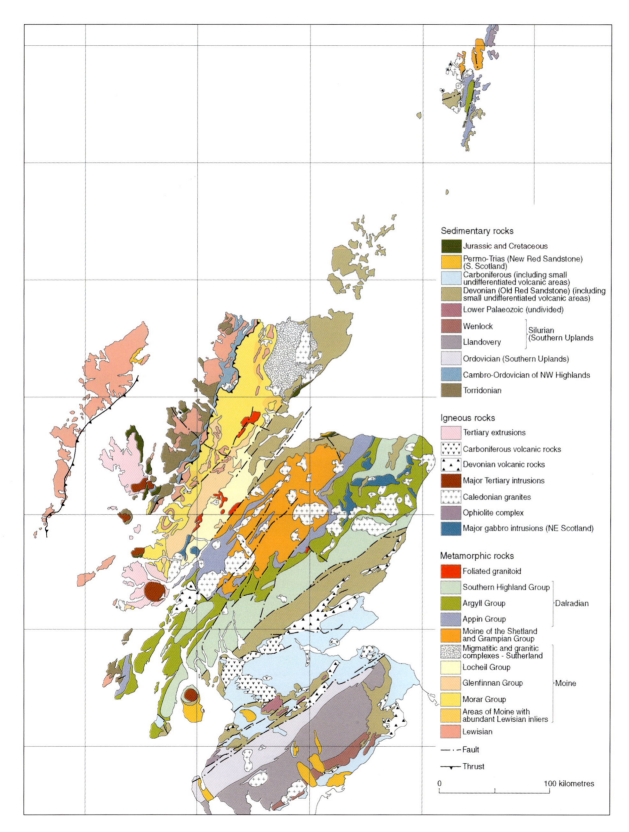

Simplified geological map of Scotland, from *Quarries of Scotland* (McMillan, 1997). Reproduced by permission of the Director, British Geological Survey. Permit Number BD/IPR/5-15 British Geological Survey. © NERC. All rights reserved.

**Sedimentary rocks**

- Jurassic and Cretaceous
- Permo-Trias (New Red Sandstone) (S. Scotland)
- Carboniferous (including small undifferentiated volcanic areas)
- Devonian (Old Red Sandstone) (including small undifferentiated volcanic areas)
- Lower Palaeozoic (undivided)
- Wenlock } Silurian
- Llandovery } (Southern Uplands)
- Ordovician (Southern Uplands)
- Cambro-Ordovician of NW Highlands
- Torridonian

**Igneous rocks**

- Tertiary extrusions
- Carboniferous volcanic rocks
- Devonian volcanic rocks
- Major Tertiary intrusions
- Caledonian granites
- Ophiolite complex
- Major gabbro intrusions (NE Scotland)

**Metamorphic rocks**

- Foliated granitoid
- Southern Highland Group }
- Argyll Group } Dalradian
- Appin Group }
- Moine of the Shetland and Grampian Group
- Migmatitic and granitic complexes - Sutherland
- Locheil Group }
- Glenfinnan Group } Moine
- Morar Group }
- Areas of Moine with abundant Lewisian inliers }
- Lewisian

— · — Fault

—▼— Thrust

0            100 kilometres

midst of a sub-plate, broad boundary shear-zone that interconnected the Alpine and North Atlantic plate boundaries. West of Britain, a large series of rifts developed into the Atlantic Spreading Centre. As Africa slid away from North and South America, rift systems developed between Greenland and northeastern North America, and then between the Outer Hebrides and Rockall Bank. Intense volcanic activity preceded the final separation, around 55 m.y. ago, of Greenland from the British Isles continental crust.

Igneous activity in western Scotland spanned 11 m.y. from 63 to 52 m.y. ago, with a peak around 58 m.y. ago. Eroded remnants of major central volcanoes occur in Arran, Ardnamurchan, Mull, Rhum, Skye, and at several submarine sites. The central masses are primarily sited on ridges of pre-Mesozoic and Precambrian rocks close to, or on, major faults. Scottish Tertiary volcanic centres appear to be located within areas of thinned crust and high heat flow. Lavas were erupted from the central volcanoes during the early phases and as the volcanoes aged, explosive activity characterised the later effusions. Subaerially erupted lavas possess reddened oxidised tops, especially in parts of Mull and Skye. Thin sediments with plant remains intercalated with lavas appear to have formed in shallow lakes or as river deposits. Separate flows range from less than 1m thick to over 50m. Many are laterally impersistent, which strongly suggests that the lavas were predominantly fissure-fed. Zeolitisation around and within the central complexes resulted from extensive convective circulation of meteoric waters within the lavas. Crustal tension accompanied magmatic activity, permitting intrusion of thousands of dykes, some of which are traceable over long distances into northern England. As much as 10% crustal extension has been measured in places.

The Tertiary central complexes represent volcanoes eroded to their roots. Characteristic features include numerous arcuate gabbroic intrusions and layered gabbros. Granites, granophyres and effusive silicic rocks are spatially and genetically associated with the central complexes.

Present-day features belittle the size of the Mull volcano, which probably formed the highest land in Tertiary Britain; it poured out a great thickness of lavas (up to 1830m thick) from a caldera 9.6km in diameter (Bailey *et al.*, 1924). By this time Britain had gently tilted downwards in an easterly direction as the European plate warped. It continued moving northwards, reaching its present position 15 m.y. ago.

Following the volcanic episode, marine erosion lowered the landscape to form the 600-900m platform of the Highlands as a great thickness of stripped-off sediments accumulated in the North Sea Basin. A second platform, at approximately 300-365m, is thought to be Pliocene in age.

As the global climate cooled, glaciers formed ice sheets which flowed radially from centres in the Highlands, Cairngorm mountains and Southern Uplands. By 18,000 years ago the ice sheet covered most of Britain and was greater than 1km thick in the Midland Valley. The ice sculpted Scottish rocks into present-day landscapes and, with the final warming, huge rivers carved large valleys. Dramatic drops in sea-level (up to 150m), due to ocean ice-sheet interchange plus crustal loading effects, became reversed upon final warming. Isostatic readjustment left numerous raised beaches around Scotland's coastline.

Over the last 10,000 years man, as a geological agent, has continued to modify the landscape.

# COLLECTORS AND COLLECTIONS

## INTRODUCTION

Advancement of mineralogy is reflected in the growth and function of mineral collections. Many collections attain a zenith then decline, yet they are the cornerstones from which our science is projected. The stimulus for people to amass mineral collections has depended not upon financial ability but purely on derived intellectual satisfaction. Enlightened self-interest is not the prerogative of any one level of society, for mineral collections in Scotland have been accumulated by academics, the aristocracy, medics, ministers, men of industry and commerce; people from all walks of life. Without their dedication, skill and generosity, national collections today would be much impoverished.

Many mineral collections expanded during periods of economic exploitation, reflecting prosperity and scientific curiosity. With the closure of metalliferous mines, traces of past mining activities obliterated or localities no longer accessible, mineral collections are now an irreplaceable part of our cultural, industrial and scientific heritage. Yet conservation, now to the forefront, has not always been uppermost in the minds of people in authority, for some historic collections have not only been despoiled but annihilated.

In the eighteenth century, great artistic, intellectual and scientific advances occurred throughout many parts of Scotland; in particular, Edinburgh attained status as a leading cultural city, not only in Scotland but throughout Europe. The 'Enlightenment' took place from around 1730 to 1790. At this time, the nobility undertook the 'Grand Tour' to Europe, considered an essential component of gentlemanly education. Most of these 'tourists' returned with statues and antiques, for no gentleman's house was complete without a 'cabinet of curiosities'. Cabinets frequently included natural history specimens and minerals. Through the generosity of cabinet owners, many mineral specimens were donated for scientific study.

A flourishing lapidary trade built up in Edinburgh and Aberdeen. One lapidary in particular, W. Somerville, who resided at numerous addresses in Edinburgh between 1813 and 1840, supplied minerals to interested parties, including Professor Jameson at the University Museum, Edinburgh. There was also interaction between the optical lapidary George Sanderson, who resided at St Andrews Square, Edinburgh, and William Nicol (*qv*). Sanderson manufactured polarising instruments of tourmaline and Iceland spar (Rose, 1847), as Nicol had revealed to him the earlier method that he had developed of preparing mineral thin-sections mounted directly on glass rather than wood (Morrison-Low and Nuttall, 1984). An early relationship was thus forged between amateur, lapidary and professional – a tradition continued to the present day, to the benefit of mineralogy.

During the early days of collecting, mineral specimens were regarded primarily as scientific commodities rather than as aesthetic entities. New chemical elements were being discovered in minerals, which stimulated further search. Developments in morphological crystallography provided a driving force to seek better crystals, and thus elevated aesthetic aspects to primacy:

> There are two sides to collecting, esthetic and scientific. Many mineral specimens are beautiful objects to contemplate, and are well worthy of collecting for that reason alone. The variety of crystal shapes and the seemingly endless variations in composition and occurrences of minerals initially excite a desire for understanding, which has given rise to the scientific side of collecting. (Bideaux, 1972)

In his foreword to the facsimile of Greg and Lettsom's (1858) *Mineralogy of Great Britain and Ireland*, Embrey (1977) writes:

> Greg and Lettsom' owes more to the amateur than to the full-time professional development of the science, and this comes as no surprise to those of us who believe that the contributions of the talented amateur collector and of the specialist dealer

have, over the years, been sadly and generally under-appreciated. Mineralogy is pre-eminently a specimen-based science, and relies for its progress on extensive collections, yet I doubt whether there is a single public collection anywhere in the world that does not owe its origin and probably most of its specimens to the efforts of the private collector. The private collector and the dealer, moreover, by drawing the attention of the professional to the unusual, have very likely been responsible for the greater part of the new species that have been described, and by their wide contacts have helped the professional in many other ways.

Nagel (1994) reports that:

Most mineral specimens, by far, are in the hands of private collectors and dealers, not museums. This might surprise collectors who have in mind the vast collections of some of the large public museums whose holdings number in the hundreds of thousands. Those large collections, however, represent only a small fraction of public mineral collections; most such museums have only some tens of thousands of pieces.

Implicit in Nagel's statement is 'a latent heritage': with diminishing resources, combined with museums' changing roles, the private collector should develop up-to-date curatorial skills to document, accurately and electronically, considerable amounts of information. Data can then be passed on to a wider audience when the collection is handed over.

Devoted and erudite amateurs are becoming more sophisticated in mineral identification, yet they encounter insurmountable problems which necessitate collaboration with museum curators who have access to modern instrumentation. Many museums are experiencing a period of unprecedented and accelerating changes, as their functions are challenged and changed, and resources subject to increasing pressure. Nevertheless the amateur:professional axis is as strong today as it has ever been, with fruitful spin-offs for both individuals and the scientific community. The amateur has become more professional in fieldwork, and in response to the mineralogical data explosion. Burgeoning suites of high-standard papers by amateurs are now appearing, thus adding their major contributions to topographical mineralogy.

Collections are latent databases offering both insights into past characters and personalities, and avenues for potential economic exploration and exploitation. In a survey by the author

of currently accepted new mineral species originally published in *American Mineralogist* from 1916-90, and *Mineralogical Magazine* from 1877-1990, a total of 632 species came to light. Well over 50% arose from economic exploitation studies, 20% directly from fieldwork and 10% originated in museum collections. (This latter figure contains a very small proportion from private and university collections.)

Immense importance is placed upon the scientific value of collections and especially of holotype minerals for which information is internationally disseminated. Information derivable from minerals is related to the technology of the period; consequently, a sample of seemingly no particular value may suddenly assume supreme importance when examined or re-examined.

## JAMES BALFOUR (1600-57)
### Historian and King's Commissioner

Sir James Balfour and his brother Sir Andrew Balfour (*qv*) of Denmill (Denmylne), Newburgh, Fife, were among the first of Scotland's scholarly scientific naturalists to create collections, for natural history object classification was then in its infancy.

*Right:* Sir James Balfour, artist unknown. Scottish National Portrait Gallery.

Sir James Balfour, antiquarian and Lyon King-of-Arms, was addicted to heraldry and amassed a large manuscript collection; his life was largely devoted to preserving the vestiges of Scotland's past. He wrote numerous tracts, including *A Treatise on Gems and the Composition of False Precious Stones* (NLS MS Adv. 33.7.9, undated). This manuscript reveals Sir James's intense interest in minerals and ranks among the earliest extant mineralogical writings in Scotland. The 106-page compilation is divided into six 'caskets': the first five each deal with 13 kinds of stones, and the sixth with ten. Each stone is described, accompanied by its synonymy, place of origin and medical uses. An alphabetical list of precious stones mentioned by the Ancient Greeks concludes the work. In his *Treatise*, Sir James frequently refers to Georgius Agricola's (1494-1555) *De Natura Fossilium* (1546). The profound influence of Agricola (the 'Father of Mineralogy') is clearly apparent, for the caskets are organised in a systematic manner. The first casket details diamond, ruby, sapphire, emerald, garnet and 'cristall', amongst others. Within the second casket, pearl, 'opall', onyx, 'sardornix' (sardonyx), 'achates' (agates) and 'chalcedonius' are described.

## ANDREW BALFOUR (1630-94)
### Botanist and medical practitioner

Sir Andrew Balfour, born on 18 January 1630 at Balfour Castle, Denmill, Newburgh, Fife, was educated at St Andrews University, where he studied natural history and medicine, and at Oxford. After further study at the eminent medical and botanical centres of Europe, he practiced medicine in London. Following a further four-year study visit to the Continent, he returned to St Andrews in 1667 after an absence of 15 years to re-commence medical practice, then moved to Edinburgh. With him, he brought a library considered to be the most outstanding in Scotland, particularly for its medical and natural history works, together with an extensive museum of medals, antiquities, materia medica, plants, animal remains and fossils, which mirrored his brother's collection (Simpson, 1982).

Upon Sir Andrew's death (10 January 1694) his executors and friends sold his collection to Edinburgh University for £400 Scots (at the Act of Union in 1707, when the Scottish currency was abolished, £100 Scots equalled £8-6s-8d,

Chalcedonius. Extract from *A Treatise on Gems and the Composition of False Precious Stones*, (undated) in the handwriting of Sir James Balfour. Note reference to quality of Scottish chalcedony and comparison in the first paragraph with 'ascahates' (agates). Courtesy Trustees of the National Library of Scotland, MS Adv. 33.7.9.

*ie* one-twelfth the Sterling value), on condition that it be retained in a 'closet' called after him. (It is apparent from the Edinburgh Town Council Minutes for 10 May 1695 that the terms of the sale were deemed favourable compared with purchasable collections from other sources.) Sir Andrew's curiosities and manuscripts were bequeathed to his long-standing friend Sir Robert Sibbald, another distinguished Scottish naturalist of this turbulent period of religious and political strife.

## ROBERT SIBBALD (1641-1722)
### Antiquary and physician

Sir Robert Sibbald was born in Edinburgh on 15 April 1641 'in a house near to the head of Blackfriars Wynd upon the east side'. His father held political office as Keeper of the Great Seal of Scotland. Sibbald's early life is carefully revealed in his autobiography (NLS MS Adv. 33.5.1), which tells of the family hardships suffered during the Civil War of the 1640s, his religious education and his decision to enter medicine.

At an early age he developed a great aptitude for study. After the family moved to Fife, he was sent to study at Edinburgh High School, then progressed to university to read theology, followed by medicine. In 1660 he attended Leyden University and graduated MD the following year, thereafter proceeding to Paris, then Angers, for his doctorate. On 30 October 1662, Sibbald returned to Edinburgh to practice medicine. Botany also assumed great importance, for he immersed himself in

Balfour's library to study whatever 'animalls, vegetables, mineralls, metalls, and substances cast up by the sea, were found in this country, that might be of use in medicine, or other artes usefull to human lyfe'. Being 'curious in searching after them and collecting them' (Simpson, 1982), Sibbald, too, assembled a substantial museum.

A plot of land at Holyrood, Edinburgh, was leased in 1670 by Andrew Balfour and Sibbald for the cultivation of medicinal plants. From this venture, around 1680, Sir Andrew Balfour founded the Royal Botanic Gardens of Edinburgh. On 2 November 1681 the Royal College of Physicians of Edinburgh, also largely founded by Sibbald, received its charter. On 30 September 1682 he was appointed physician to Charles II

*Right:* Sir Robert Sibbald, aged 80, oil on wood, 76 x 66cm, by John Alexander (1686-c.1766) who studied in Italy (Macmillan, 1990). Reproduced by kind permission of the Royal College of Physicians, Edinburgh.

*Far right:* Birthplace of Sir Robert Sibbald – Blackfriars Wynd, one of the most important and ancient thoroughfares diverging from the High Street – looking north from Cowgate, Edinburgh. From 1230 for five centuries the dwelling places housed some of the most aristocratic families, and highest ecclesiastics, in Scotland. The Wynd was renamed Blackfriars Street after the Edinburgh Improvement Act passed in 1867. From Grant (1882).

and on 30 December of the same year he assumed the post of Geographer Royal of Scotland. This prestigious appointment enabled him to assemble natural history material for a geographical and statistical account of Scotland, a most elaborate work which appeared as *Scotia Illustrata sive Prodromus Historiae Naturalis* in 1684. In the previous year, Sibbald had received memoranda from Colonel Borthwick relating to Scottish mines and minerals, in particular gold, silver, copper and lead (Cochran-Patrick, 1878). Sibbald presented his natural history collection to the University of Edinburgh during 1697, with a catalogue (printed at the expense of the university) to supplement that of Sir Andrew Balfour's collection. The Town Council minutes for 19 May 1703 record that the common hall in which the museum was housed contained an additional cupboard 'on the north side of the hall with Doctor Sibbald's name upon it having putt some fossills vegetables and Animalls skeins whereof he keeps the key and promises to fill it up'. Defoe's (1748) account of the University Museum relates:

> In this higher Common Hall, which is a very spacious Room, are placed such Books as have been bought by, or given to the College, since the Library below was full; and in the South-end of it is a curious and noble Museum, collected by the very eminent Sir *Andrew Balfour*, who was once Tutor to the famous *Earl of Rochester*. It contains a vast Treasure of Curiosities of Art and Nature, domestic and foreign, from almost all Parts of the World; and is greatly valued by the Virtuosoes, containing some Rarities that are not to be found, either in those of the *Royal Society* at *London*, or the *Ashmolean* at *Oxford*. Sir *Robert Sibbald*, having a mind to engraft his Name and Merit on that of the celebrated *Balfour*, made a Present of a great Number of Shells, and other Curiosities, to the College, on Condition the Magistrates would print the Account of it, called, *Auctarium Musei Balfouriani e Museo Sibbaldiano*.

de Fossilibus rarioribus.    71

'Tis found in several places in this Coun-
trey.
Copper Ore from Devennen in the West-Coun-
trey.
A piece of the Copper made of this Ore.
Copper Ore from Linton.
Copper Ore from Braids Craigs.
Copper Ore from Arthrie, with green spots and
yellow sparks.
Copper Ore from Malainie.
Copper Ore with blue spots, and blue Veins.
Copper Ore, green upon one side, and of a russie
colour on the other side.
Copper Ore from Caithness.
Minera Ferri. Iron Ore.
Iron Ore of a gray colour brought to me from
England.
Lapis ruber, nostras, ex quo ferrum confi-
citur. Iron-stone of a red-colour, band-
ed with white Lines arising above the
surface found in this Countrey. The Fi-
gure of it may be seen in Scotiâ Illustra-
ta, Tab. 22. Fig. 2.
Plumbum nigrum; simpliciter etiam Plum-
bum vocatur. Nobis Lead. In multis locis
hujus Regionis reperitur.
Lead, from Hoptons Lead-hills.
Lead Ore, from South Ronaldsa, one of the
Orkney Isles.

Lead

72    Liber Primus,

Lead Ore, from the Isle of Sanda, there.
Lead Ore, from the Lord Duffus his Ground.
Lead Ore, from the Duke of Queensberry his
Mine.

CAPUT TERTIUM.
De Semi-Metallis.
Of Imperfect Metals.

Imperfecta Metalla, seu semi-Metalla, ea
sunt quæ affinitatem cum Metallis ha-
bent; sed non sunt ductilia; sed friabilia
aut fluida. Cujusmodi hic habentur
Bismutum, Marcasita Argentea Cæsalpini;
Germanis, Bismut. Differt ab utroque Plumbo, colore & duritie; nitet quandoque Ar-
genti colore, interdum purpureo diluto;
Simile Stibio, seu Antimonio, quoàd Figu-
ram, sed colore præstantius, Stellatum An-
timonii Regulum referens.
Antimonium; quibusdam Stimmi;Stibium.
Nobis Antimony; said to be found in
the North of Scotland.
Argentum vivum. Nobis Quick-silver.
Cinnabaris nativa rubra striata Nobis Na-
tive Cinnabar. 'Tis of a Scarlet colour.
Cinnabaris nativa rubra lævis; This is of a
darker red colour; and is ponderous.
Finis Libri Primi.

LIBER

4

a Catalogue
of Minerall an figur'd
Stones found in Scotland
in MRW cabinet.
1 sho the
a piece of fosil Stone, halff a foot
in Length. 5 inch in breadth,
with an impression on it from
the Banks of Kelvin.
2 two more small pieces, much of
the same kind
another of 4 inch in Length
and 3 in breadth with an
impression in form of Cockles
4 a piece of the Lapis undulatus
from the West Indies of the
coralline Kinde
5 a piece of Stone with an auxiliar
and round stone sticking to it
got among the same coralline
substances

*Above:* Catalogue pages showing entries for predominantly Scottish ore specimens, from *Auctarium Musaei Balfouriani e Musaeo Sibbaldiano* (Sibbald, 1697). NMS Library archives.

*Above, right:* A page in Sibbald's handwriting, from 'A Catalogue of Minerals and figured Stones found in Scotland in MRW cabinet'. The cabinet belonged to Robert Wodrow (1679-1734), ecclesiastical historian, and the catalogue dates from 1703. Courtesy of Trustees of the National Library of Scotland, NLS MS Adv. 13.2.8.

The catalogue (Sibbald, 1697) reveals a four-fold division into Fossils, Vegetables, Animals and Artificial Rarities. Sibbald's mineralogical *modus operandi* drew upon *De Natura Fossilium* (Agricola, 1546), for within Fossils we see represented one of the earliest examples of systematic mineralogy applied to a Scottish collection. Sibbald's catalogue illuminates interest in minerals in their own right: morphology, colour, lustre and locality details are stated for many specimens. Not only is systematic mineralogy foremost, but also regional surveying for economic minerals. The catalogue details approximately 250 geological specimens, mainly minerals.

Sir Robert Sibbald's later days passed almost unnoted. He died in Carruber's Close (in close proximity to his birthplace) on 9 August 1722, aged 81 years, and was buried on 12 August 1722 within the burial ground of the Phesdo family in Grey-friars Churchyard, Edinburgh (Craig, 1976).

Even prior to Defoe's 1748 account, the Balfour-Sibbald museum had remained neglected, with objects decaying, for many years, yet in 1750 it still remained a sizeable collection. In 1782, three years after his appointment as Regius Professor of Natural History, Dr John Walker (*qv*) rescued many valuable objects and placed them in the best order possible (Eyles, 1954). Yet, just less than a century after acquisition of the Balfour-Sibbald collections, it is a tragedy that not a trace remained in Edinburgh University (Waterston, 1972; Fraser, 1989).

## WILLIAM HUNTER (1718-83)
### Natural historian and physician

William Hunter was born the seventh of ten children of John and Agnes Hunter on 23 May 1718, at Long Calderwood, East Kilbride, Lanarkshire. After attending grammar school in East Kilbride, at the age of 14 he went to Glasgow University where he studied Greek, natural philosophy and theology for five years. William Hunter was originally destined for the Church, but declined; upon becoming friendly with the physician William Cullen (1710-90) and eventually becoming his assistant, he entered the medical profession. On Cullen's advice, Hunter attended anatomy classes given by Professor Alexander Monro at the University of Edinburgh. On 25 October 1740 he set sail for London, where he stayed with the obstetrician William Smellie and pursued his classical and medical studies. Prospects at that time were highly favourable to Hunter, who decided to stay in London. He advertised in the London *Evening Post* a course of lectures on anatomy to commence on 13 October 1746, for a charge of four guineas. The lectures were given in his Covent Garden home (1746-60), at Litchfield Street (1763-67), and in Windmill Street from 1768 until his death in 1783. Hunter prepared numerous anatomical dissections, and corrosion casts to demonstrate ramifications of even the smallest vessels, and also documented diseases and mid-eighteenth-century accidents. The foundations of an anatomical and pathological museum were laid down and in 1765 Hunter began entertaining thoughts of assembling a great museum 'for the improvement of anatomy, surgery and physik'.

Having obtained an MD degree from Glasgow University on 24 October 1750, and been admitted licentiate of the Royal College of Physicians of London on 30 September 1756, Hunter had now become a highly successful obstetrician and was consulted in 1762 by Queen Charlotte. Five years later he became a Fellow of the Royal Society and the following year (1768) was appointed the first Professor of Anatomy at the newly founded Royal Academy.

During 1765 Hunter developed his project for a museum and offered £7000 for the building if a plot of ground was granted to him. The request was refused, but Hunter purchased a plot of land in Great Windmill Street, London, on which he built a house (occupied in 1770), lecture theatre, dissecting

William Hunter by Allan Ramsay (1713-84), oil on canvas, 96 x 75cm. © Hunterian Art Gallery, University of Glasgow.

room and a large museum. To his anatomical collections he now added coins, medals, corals, shells and minerals. In his Will, Hunter left the museum collections, which he calculated in 1783 had cost him £20,000, to three trustees for 20 years, after which they passed, in their entirety, to the University of Glasgow, where they are now lodged in the Hunterian Museum (see University of Glasgow).

William Hunter commenced mineral collecting around 1765, but between 1771 and 1782 he spent £1500 on mineral specimens (c.£130,000 at present-day values). Purchases came from two main sources: Adolarius Jacob Forster (1739-1806) and Peter Woulfe (1727?-1803) (Durant and Rolfe, 1984). The former dealt actively in minerals for about 45 years, from premises in Covent Garden and Soho, London, in Paris, and in St Petersburg (Frondel, 1972). Forster supplied Hunter principally with European minerals, many from old mining districts and classic localities of the Harz, Bohemia, Saxony, Thuringa, Kremnitz and Transylvania. Hunter's most

Catalogue prepared by Jameson 1809-10 of the Hunter Collection of Minerals and Fossils, Glasgow. Note white lead ore (cerussite), green lead ore (pyromorphite), and lead vitriol (anglesite), descriptive names used prior to assignment of current nomenclature (cerussite, Haidinger, 1845; pyromorphite, Hausmann, 1813; and anglesite, Beudant, 1832). MS no. 3, Courtesy of Hunterian Museum, University of Glasgow.

Black Lead Ore.

1095. Black lead ore with white lead ore —
1096. Black lead ore, with white lead ore, lead glance

White lead ore.

1097. 1104
1108. 1109. White lead ore
N 908 - 909. white lead ore ? —

Green Lead Ore

1110. 1116
N 910 to 916. Green lead ore

Lead Vitriol

1117. 1120
N 917 to 920. Lead vitriol —

Tin Stone

1121. 1145
N 921 to 945. Tin stone —

Yellow Blende

1146 to 1150
N 946 to 950. Yellow blende

Brown Blende

1151 — to 1161
N 951 to 961. Brown Blende —

Calamine

1162 to 1170
N 962 to 970. Calamine

Native Antimony

1171 to 1
N 971. Native antimony —

Compact Grey antimony ore.

1172 to 1173
N 972 - 973

expensive specimen, 'A very fine and rare specimen of Needle Antimony: from Hungary', cost 30 guineas. This specimen was mentioned by John Walker in his lectures towards the end of the eighteenth century (Scott, 1966, p. 12). The Forster-Hunter axis reveals the latter's association with European royalty, for one of Hunter's specimens, 'A large and rare piece of yellow crystallised sulphur from a mine, now lost near Cadiz in Spain', was bought from Forster for eight guineas. Forster collected from the mine upon grant of a special licence from the King of Spain (Russell, 1952). Upon Forster's death, five or six boxes of Russian specimens worth £1000 and collected for the King of Spain were disposed of to his wife and relatives, whereas his main collection was bequeathed to his brother, and to his nephew John Henry Heuland. Hunter's second main supplier, Peter Woulfe, was a chemist and early mineral analyst from whom Hunter received 1000 specimens between 1774 and 1778 (Durant and Rolfe, 1984). Other suppliers included Jan Ingen Housz MD (1730-99), Physician to Their Imperial Majesties in Vienna, who secured a series of Transylvanian gold specimens. John Jeans sold 20 Scottish minerals for £6-7s-9d in 1774 and 1777. Lady St Aubyn, wife of Sir John St Aubyn (1758-1839), a renowned Cornish collector, in 1764 or 1765 forwarded a collection of minerals to Hunter (Wilson, 1994).

Some of William Hunter's mineral specimens, including a 5 x 5 x 5cm sample extremely rich in dendritic silver from Alva, are on display in the Hunterian Museum. In the mineral collection is a Hunter Freiberg argentite, plus a silver button smelted from it by George Fordyce (1736-1802). The latter, one of Hunter's trustees, together with Hunter's other trustees D. Pitcairn (1749-1809) and C. Combe (1743-1817), compiled a catalogue during 1794-97 (Hunterian ms. no. 1) of Hunter's minerals before they arrived in Glasgow: 'This we believe to be a correct catalogue of the minerals in Dr. Hunter's Museum.' It reveals up to 38 sub-divisions, from no. 1 Gold, no. 2 Silver and no. 3 Mercury, through to no. 38 Sulphur. George Fordyce, physician and chemist, used Hunter's collection for his work on the chemical composition of minerals. He evolved a rational system of mineral classification which he applied in arranging and cataloguing Hunter's minerals for him, and published a paper on 'An Examination of various Ores in the Museum of Dr. William Hunter' (Fordyce and Alchorne, 1779).

William Hunter was a man of elegance and spirit, and slender build. He possessed a good memory, quick perception and sound judgement. His mode of life was very frugal. Hunter epitomised the polite scholar, accomplished gentleman, complete anatomist and highly talented demonstrator and lecturer. (For additional details of Hunter's life, see Brock, 1983.) He died, unmarried, on 30 March 1783, aged 64, and was buried privately at 8.00pm on 5 April in the vaults of St James's, Piccadilly, London.

## REV. JOHN WALKER (1731-1803)

Minister, Professor of Natural History,
'Father of Geological Education'
and 'Father of Scottish Mineralogy'

The Rev. Dr John Walker was born in 1731 in the Canon-gate, Edinburgh, where his father was rector of Canongate Grammar School. A gifted child, he possessed a great propensity for learning, studied Latin and Greek, and had a leaning towards botany. After finishing studies at his father's school, Walker entered the University of Edinburgh to prepare for the ministry in the Church of Scotland. As an undergraduate at the age of 15, he

> began to collect minerals in the year 1746, when attending the Natural Philosophy Class, and was first led to it by the perusal of Mr. Boyle's works, and especially his Treatise on Gems … and visited the quarries and coalleries near Edinburgh; but had no book at the time, to direct us concerning the species of minerals, but Woodward's Catalogues [two volumes, 1728 and 1729; Brewster and Jameson, 1822].

THE

# NATURALIST'S LIBRARY.

ORNITHOLOGY.

VOL. XII.

BIRDS OF GREAT BRITAIN
AND IRELAND.

PART III.

RASORES AND GRALLATORES.

BY

SIR WILLIAM JARDINE, BART.
F. R. S. E., F. L. S., &c. &c.

EDINBURGH:
W. H. LIZARS, 3, ST. JAMES' SQUARE,
S. HIGHLEY, 32, FLEET STREET, LONDON; AND
W. CURRY, JUN. AND CO. DUBLIN.
1842.

*Left:* Title page to Sir William Jardine's *The Naturalist's Library*, (1842).

*Below, left:* Frontispiece to *The Naturalist's Library* – John Walker, with the spire of St Giles Cathedral, Edinburgh, in the background.

He was inspired by the teaching of Professor Cullen, whose chemistry courses he attended. Cullen's interest in minerals was contagious, for the young Walker became deeply attached to the subject, and to geology in general. The University Museum at this period, with its fine natural history collection (Balfour's and Sibbald's), aroused Walker's scientific interests:

> I may well remember it for it was the view of it which first inspired me with a taste for Natural History when a boy at this College. (Scott, 1966)

In 1754 Walker was licensed to preach by the Kirkcud-brightshire Presbytery, and four years later became minister at Glencorse, Edinburgh. There he undertook botanical and mineralogical excursions into the nearby Pentland Hills. During the latter part of 1758, Walker became acquainted with Henry Home, Lord Kames, a prominent figure of the

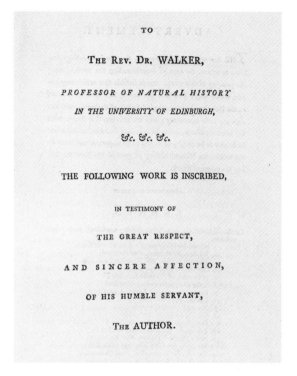

TO

THE REV. DR. WALKER,

*PROFESSOR OF NATURAL HISTORY
IN THE UNIVERSITY OF EDINBURGH,*

&c. &c. &c.

THE FOLLOWING WORK IS INSCRIBED,

IN TESTIMONY OF

THE GREAT RESPECT,

AND SINCERE AFFECTION,

OF HIS HUMBLE SERVANT,

THE AUTHOR.

Edinburgh Enlightenment, and a like-minded man with a great thirst for knowledge. The relationship flourished, surviving Walker's move in 1762 to become Minister of Moffat, Dumfriesshire, and it changed Walker's entire course through life, for it provided the opportunity to make his great educational and scientific contributions. Lord Kames became influential in commissioning Walker to undertake a survey of the Highlands and Hebrides: the journey (5082km; McKay, 1980) lasted from May to December, 1764. The trip, an enormous one-man task, was not Walker's first to the Highlands and Hebrides, for he visited them in 1761 and 1762 respectively.

After returning to Moffat in 1765, Walker was granted degrees at Edinburgh University (Doctorate of Divinity, diploma dated 11 April 1765) and Glasgow University (Honorary Doctorate of Medicine). His insatiable appetite for knowledge, combined with a restless inquisitiveness for collecting natural history specimens, including 'weeds', earned him the nickname 'the mad Minister of Moffat':

> During my long residence at Moffat [until 1783], I collected, in a number of short tours, all the remarkable minerals in Dumfries-shire, the Forest of Selkirk, Tiviotdale, Ayrshire,

and Clydesdale. I visited the lead-mines at Mackrymore[1], the copper mines at Covend[2], and the mines of antimony in Eskdale[3]. Leadhills and Wanlock being within a forenoon's ride, I frequently visited the mines at these places, and went down in them to the greatest depths .... Between the years 1761 and 1764, I found in those mines the Strontianite[4]; the Ore, and the Ochre of Nickel; the Plumbum pellucidum [cerussite] of Linnæus; the Plumbum decahedrum [cerussite] and cyaneum [linarite], both undescribed; the Saxum metalliferum [metallic stone] of the Germans; the Ponderosa aërata [witherite] of Bergman; and the Morettum, which afterwards appeared to be a peculiar sort of Zeolite [brewsterite]. All these were here, for the first time, discovered in Britain. (Brewster and Jameson, 1822)

During his spare time Walker pursued studies in botany and geology. In 1778, on the death of Dr Ramsay, Professor of Natural History at the University of Edinburgh, Dr Walker made application to the Crown for the vacant Chair (Kay, 1842). 'There is no evidence that Ramsay ever gave lectures, held class meetings, or prepared syllabi for students' (Scott, 1966). Most of Walker's geological and botanical papers were prepared, or published, between 1758 and 1781 (ibid). On 15 June 1779, Walker was appointed Regius Professor of Natural History at Edinburgh while retaining his clerical post at Moffat some 80km away. Naturally his parishioners expressed concern. His move to become Minister at Colinton, Edinburgh, in 1783 alleviated the situation. Walker enrolled his first class at least as early as 1781. In his initial lecture to this class Walker said, 'I am to teach a science I never was taught' (Scott, 1966, p. 1). Thus, in contrast to Ramsay, Walker began to deliver vigorous lectures and actively pursued a teaching career for virtually the remainder of his life. For his profound influence in establishing geology as a science, and as a discipline within higher education, Walker is accredited with the title 'Father of Geological Education' and 'Father of Scottish Mineralogy' (Scott, 1966, p. xliv):

> The modern character of Walker's scientific vocabulary is impressive. The first use of several geological words and phrases should be credited to him. In other instances, geological terms had been used before but Walker employed them in a more precise manner and related them to definite geological phenomena. He may be credited with using most of them for the first time in a classroom, thus establishing their status and giving them respectability in the new science. (Scott, 1966, p. 263)

Walker became an eminent, stimulating lecturer; even professors enrolled for his classes. With his encyclopaedic scientific knowledge, art of demonstration, originality of thought and search for the truth, he inspired many well-known scientists. Examination of Walker's class lists reveals Sir James Hall ('Father of Experimental Petrology'), the Rev. John Playfair, who was appointed Professor of Mathematics of Edinburgh University in 1785, and 18 year-old Robert Jameson (later Professor Jameson).

Walker worked with polished rock slabs, reflected light, a microscope and chemical methods. In the 1780s he formulated a hardness scale, albeit slightly different, some 40 years before that of Mohs. Although Mohs published his scale in 1820, Staples (1964) reported that Kirwan had utilised a numbering system in 1794. (Kirwan's scale also had ten numbers of hardness defined by the relationships to the hardness of chalk and a knife.)

Of mineralogy Walker commented: 'This seems at present to be the most cultivated part of Natural History in every part of Europe, and indeed it is the part most behind' (Scott, 1966). Walker, strongly influenced by Linnæus, classified minerals by their external and internal properties, *ie* appearance and chemical composition. An extensive collection of minerals was amassed:

> the catalogue of my collection [1787] contains 1569 species [323 genera] and varieties of minerals … that the number of specimens may probably amount to above 3138. These, however, do not form the whole of my collection. (Brewster and Jameson, 1822)

By the 1770s Walker had visited most of Scotland and records how 'At different Times also, I had occasion to traverse most of the counties in England' (Withers, 1991). It is known that he visited Strontian, Skye, Tiree, the Inner and Outer Hebrides, Glenelg and Assynt, and the north of Ireland, areas which could provide ample specimens 'and in large masses, as their conveyance home by the cutter [one of the King's cutters] was so easy' (Brewster and Jameson, 1822).

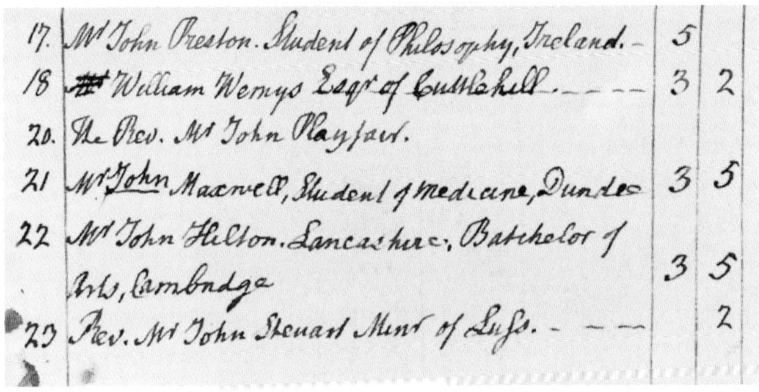

Walker became the first secretary of the Physical section of the Royal Society of Edinburgh (founded 1783) and was one of the earliest Fellows of the society. In 1782 he became influential in organising the Natural History Society of Edinburgh; in the same year he formed the Edinburgh Agricultural Society; and meantime he became a long-term member of the Highland

List of Students
at the
Class of Natural History,
in the
University of Edinburgh.
Second Session.
Novr 1782.

1. Sir James Hall of Dunglass Bart. — 5 5
2. Mr John Porter of Waterford in Ireland — 3 5
3. Mr Thomas Hurst of Portsmouth, Hampshire, Student of Medicine — 3 5
4. Mr James Edward Smith of Norwich, Student of Medicine. Second Course. — 3 5
5. Right Honble the Earl of Ancrum. 2d Course — 5 5
6. Mr Thomas Horner Junr, of Mells-Park, Somersetshire. — 3 3

17. Mr John Preston. Student of Philosophy, Ireland. — 5
18 Mr William Wemyss Esqr of Cuttlehill — 3 2
20. The Rev. Mr John Playfair.
21 Mr John Maxwell, Student of Medicine, Dundee — 3 5
22 Mr John Hilton. Lancashire, Batchelor of Arts, Cambridge — 3 5
23 Rev. Mr John Steuart Minr of Luss. — 2

List of Students May 1st 1793.

1. John Barnett. M. S. Staffordshire.
2. Michael Bulkely. M. S. Cork, Ireland. Not paid.
3. John Foster. M. S. Kingston upon Hull, Yorkshire.
4. John Innes Crawford. Ph. S. Jamaica. } Dr Wright.
5. William Webb of Alton, Hampshire. Ph. S. }
6. Patrick Walker. Ph. S. Edinr.
7. The Honble Archibald Fraser of Lovat.
8. Robert Jameson. Ph. S. Leith. 2d Course.
9. John Rogers. A. M. Preacher.
10. Robert Johnson of Byker near Newcastle. Ph. S.
11. Capt. West. Edinr. 3d Course.
12. Richards Brown. Stamford. M. S.
13. Hugh Ferguson. Dublin. M. S.
14. Henry Scot Alves. Dalkeith. Ph. S.

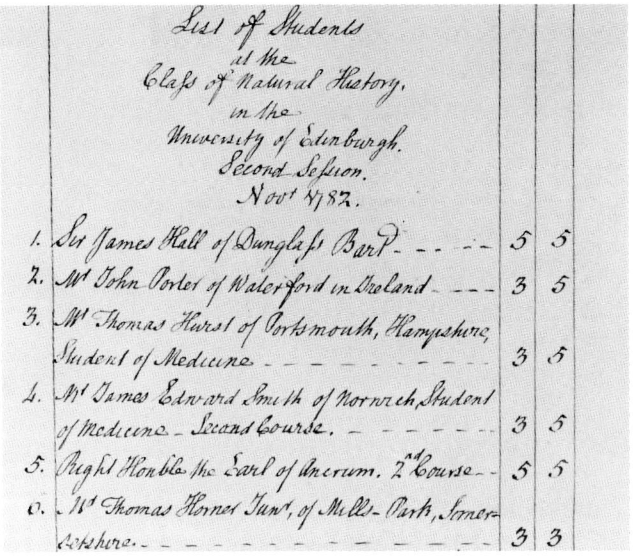

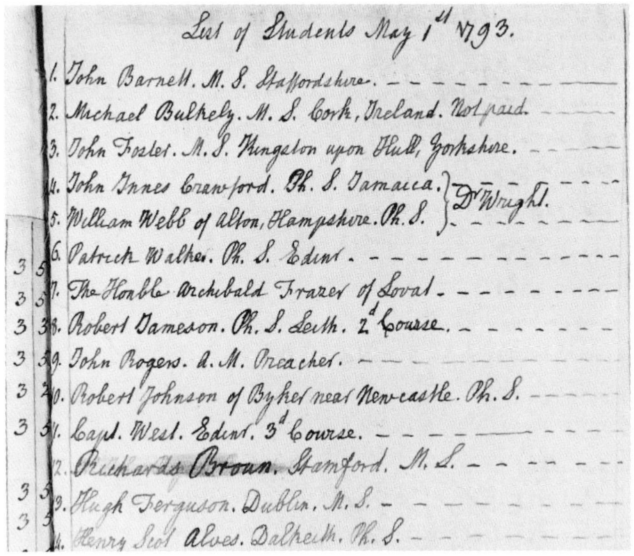

Walker's registration lists. The 1782 list shows Sir James Hall and the Rev. John Playfair; and Robert Jameson as the eighth registrand in 1793. The lists are in Walker's handwriting. Edinburgh University, Special Collections Dcl.57.

Polished specimens of Italian ornamental marbles from the geological collection presented to the Royal Society of Edinburgh by the third Earl of Hopetoun, c.1785, probably from 'a box of 100 different kinds of marble' purchased by Lord John Hope while in Italy on the Grand Tour between 1725 and 1727. Each marble bears a label, in John Walker's writing stating genus, species and variety according to Walker's usage. Note 'ME' means *Musæum Edinensis*, followed by a number allocated according to the order in which the items were deposited in the University Museum (Waterston 1997). Top row NMS G 1993.34.97 and 1993.34.94. Bottom row NMS G 1993.34.20 and 1993.34.31.

Society of Scotland. Most of Walker's fieldwork was done, and his manuscripts written, between 1758 and 1781, and therefore predate Hutton's dissertation, read in 1785, and his *Theory of the Earth* (in two volumes), published in 1795 (Scott, 1966). Both Walker and Hutton were Fellows of the Royal Society of London (Walker probably being elected in 1794; Sweet, 1972).

A specimen of zeolite, named wollastonite by Thomson, 'was found by one of our best geological observers, General Lord Greenock' at Corstorphine Hill, Edinburgh, and was analysed by Walker (Jameson, 1833). Walker's analysis yielded $SiO_2$ 54%, CaO 30.79%, $Na_2O$ 5.55%, MgO 2.59% and $H_2O$ 5.43%. Heddle (1880) analysed a similar specimen from the same locality, found in the Jameson-Torrie Collection that he had purchased. His work revealed a two and a half-fold magnesia increase over Walker's result and he surmised that the latter 'had thrown down much of his magnesia along with his lime'. Heddle concluded that the yellowish-white radiating mineral analysed by Walker was a new species and appropriately named it walkerite. A logical explanation for Corstorphine Hill pectolite apparently high in magnesia was proffered by Davidson (1989), who showed that stevensite occurred intimately associated with the pectolite. For Walker's many diligent achievements and discoveries in mineralogy (some later accredited to others), Heddle appropriately referred to him as the 'Father of British Mineralogy'.

When Walker undertook a second position, that of Keeper of the University Museum (1779-1803), without salary, the extensive collections of Balfour and Sibbald had, inexcusably, been previously decimated. With characteristic zeal, he endeavoured to rectify the situation by trying to establish a public museum within the college, commenting, 'I have a commission from the King, and another from the City of Edinburgh to be Keeper of the Museum; but when I settled here there was nothing to keep' (Scott, 1966). The charter of the Royal Society of Edinburgh (1783) decreed that all natural history objects gifted to the society should be deposited in the museum of the University of Edinburgh. For six years Walker endeavoured to establish a public museum in the college, but sufficient financial support never materialised. A frustrated Walker resorted to using his own substantial private collection of natural history specimens, rocks and minerals. The famous French geologist Faujas Saint-Fond (1741-1819) visited Edinburgh in 1784 and recorded that:

> The cabinet of natural history, in the university, is under the direction of Doctor Howard. The examination of this collection gave me great pleasure, and interested me much more than that of the British Museum, in London, though it was far less considerable; but, the objects which compose it, are in a more methodical order, particularly the stones and minerals: Besides, the managers of this museum have very properly taken care to collect all the productions of Scotland they have been able to procure .... The place allotted for it ought to be larger and decorated with more taste. The classification should also be extended to the other parts as well as the minerals .... Lithology, and the study of minerals, have as yet made little progress in Scotland. There are therefore few collections of these objects. Doctor James Hutton is, perhaps, the only individual in Edinburgh who has placed in his cabinet some minerals and a number of agates chiefly found in Scotland. (Faujas Saint-Fond, 1799)

Specific instructions to collectors of natural history specimens were issued by Walker in 1793 (reproduced in full by

Sweet, 1972), which assisted the College Museum in acquiring foreign material. A memorial dated 1793 regarding the New College states:

> As the Hall alotted for the Museum, in the New College, is of great Extent, and as it is to be hoped may become one of the greatest ornaments to the University and to the City: He [J.W.] is anxious, while the building is going on, to provide as far as possible furniture suitable for such a noble Repository.
>   This Hall is 160 Feet long, 35 Feet wide and 25 Feet high. It would require a very great Collection to form in it. (EDU Special Collections La 352/2)

John Walker died in Edinburgh on 31 December 1803 and his collection, according to a footnote by Jameson (1822) '... will soon be arranged for public sale by the Trustees of Dr. Walker'. John Walker's collection has largely disappeared.

He was buried in his own ground in the Canongate Churchyard, close to the eastern side of Canongate Church. A tablet to his memory is on the wall of the large burial enclosure a few yards east of the church wall. John Walker bequeathed to the University of Edinburgh the proceeds of the sale of his house in the Canongate, to be applied to 'The College Museum and the Library'. Walker's work, apart from that included in Scott (1966) and McKay (1980), remains unpublished.

## Footnotes

1  Newton Stewart area. Mines opened up shortly after 1763.
2  Sinclair (1796) reports: 'About 25 years ago a copper mine was opened in this parish, near the rocky shore. A considerable quantity of ore was dug up.'
3  Glendinning, Louisa Mine.
4  Strontianite, a mineral not known from the Leadhills-Wanlockhead orefield. This refers to a pale-green radiating mineral (stronites) which is a strontium-bearing plumbo aragonite (see Aragonite).

### ALEXANDER WEIR (1735-97)
Heraldic painter, sign writer and museum owner

Alexander Weir was born on 8 February 1735, the son of an Edinburgh stabler. He undertook an apprenticeship with James Miller, a painter in Edinburgh, on 17 June 1747 and on 9 March 1757 was elected a Burgess of the City of Edinburgh,

South Bridge looking north, Edinburgh. Amateur sketch c.1800. Alexander Weir's museum properties, nos. 27 and 31, would be situated approximately at the junction of the two darker blocks. Edinburgh University, Special Collections.

being sponsored by his master James Miller (Taylor, 1992). On the death of the Marquiss of Lothian, 22 July 1767, Alexander Weir was commissioned to execute the funeral paintings. He worked from various premises in Edinburgh, including Baillie Fife's Close (High Street, 1773-74), Toddricks Wynd (High Street, 1774-82) and St James' Square (1784), after which he returned to Toddricks Wynd until his death.

From around 1760 Alexander Weir had been deeply engrossed in discovering, and employing, the best taxidermic preservation methods. As the foundations of modern science were being laid down in the city during the Enlightenment, Alexander Weir's talents soon attracted notice. Dr John Walker strove during the 1780s to re-establish the devastated natural history museum within the university. At this period, various august Edinburgh institutions were openly feuding with William Smellie (1740-95), the curator of the collections owned by the Society of Antiquaries of Scotland. Smellie proposed to lecture on natural history, thus potentially threatening Walker's position. In 1781 Smellie had discussed with the Society of Antiquaries of Scotland the desirability of forming a natural history museum. Weir and Smellie were acquainted, although Weir distanced himself from the argument by proposing, in November 1782, to establish his own museum (Taylor, 1992). Weir's museum did not open until 31 January 1789, at 31 South Bridge Street, Edinburgh. (Three years previously, on 15 February 1786, Weir had sold a collection of natural history objects for 200 guineas to John Walker, the money being raised by subscription. A printed document of 113 subscribers, dated Edinburgh December 1784 (EDU La III 352/2), details amongst others Professor Charles Hope (see Sir David Brewster) and Mr Ferguson of Raith, who contributed two guineas.)

Rapid expansion of the Weir collections created intense pressure upon space, and in July 1790 Alexander Weir announced a move from 31 to 27 South Bridge Street. In January 1791 he proposed to 'fit up the spare room adjoining to the Museum with a Collection of Productions of Scotland in Minerals, Fossils and Insects, properly arranged' (Taylor, 1992). An article in the *Edinburgh Evening Courant* for 28 February 1791 highlights mineralogical additions under the banner:

> NEW BRANCH of NATURAL CURIOSITIES among which is the MATRIX of the OPAL, a PETRIFIED CRAB, from Calabria; a fine specimen of IRON ORE, from Elba; and that most extraordinary phenomenon of the Fossile Kingdom, the ELASTIC STONE, etc, etc.

By Saturday, 21 July 1792, the museum had

> removed to elegant appartments fitted up for the purpose, No 16, PRINCES STREET and which will be ready for the reception of Company on Monday next.

In a testimonial letter (*Edinburgh Evening Courant*, 16 March 1795) written by William Smellie to the Hon. Henry Erskine (Chairman of a meeting of Mr Weir's patrons), he records:

> With regard to minerals, Mr WEIR has amassed above sixteen hundred specimens, among which, beside the ones of silver, copper etc he possesses a number of crystals and spars, with petrefactions of various kinds of wood, converted into beautiful agates etc ....
>
> One deserves particular notice, because I believe, it is the only example to be found in this country – it is an elastic stone from Brazil, which was presented to Mr WEIR by the late ingenius Lord GARDENSTONE – by a simple shake of the hand, it vibrates quickly from side to side like the pendulum of a clock.

A further letter in the *Edinburgh Evening Courant* of 24 July 1797 mentions 'beautiful specimens of Mineralogy'.

Inevitably, decline set in; following his death on 14 October 1797, museum management devolved upon his son Walter, but representatives of Alexander Weir offered the collection and property for sale to the town (Town Council Minutes, 3 December 1800). The offer was declined and disposal took place by auction on 11 February 1801. A revival appears to have occurred, for in 1802 the museum's last known address was 62 South Bridge Street. Again disposal was by auction, in March 1802; the whereabouts of extant specimens, if any, is unknown. A letter in the Robert Ferguson documents held within the NMS Geology and Zoology Department archives, dated 'Edinburgh, 2 November 1802' from William Burrell to Robert Ferguson, states:

> I have several very fine large specimens from Sweden, Norway, Denmark, Iceland, the West Indies, Russia and Siberia. I bought likewise the whole of Weir's Museum, with the fine specimens put in by Lord Gardenston particularly the elastic stone. Next May I propose filling up Clivers (?) Chapel (next door) as a Public Museum.

From the letter it is obvious that William Burrell was a mineral collector and dealer and he had forwarded specimens to Haidinger in Vienna. He was acquainted with Black, Hall, Hutton, Jameson, Mackenzie, Laing, Playfair and Walker. He also complained that vast numbers of foreigners for some years had collected in Scotland and fine specimens were becoming much scarcer.

Alexander Weir is interred in the Calton Hill Burial Ground, close to the northern side of Mr James Brown's ground (d.22 March 1796; Taylor, 1992).

## NINIAN IMRIE (? -1820)
### Soldier and geologist

Towards the end of the eighteenth century, Lt. Col. Ninian Imrie became a familiar figure within Edinburgh's geological fraternity, and a good friend of Professor Jameson. He served in the Royal Scots from 1768-99. During 1786-87 he toured Cyprus and Greece with Professor John Sibthorp (1758-96; Professor of Botany at Oxford) and John Hawkins FRS (1761-1841; an early mining geologist; Russell, 1954).

In 1808 Thomas Allan accompanied Imrie to view Giesecke's collection housed at Leith (see Thomas Allan; Nairn Literary Institute). Imrie and Allan purchased Giesecke's material for £40. In his cryolite account, Jameson (1809) wrote:

> Through the politeness of Colonel Imrie, who presented me with several specimens of this rare mineral I have been able to draw up a description of its external aspects.

At the request of Lord Gray (see Francis Gray), the Perthshire Literary and Antiquarian Society received (c.1827) 91 Imrie specimens, 31 specimens from Greece being self-collected. Additionally the donation included 23 Scottish minerals and 19 from Greenland, the latter group including a blue quartz-garnet rock. Although this collection is no longer traceable, the donated blue quartz-garnet rock, cryolite and sodalite are duplicated within Giesecke material at Nairn Literary Institute Museum.

Imrie became a Fellow of the Royal Society of Edinburgh in 1797 and was elected on 10 March 1810 to the Wernerian Natural History Society. During 1791 he wrote numerous letters to John Hawkins, many requesting guidance in geological matters. By the turn of the century his geological knowledge was considerable: a paper, mineralogical in nature, about the Rock of Gibraltar, its strata and caves, appeared in 1798 (Imrie, 1798). (He served at Gibraltar as an *aide-de-camp* to General Skene.) Three papers followed over the next few years, detailing geological features in the Grampians (Imrie, 1810), Campsie Fells (Imrie, 1812) and Kincardineshire to the Grampians (read 1804; Imrie, 1812a). During 1811 he served on the Royal Society of Edinburgh's committee on 'Edinburgh Geology'.

Throughout his travels Imrie collected extensively, and also obtained specimens from contacts; donations included 167 rocks and minerals to Edinburgh University (Imrie, 1817). Eighty rocks and minerals were presented by G. Govan, Esq. 'from the collection of the late Colonel Imrie' to the Industrial Museum of Scotland. Some specimens, registered under NMS G 1862.38. 1–80, are traceable, including Elba hematite and unlocalitated galena, graphite and native copper.

On 13 November 1820 Imrie died at his home, 63 Queen Street, Edinburgh.

Lord Gray, artist unknown (print, possibly 1830s). In a private Scottish collection.

## FRANCIS GRAY (1765-1842)
### Public figure, Postmaster General

Edinburgh-born Lord Francis Gray followed a public career. For many years he acted as one of Scotland's representative peers and served as Postmaster General for Scotland from 1807-10. He became a co-founder of the Perthshire Literary and Antiquarian Society and in c.1827 persuaded Lt. Col. Imrie to donate a collection of 91 rocks and minerals, including some of Giesecke's Greenland minerals, to the society (see Ninian Imrie; Nairn Literary Institute). Lord Gray gained a reputation as an agate and mineral collector. Lord Panmure purchased Lord Gray's agate collection of 875 stones at the sale of his effects, and in 1844 gifted them to Montrose Museum, where

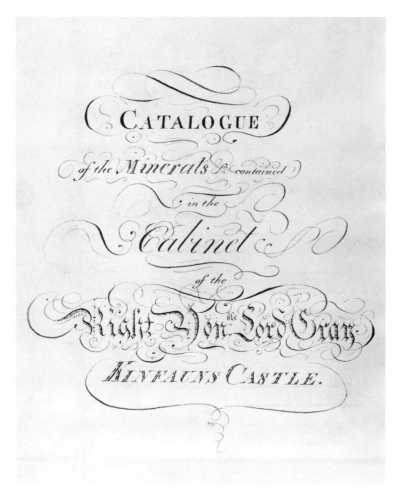

Title page to Lord Gray's
manuscript catalogue; seven
pages are watermarked 1818,
and eleven 1819. In a private
Scottish collection.

Lord Gray's mineral cabinet c.1805, constructed in ebony-inlaid walnut, originally
contained 116 drawers of which 14 are now missing. A particularly interesting feature
of the glazed doors is the narrowness (0.8cm) of the astragals, unusual for that period
of construction. The cabinet measures 2.92m high, 2.68m wide and 0.56m deep. In a
private Scottish collection.

they were publicly displayed for the first time in 1935 (*Montrose Standard*, 25 January 1935).

Lord Gray, the builder in 1820 of the present-day Kinfauns Castle, Perthshire, constructed a museum in the elaborate development (Lord Moray, *pers. comm.*, 1994). His mineral collection, together with a range of natural history objects, probably amassed close to the turn of the eighteenth century and opening years of the nineteenth century, is housed in its own large, original, glazed wooden cabinet. Some drawers contain paper base sheets watermarked 1803 or 1807. Two brown, leather-bound catalogues (21 x 13.3cm) bearing the Gray monograph in gold leaf ('G' under a coronet) relate to Lord Gray's mineral collections. One work, dated 1815 and in the same distinctive hand, which refers to the Nairn Literary Institute's 120 rare minerals, is entitled 'A Descriptive Cat-

alogue of Lord Gray's Collection of Minerals contained in the first 10 volumes of His Lordship's cabinet portatif with an appendix containing tables of hardness and specific gravity compiled for the Right Honourable Lord Gray by Lt. Colonel Imrie'. The whereabouts of the collection detailed in this catalogue is unknown.

A total of 2085 specimens, mainly minerals, is detailed in the second catalogue, which relates to extant specimens in the cabinet. Approximately 50% of the collection remains and many excellent, numbered specimens can be directly matched to catalogue entries. The lower cabinet drawers, numbers 15, 16 and 17, housed 'A Collection of Rare Minerals containing 120 Specimens presented by Lt. Col. Imrie'. It is highly probable that the collection of 120 specimens was gifted from these drawers by Lord Gray to Elizabeth, Duchess of Gordon in 1828,

Drawer 31st continued ———— 1028

770 Compact Mallachite, Shetland
771 Do ——— Do ——— Do
772 Foliated ... Do ——— Do
773 Do ——— Do with Copper Pyrites Do
774 Earthy Mallachite ——— Do
775 Do ——— Do with Copper Azure — Leadhills
776 Do ——— Do ——— Do ——— Do
777 Do ——— Do ——— Do ——— Do
778 Mallachite with Quartz, Leadhills
779 Do ——— Do ——— Do
780 Do ——— Do & Pyrites
781 Do with Grey Copper Ore, Cornwall
782 Iron Shot Copper Green ——— Siberia
783 Ditto ... with Copper Azure - Do
784 Do ——— with Quartz
785 Fibrous Phosphate of Copper, Siberia
786 Do ——— Do ——— Do
787 Do ——— Do ——— Do
788 Copper Pyrites ——— Cornwall
789 Do ——— Do ——— Do
790 Do ——— Do with Grey Copper, Siberia
791 Do ——— Do ——— Do ——— Do
792 Ruby Copper ——— Cornwall
793 Copper Ore with Quartz
794 Do ——— Do

1068

# ROBERT FERGUSON (1767-1840)
### Advocate and politician

Robert Ferguson was the eldest son of William and Jane Ferguson, who resided in the family home, Raith House, some 3km west of Kirkcaldy, Fife. At Edinburgh University Robert Ferguson's tutor was John Playfair, who inspired his pupil with a deep interest in literature and science. In 1791 he entered the Faculty of Advocates but never practised and in the following year was elected to the Royal Company of Archers.

Ten extant diaries in Robert Ferguson's hand span the period 22 August 1795, when he was in Zurich, until 4 August 1805. An entry for the last date states, 'Newcastle – I am again on my return to Raith from my visit to London'. He spent many years on the Continent and associated with aristocrats and men of letters, travelling through numerous towns and cities, including Verona, Florence, Pisa, Naples, Rome, Genoa, Berne, Leipzig, Hamburg and Dresden. He was detained in France by the Revolutionary Government and spent several years in Paris during the career of Napoleon Bonaparte. Robert Ferguson was a prisoner-of-war in France during 1803 but was free in June 1804 (Bank's letters; Dawson, 1958).

On 7 March 1805 Ferguson was elected a Fellow of the Royal Society, becoming a Fellow of the Royal Society of Edinburgh in 1806. Upon returning to Scotland, he was elected a Member of Parliament for Fife (1806-7) and remained in politics for the rest of his life. After the Reform Bill (1831) he represented the Kirkcaldy District Burghs (1831-32, and 1837) and Haddingtonshire (1835-37). During August 1837, he took the office of Lord Lieutenant of Fife, which he held until 1840. He was a staunch supporter of the Whigs. When in Paris he counted among his most particular friends Baron Cuvier (1769-1832), the famous French geologist, whom he accompanied on his geological excursions in Montmartre, Paris. Cuvier discovered fossil remains which first clearly disclosed the existence of extinct animals of former ages (Conolly, 1866).

Soon after the formation of the Geological Society of London on 13 November 1807, Robert Ferguson was elected to membership in February 1809. During 1810 he became a Trustee of the society and Vice President, remaining in the latter capacity for four years. Additionally he provided funding for Count de Bournon's (1751-1825) work which was published in 1813 (Haüy, 1813).

of which remnants still exist at Nairn (Livingstone, 1993b). The catalogue gives non-specific localities from which Lord Gray's specimens were obtained, including Derbyshire, Cumberland, Cornwall, Leadhills, Shetland, Andreasberg, Freiberg, Schellenberg, Saxony, Moravia, Transylvania, Siberia, Greenland, Norway, Brazil, Dauphiné and Vesuvius. The remnants of Lord Gray's mineral cabinet represent one of the oldest surviving collections in Scotland in private hands.

When Lord Gray died it was recorded in *The Perthshire Constitutional* (August, 1842) that he was 'Distinguished by a delicately refined taste, sound common sense and affable engaging manners, he was esteemed and respected by all who knew him'. He is buried at Fowlis, Perthshire (Speake, 1982).

On his European Grand Tour, Robert Ferguson vigorously pursued a policy of mineral purchase; he acquired a large mineral collection in excess of 4000 specimens from Baron Block of Dresden for £600. (Included in this purchase was a collection from the cabinet of Dr Titius. A catalogue of Block's collection was found amongst Robert Ferguson's mineralogical papers, NMS Geology and Zoology Department archives; see Wilson, 1994, pp. 205, 226.) His diary of 21 January 1801 states:

> I shall then possess a most complete and distinguished cabinet of minerals … with what I already possess added to it and what I shall now and then continue to pick up, I believe I shall have a very valuable collection … well worth £12 or 14 hundred pounds.

Between 1799 and the end of 1801, he spent £385-10s on mineral purchases (excluding Block's collection); the sum includes a purchase of £100 for 'Hutton's minerals' (possibly part of James Hutton's collection, for Hutton died in 1797). Robert Ferguson's collection was sited in a purpose-built room in the tower block of Raith House and housed in drawers, glazed desks and wall cases, of which the latter had several shelves. Shortly after Robert Ferguson's death Alexander Rose compiled a 19-page report, dated March 1841, regarding the collection. The following specimen descriptions, and sections of the collection, are taken from Rose's report. (The report and valuation was drawn up for probate under the Will of Robert Ferguson; letter from Raith House to J. G. Goodchild, 20 December 1899; NMS Geology and Zoology Department archives.)

Large, high-quality specimens included beryls from Siberia, tourmaline, aragonite, galena and rock crystal groups. The beryls and tourmaline were of outstanding appearance, one beryl crystal measured 16.5 x 6cm in diameter. A beautiful rose-coloured fluorite displayed octahedral form.

The report comments upon:

> Red Silver probably unparalled in any other collection. The mass is 9x6 inches, the greater part of which is covered by crystals of large size and beauty. There is also another similar specimen of Red Silver, alike in crystallisation 3x6 inches and is almost entirely a mass of ore. These are from the Hartz.

Two magnificent specimens of Kongsberg (Norway) silver were also displayed.

Various sections concentrated upon cut stones: diamonds, lapis, quartz, agates, opal, amethyst, sapphires, amber and numerous marbles. In the gem section were exhibited over 60 sapphires and rubies, either massive or cut, faceted green zircon and large cut emeralds, beryls and gem garnets. Additional highlights in this region of the collection included axinites, topaz (small cut stones), tourmalines (rubellite), chrysoberyls and olivine. Further sections concentrated upon sulphates, fluorites, carbonates, zeolites, metals and ores.

A specialist feature of Robert Ferguson's collection centred around native gold and silver:

> The collection of Gold and Silver may literally be termed Extraordinary. Of Gold of its few ores, there are above 40 specimens, presenting beautiful crystallised forms and other shapes. The Native Silver, 44 in number, are perhaps still more valuable for beauty and variety, there are likewise fine sulphurets of Muriates of Silver, about 30 specimens, and a superb collection of 30 Red Silver.

Of other species, Rose's report continues:

> There are great and beautiful varieties of Blue Carbonate of copper from Chessy and other localities and very fine malachites …. Amongst the irons is a large mass of Pallas's

meteoric Iron and several other meteorites of native Iron; other ores of Iron, ... Goethite, Phosphates and carbonates of Iron &c.

A valuable insight into the tenor and condition of Robert Ferguson's collection at his death may be gained from Rose's concluding statements:

It is but justice to add that this splendid collection is at present seen under very great disadvantages. The show specimens in order to be properly displayed, would require about double the space they now occupy. The specimens are crowded together, and many fine things must escape observation. Several of the large specimens above the glass cases, being unprotected, are covered with dust, by which their beauty is concealed, and their quality can be judged of by such persons only as have some experience of minerals.

The Collection in the drawers, too, is seen in a very unfavourable condition. Arrangement is not, in general, adhered to. From careless movement of the drawers, the specimens are often huddled together, and dust conceals their beauty. Many are not named, of those which are, have the names in different languages, in English, French, Italian and German, and not always in very legible characters. These observations are made with the view of presenting a false estimate being formed of the real value and quality of this Collection. [Numerous labels in the collection bear symbols, commonly used at that period, to denote mineral chemistry *eg* ♃ lead, ♀ copper; Wilson, 1994, p. 60.]

Rose concluded his report by valuing the whole collection at £1646.

Robert Ferguson clearly possessed an interest in crystal morphology, for an extant series of crystal models, carved from white gypsum, is similar to those figured and described in *Cristallographie ou Description des formes propres à tous les corps du regne mineral. Avec figures et tableaux synoptiques de tous les cristaux connu* by Romé de L'Isle (1783). Furthermore, a wainscot cabinet housed 600-700 named, wooden crystal models, several of which were composite and made of several pieces of wood. This collection of crystal models was arranged to illustrate the crystallographic systems of Haüy (1809). Surviving mineralogy books, many inscribed with Ferguson's signature, include *Traité Élémentaire de Minéralogie*, Brongniart (1807); *Traité de Minéralogie*, de Bournon (1808); *Theory of the Earth*, Cuvier (1815); *Treatise on Mineralogy*,

INSCRIBED

TO

ROBERT FERGUSON, Esq.

OF RAITH,

FELLOW OF THE ROYAL SOCIETIES OF LONDON AND EDINBURGH, MEMBER OF THE GEOLOGICAL SOCIETY, OF THE WERNERIAN NATURAL HISTORY SOCIETY, &c.

IN TESTIMONY

OF HIS

DISTINGUISHED TALENTS AS A MINERALOGIST,

BY HIS

FAITHFUL AND SINCERE FRIEND

THE AUTHOR.

Haidinger (1825); together with works by Jameson, Allan and Bergmann.

The extant collection contains around 5000 specimens. Many fine minerals remain, including a Russian aquamarine crystal 16.5cm long and 5cm across prism faces. Excellent fluorite specimens range from inky-blue cubes (3cm edge) and a deep salmon-coloured octahedron (3cm edge) in matrix, to classic green Weardale material. Silver minerals are represented by Kongsberg, Norway, with native silver loops 4cm across on quartz, argentite octahedra (to 1cm) that form a base from which a 7cm column of smaller octahedra projects, and an outstanding pyrargyrite which completely covers a crudely rectangular area some 20 x 5cm, with crystals to 1.5cm. Andreasberg (Harz mountains) trigonal calcite prisms which attain 2.5cm across; a 17cm high by 16 x 14cm Siberian amazonite crystal, azurite from Chessy, France, a group of polished framed chalcedonics/agates, and around 1000 carved crystal models after Romé de L'Isle's (1783) figures, reflect a fraction of this once-great, versatile collection. A 1.55kg

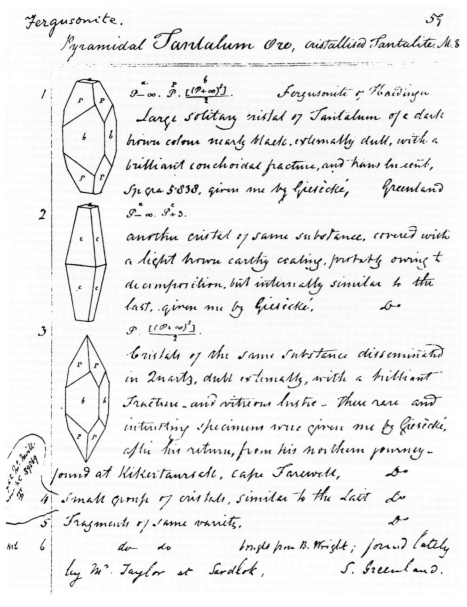

Fergusonite entry from the Allan-Greg catalogue, note Giesecke – Greenland – Haidinger association. Entries are by Thomas Allan and lists of forms by Robert Allan, c.1825. Courtesy of Mineralogy Department, Natural History Museum, London.

portion of the L'Aigle meteorite remains in the collection. This specimen, part of a shower of stones, fell after the appearance of a fireball followed by a detonation, at 1.00pm, on 26 April 1803. The detailed report of the phenomena (Biot, 1806) first established beyond doubt that the fall was from outer space. As a measure of the collection's quality, the residue, including a number of antiquarian books, was valued at £112,000 by Brian Lloyd during 1997. (Compare Alexander Rose's 1841 valuation

at £1646 (with the contemporary equivalent being £74,684; S. Langdown, *pers. comm.*, 1997.) The collection was purchased in 1997 by Brian Lloyd, mineral dealer, and during May 1998 a significant portion of it was purchased by Trinity Mineral Company and The Arkenstone of La Jolla, California, for sale via the internet. Sadly the opportunity to undertake detailed research on this historically significant collection has now been lost. For further details of Ferguson's travel journals, extant collection and photographs of some of his specimens, see Lloyd and Lloyd (2000).

Robert Ferguson, distinguished as a politician, became an eminent mineralogist of his era. He communicated with foremost mineralogists of whom may be mentioned Haüy, Crichton, Heuland, Chenevix, Wollaston, Allan, Jameson and Count de Bournon. His house became a centre of intellectual and artistic society, for it contained 'a cabinet of minerals, which, for richness and extent, is surpassed by few private collections of this sort in the Kingdom' (Murray, 1845). He was revered by leading mineralogists and served on a committee formulated to appraise Charles Francis Greville's (1749-1809) mineral collection (20,000 specimens) as to its value and suitability for purchase by the British Museum. The committee assessed the collection at £13,727 and published a document 'Report to the Committee of the Honourable House of Commons, on the Petition of the Trustees of the British Museum; respecting the purchase of Mr Greville's Collection of Minerals, London, May 9th 1810'. (The report was signed by William Babington, L. Compte De Bournon, Richard Chenevix, Humphry Davy, Robert Ferguson, Charles Hatchett and W. H. Wollaston.)

A specimen in the cabinet of Thomas Allan, which contained a dark brownish-black mineral occurring as either embedded groups or single crystals in white quartz, was found by Giesecke at Kikertaursuk, Cape Farewell, Greenland. Brooke (1823) and Phillips (1823) ascribed a rectangular four-sided prism to the mineral, which they classified as allanite. Haidinger (1826) recognised the mineral as a new species which, at the suggestion of Thomas Allan, he named fergusonite 'in honour of a gentleman too well known to the mineralogical world at large, and to the members of this society (Royal Society of Edinburgh) in particular, to require, in the present place, a more detailed acknowledgement' (Haidinger, 1826). (Currently there are six species of fergusonite based upon

monoclinic (3), tetragonal (3) and dominant rare-earth characteristics [Fleischer and Mandarino, 1995].) Fergusonite is not
named after the Scottish physician Robert Ferguson (1799-
1865) as incorrectly reported in Dana's *System of Mineralogy*
(Palache *et al.*, 1966, 7th edn, vol. 1, p. 762). This error is perpetuated by Mitchell (1979), although corrected later (Mitchell,
1988).

Robert Ferguson was a man of great compassion. Of
generous mind, he was an undeviating friend of civil and
religious liberty. As a landlord he possessed many qualifications
which endeared him to his numerous tenantry. His tenants not
only respected him; he was beloved by them. Even to the last his
charitable feelings were uppermost in his mind. Only a week or
so before his death, his attention was drawn to a destitute who
had partaken of his bounty and with his own hand he instructed
his agents to provide for the individual's necessities (Conolly,
1866). On 3 December 1840 Robert Ferguson died at his home
in Portman Square, London. His obituary in *The Globe* is
quoted by Lloyd and Lloyd (2000).

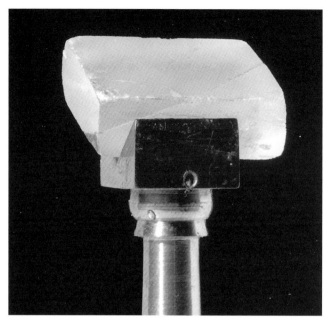

Nicol prism (16 x 7 x 7mm, supported on a brass pillar 55mm high) made
by William Nicol at the age of 80, now in the Royal Museum, Edinburgh,
NMST 1856.54. Courtesy of Trustees of the National Museums of Scotland.

## WILLIAM NICOL (c.1771-1851)
Inventor of the Nicol prism,
lecturer, natural philosopher and collector

William Nicol's early life is shrouded in obscurity and his exact
date of birth is unknown, for he was born, probably in Humbie
outside Edinburgh, prior to official registration of births and
deaths within Victorian Scotland in 1855 (Morrison-Low,
1992). Around the age of 15 in 1786, he became the blind Dr
Henry Moyes's (1749-1807) assistant. Moyes was a remarkable
public lecturer on science, and after his death in December 1807
(at Doncaster while on a lecturing course), Nicol eventually
acquired his manuscripts and apparatus. Nicol returned to
Doncaster in 1808 to complete Moyes's lecture course; this was
the commencement of his itinerant lecturing career (Morrison-
Low, 1992).

William Nicol associated with the Edinburgh scientific
community and became a Fellow of the Royal Society of
Edinburgh in 1838. He commenced publishing *c*.1826 and
throughout the following years twelve papers appeared in the
*Philosophical Journal* on such diverse subjects as greenockite,
fluid inclusions and fossilised wood. Prior to 1826 he had

accumulated an appreciable mineral collection which was
consulted by others with similar interests, including Sir David
Brewster. He published a paper describing the use of 'Iceland
[calcareous] Spar' for generating only one image – the Nicol
prism (Nicol, 1829). Consequences of the discovery were not
immediately realised, although subsequently it had a profound
impact upon optical mineralogy, petrology and palaeobotany.
Ten years later he announced details of an improved version
(Nicol, 1839).

William Nicol's mineral collection (Jameson, 1829) is
mentioned in his Will (SRO 70/4/17.959), which states:

> Assign and make over to the aforesaid Alexander Bryson
> [friend and fellow geologist] and his heirs ... all my books,
> Philosophical apparatus, minerals, fossils, shells, optical
> instruments.

Nicol's work was probably known by Brewster who used
minerals from Nicol's collection; it seems probable that in a
closely knit scientific community two people with common
interests would be in contact (Brewster produced polarised light

via plates of cemented agate and tourmaline). Thin-section techniques, first used in Edinburgh and developed by Nicol who was the first to use minerals mounted directly on glass instead of wood (Morrison-Low and Nuttall, 1984), were utilised by Henry Clifton Sorby (1826-1908) in his early geological papers, which passed unappreciated by the geological world at that time. Although Sorby independently developed thin-section methods (Eyles, 1951), he presented a paper to the Geological Society on the microscopic structure of fluid inclusions in rock-forming minerals, in which he referred to Alexander Bryson's collection (Sorby, 1858). Bryson and Nicol's joint collections were sold on Bryson's death: part of these collections are now in the mineral department of the Natural History Museum, London (Edwards, 1951).

William Nicol died at his home, 4 Inverleith Terrace, Edinburgh, on 2 September 1851. A white marble wall-mounted memorial tablet in Warriston Cemetery, Edinburgh, bears the inscription: 'In Memory of William Nicol 1766-1851 Inventor of the Polarizing Prism everywhere known by his name.'

## THOMAS THOMSON (1773-1852)
### Professor of Chemistry and mineralogist

Thomas Thomson, the first Regius Professor of Chemistry at the University of Glasgow, was born on 12 April 1773 at Crieff, Perthshire. In 1786, in his thirteenth year, he was sent to the burgh school in Stirling, acquiring a classical education (Thomson, 1853, *ie* R. D. Thomson, 1810-64, nephew of Thomas Thomson). He obtained a bursary at St Andrews University which enabled him to study for three years. During 1795 he commenced medical studies at Edinburgh University, attending chemistry lectures delivered by Professor Joseph Black. Thomas Thomson graduated MD in 1799 and during this early period contributed articles to the third edition (1788-97) of *Encyclopaedia Britannica* on chemistry, mineralogy, and 'Vegetable, Animal and Dyeing Substances'. A course of lectures in Edinburgh instituted by Thomson in 1800 delivered practical instruction in chemistry. The lectures and practical classes ran until 1811, in a chemical laboratory thought to have been the first in the United Kingdom opened for the purpose of instruction. On 28 March 1811 he was elected a

Thomas Thomson, mezzotint engraving 1853, 52 x 38cm, by James Faed after a painting by J. Graham Gilbert. © Hunterian Art Gallery, University of Glasgow.

Fellow of the Royal Society of London. Thomson became a member of numerous scientific societies in Great Britain, and also of many international academies of sciences.

In the article on mineralogy for *Encyclopaedia Britannica* Thomson became the first person to introduce the use of symbols into chemical science (Thomson, 1853). He arranged minerals into genera, the first genus being A (alumina) which contained two species: topaz and corundum. The second genus is AMC, *ie* spinel, designated from alumina, magnesia and chrome-iron. Thomson was thus the first chemist to bring mineralogy systematically within the sphere of chemistry. Berzelius in 1814 adopted, with modifications, Thomson's system of symbols (Thomson, 1853). (In 1804 Thomson introduced his nomenclature of oxides – protoxide, trioxide, *etc* – and later used Greek and Latin prefixes for the numbers of atoms of the radicals and oxygens respectively.) During 1817 he was appointed Lecturer on Chemistry at the University of Glasgow, and Professor from 1818-52. Soon after his appointment he commenced research into the atomic constitution of chemical compounds. His atomic numbers, derived from a vast series of experiments, resulted in atomic weights being multiples, by a whole number, of the atomic weight of hydrogen. After the publication of this work, he devoted himself to the inorganic kingdom of nature:

> purchasing and collecting every species of mineral obtainable, until his museum, which he has left behind him, became not only one of the noblest mineral collections in the kingdom, but a substantial monument of his taste and of his devotion to science. (Thomson, 1853)

Results of his investigations into minerals appeared in a two-volume work: *Outlines of Mineralogy, Geology, and Mineral Analysis* (Thomson, 1836). This work contains an account of nearly 50 'new minerals' which he had discovered in a period of little more than ten years. Of Thomson's 'new minerals' the following are currently accepted species: allanite, sodalite and ferrotantalite (Thomson, 1810, 1810a, 1836a). Bytownite was also named by Thomson (1836) after Bytown (now Ottawa), Canada (Steacy and Rose, 1982), although recently it has been downgraded from full species status to being a member of the albite-anorthite series (Fleischer and Mandarino, 1991, 1995). He also launched the name 'peristerite' (Thomson, 1843).

The *Catalogue of Scientific Papers of the Royal Society* (1871) lists 201 papers by Thomson, of which 41 are devoted to mineralogy, ores and geology. Thomson (1840), in a comprehensive paper describing the mineralogy of the Glasgow environment, remarks:

> Perhaps there is no part of Great Britain, not even excepting Cornwall, or the mining counties of the north of England, so rich in mineral species as the neighbourhood of Glasgow, if we include under that appellation Lead Hills and Wanlock Head, and the ranges of mountains on both sides of the Clyde, constituting the Kilpatrick hills, and the ranges of hills behind Greenock and Port Glasgow, extending as far as Kilmacolm.

The constitution of stellite seems to be
    4 atoms bisilicate of lime,
    1 atom bisilicate of magnesia,
    1 atom silicate of alumina,
    $2\frac{1}{2}$ atoms water.
The formula is $4CS^2+MS^2+AlS+2\frac{1}{2}Aq.$

Sp. 2. *Thomsonite*.
Mesotype of Hauy in part—needle zeolite of Werner in part.

This mineral was first recognised as a peculiar species by Mr. Brooke, who determined its characters. It occurs at Lochwinnoch, and near Kilpatrick, both of which localities are within a few miles of Glasgow, and in both places it is imbedded in an amygdaloidal rock.

Colour snow-white.

Occurs usually crystallized; but there is a brown coloured amorphous mineral found at Ballimony, in Ireland, and having a specific gravity of 2·289, which proved upon analysis to be an impure variety of the same species. The primary form of its crystal, as Mr. Brooke has shown,* is a right rectangular prism, with a square base. The usual crystal is that represented in the margin, in which P, M and T are the primary faces. The faces *a, b* are tangents to the lateral edges of the prism, and the faces *c* are the result of a

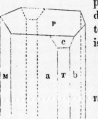

decrement of one row on the edge of the terminal plane. The height of the prism is equal to about four times its breadth.

M on a 135° 30'
a on a 90° 40'

Structure foliated; cleaves readily parallel to the faces M, T; fracture uneven.

Lustre vitreous, inclining to pearly.

From transparent to translucent.

Hardness 4·75; brittle.

Specific gravity from 2·29 to 2·36966.

Before the blowpipe it swells up like borax, and becomes opaque and snow-white, but does not melt. When exposed to a red heat it becomes opaque, very white and shining like enamel. The edges are rounded, but it does not altogether lose its shape.

The following table exhibits the constituents of this mineral, according to the best analyses hitherto made:

* Annals of Philosophy, xvi. 194.

In the 1840 paper chemical names are stated and analyses given for linarite, leadhillite and susannite, and chemical names for pyromorphite and vanadinite. The remainder of the paper deals mainly with zeolites and greenockite.

Thomson's mineral collection, upon which he drew extensively for his two-volume work published in 1836, was prior to

Description of thomsonite from *Outlines of Mineralogy, Geology, and Mineral Analysis*. Thomson (1836).

Species

10. **Hydrous trisilicate of alumine,** $Al^3\dot{S}+5Aq.$
11. **Bucholzite,** $Al\dot{S}.$
12. **Gilbertite,** $7Al\dot{S}+(\frac{4.75}{10}Cal+\frac{5}{10}Mg+\frac{2.25}{10}f)\dot{S}^2+1\frac{1}{2}Aq.$
13. **Hydrous bucholzite,** $5Al\dot{S}+1Aq.$
14. **Halloylite,** $2Al\dot{S}+Al\dot{S}^2+4Aq.$
15. **Pholerite,** $1\frac{1}{2}Al\dot{S}+Aq.$
16. **Worthite,** $5Al\dot{S}+Al\bar{A}q.$
17. **Cyanite,** $Al^{1\frac{1}{4}}\dot{S}.$
18. **Allophane,** $2Al\dot{S}+Al^2\dot{S}+10Aq.$
19. **Tuesite,** $3Al\dot{S}+2Al\dot{S}^{1\frac{1}{2}}+3Aq.$
20. **Nacrite,** $Al\dot{S}^2.$
21. **Fullers' earth,** $Al\dot{S}^2+2Aq.$
22. **Davidsonite,** $Al\dot{S}^{2\frac{1}{2}}.$
23. **Lenzinite,** $Al\dot{S}+1Aq.$
24. **Quatersilicate of Alumina,** $Al\dot{S}^4.$

### III. *Double anhydrous aluminous Salts.*

Sp.
1. **Cryolite,** $Al\dot{F}l+N\dot{F}l.$
2. **Topaz,** $3Al\dot{S}+Al\dot{F}l.$
3. **Pycnite,** $6Al\dot{S}+Al\dot{F}l^2.$
4. **Ambligonite,** $2Al^2\dot{P}h+L^2\dot{P}h.$
5. **Fibrolite,** $2Al\dot{S}+Al^2\dot{S}.$
6. **Nepheline,** $3Al\dot{S}+N\dot{S}.$
7. **Sodalite,** $2Al\dot{S}+N\dot{S}.$
8. **Idocrase,** $Al\dot{S}+Cal\dot{S}.$
9. **Grossularite,** $Al\dot{S}+Cal\dot{S}.$
10. **Melanite,** $Al\dot{S}+(\frac{23}{46}Mg+\frac{9}{46}f+\frac{8}{46}Cal+\frac{6}{46}mn)\dot{S}.$
11. **Garnet,** $Al\dot{S}+f\dot{S}$ and $Cal\dot{S}+f\dot{S}.$
12. **Essonite,** $\underline{f}\dot{S}+4Cal\dot{S}+4Al\dot{S}.$
13. **Brown manganese garnet,** $Cal\dot{S}+Al\dot{S}+\underline{f}\dot{S}+\underline{mn}\dot{S}.$
14. **Pyrope,** $10Al\dot{S}+5(\frac{10}{23}Mg+\frac{8}{23}Cal+\frac{5}{23}Chr)\dot{S}+3(\frac{3}{4}f+\frac{1}{4}\underline{mn})\dot{S}^2.$
15. **Zoisite,** $2Al\dot{S}+Cal\dot{S}.$
16. **Meionite or scapolite,** $2Al\dot{S}+Cal\dot{S}.$
17. **Prehnite,** $2Al\dot{S}+Cal\dot{S}^{1\frac{1}{2}}+\frac{1}{2}Aq.$
18. **Anhydrous scolezite,** $3Al\dot{S}+Cal\dot{S}^3.$
19. **Iolite,** $9Al\dot{S}+3Mg\dot{S}^2+f\dot{S}.$
20. **Hydrous iolite,** $3Al\dot{S}+1(\frac{3}{4}Mg+\frac{1}{4}f)\dot{S}^2+2Aq.$
21. **Staurotide,** $4Al\dot{S}+f^6\dot{S}.$
22. **Gehlenite,** $3Al^{1\frac{1}{2}}\dot{S}+3(\frac{8}{9}Cal+\frac{1}{9}f)^{1\frac{1}{2}}\dot{S}+Aq.$
23. **Trollite,** $2Al\dot{S}^2+1\frac{8}{11}Mg+\frac{2}{11}K+\frac{1}{11}f)\dot{S}^2+\frac{1}{2}Aq.$
24. **Fahlunite,** $3Al\dot{S}+1\frac{6}{10}Mg+\frac{5}{10}f+\frac{5}{10}mn)\dot{S}^2+2Aq.$

1857 in the possession of Dr R. D. Thomson at St Thomas's Hospital, London. The collection 'has been munificently accommodated by the authorities of that Royal Institution' (anon., 1857). (Thomas Thomson's daughter married her cousin R. D. Thomson, who was Professor of Chemistry at St Thomas's Hospital.) The English mineralogist John Frederick

Calvert (d.1897) purchased Thomson's collection, culled it, then sold the remainder to the Australian Victoria Mines Department between 1861 and 1866. The purchase price is unknown. The Simons (1866) catalogue of foreign minerals in the department reveals that the purchase was not the full extent of Thomson's collection. Specimens listed in the Simons catalogue were transferred from the Victoria Mines Department to the Industrial and Technological Museum (founded in 1870 and now renamed the Science Museum of Victoria) in 1887. The Thomson Collection is now in the mineral collections of the National Museum of Victoria. Approximately 340 Thomsonian specimens, mostly minerals, are extant; all are approximately 10 x 8cm, although no original labels survive (Birch, *pers. comm.*,1995).

To strangers Dr Thomson was somewhat remote initially, yet all who knew him intimately unanimously agreed that his kindhearted disposition, combined with truthfulness, always came to the fore. A most friendly and benevolent man, he contributed to numerous charitable institutions within Glasgow. Thomas Thomson died on 2 July 1852 at Kilmun on the shore of Holy Loch, near Glasgow. He was interred in a private burial ground on 8 July in the Necropilis, Cathedral Square, Glasgow.

## THOMAS BROWN (1774-1853)
### Medical doctor, mineral and fossil collector

Thomas Brown, the fourth youngest of five children and one of four sons and a daughter, was born into the medical profession. His parents, Thomas and Martha Brown, resided at Langside in the parish of Cathcart, Renfrewshire. His father had been a successful London surgeon who returned to Glasgow and entered banking. Thomas Brown (son) studied at Edinburgh University under Daniel Rutherford, the Professor of Botany and Medicine. In May 1799, he was appointed substitute Lecturer on Botany at Glasgow University by James Jeffray, to cover for him in the summer lectures. A permanent appointment followed a year later, although Brown's salary had to be met from fees as no claim on college funds could be made. Attendance records suggest he was a popular lecturer, for he continued in post until 1815 when he resigned. He became

a well-known and popular doctor in Glasgow. In 1829 Thomas Brown succeeded his 'cousin-german' Nicol Brown to the estates of Waterhaughs and Lanfine, Ayrshire. The following year he was elected a Fellow of the Royal Society of Edinburgh.

During the spring of 1803 Thomas Brown acquired, by donation, a collection of some 200 Egyptian and Indian minerals from Major Thomas Wilson. He records field-collecting as early as 1816, when he found 'blende' (later identified as greenockite) from Kilpatrick. Brown gradually accumulated a large mineral collection housed in some 260 drawers, and a fossil collection which occupied 77 drawers. He commenced a catalogue (Hunterian Museum archives) of his minerals at Lanfine on 14 November 1832, which is enlightening regarding his mineral activities: he bought and exchanged minerals until two years before his death. Interactions occurred with numerous collectors, including among others Dr John MacCulloch, Sir William Hooker, Professor William Buckland, Gideon Mantell, William Nicol, Archduke Maximilian, Alexander Rose, Thomas Allan, Henry Heuland, Dr Traill, Sir Charles Giesecke, William Somerville, Dr Anderson (of Leith),

Dr Alexander Crichton, Robert Hay Cunningham, Alexander Bryson, Professor M. F. Heddle and Krantz of Berlin.

In a letter of 1 May 1835 (EDU Special Collections Gen. 1996/8/13) by T. Jameson Torrie to Professor Jameson, the writer enthusiastically praises Brown's collection:

> My opinion of our Scotch minerals has really been considerably elevated by the examination of Mr. Brown's series of Zeolites, Prehnites, Leads, Agates etc. From the faint recollection I have of Edington's collection I should think that though it may contain some finer than Heulandites, Prehnites, Analcimes etc. yet that as a whole Mr. Brown's Scotch minerals are superior.

Of Brown's collection, Heddle (1893) remarks:

> About the year 1849 I had the pleasure of making the acquaintance of Dr. Thomas Brown of Lanfine, Ayrshire, then in about his seventy-seventh year. Dr. Brown had then far and away the largest collection of minerals in Scotland; he was the contemporary of Macculloch, Thomas Allan, Fergusson of Raith, Sir Charles Geseikie, Thomas Thomson, and Robert Jamieson; – that coterie which was the parent of British mineralogy.

*Above, left:* Thomas Brown by Colvin Smith (1795-1875), oil on canvas 1840, 127 x 102cm. © Hunterian Art Gallery, University of Glasgow.

*Above:* Entries from Thomas Brown's catalogue, in his own handwriting. Courtesy Hunterian Museum, University of Glasgow.

33

On 23 July 1851, Alexander Rose valued Brown's collection of approximately 5500 minerals and 1600 fossils at £2001, which included £250 for the fossil collection (Hunterian Museum archives; in today's terms approximately £90,000). Highlights included:

> A very fine specimen of Emerald, finely crystallised, in Limestone £15, 20 Specimens of Vanadiate of Lead; an unique series £40, 21 Specimens of Red Stilbite, a most magnificent Collection – perhaps unparalled £30, 11 Specimens, some with several crystals of Greenockite of uncommon size and splendor £25, and Meteoric Irons and meteorites £39.

Thomas Brown added a note to the valuation that:

> These minerals and fossils are contained in 320 drawers (some drawers have double numbers) as well as in a Show cabinet &c. &c. – Lanfine July 26 1851.

Following Brown's death, his daughter Martha Brown, through an agent, Robert Gairdur, donated half of her father's collections to Glasgow University Hunterian Museum in 1873, 1886, 1896 and 1897, and the remaining half to Edinburgh University in 1874-75. During 1886, she presented the National Museums of Scotland with 220 cut and polished agates, mainly Scottish, from her father's collection (NMS G 1886. 515. 1-220).

## ROBERT JAMESON (1774-1854)
Professor of Natural History,
Father of Modern Natural History and
eminent mineralogist

Robert Jameson, the third son of a wealthy soap boiler, was born in Leith on 11 July 1774, and attended Leith Grammar School and Edinburgh University. During his early years no evidence emerged of the outstanding academic career that followed, for he became a notorious truant. At the Grammar School 'he made such progress as boys who have no particular love of letters', and one of his father's servants 'had more than once to accompany him by force to the High School' (Jameson, 1854, ie Laurence Jameson 1813-67, nephew of R. Jameson).

Early maritime ambitions were driven by a strong desire to study nature in faraway places, an ambition kindled by reading Defoe's *Robinson Crusoe*, Boreman's *History of the Three Hundred Animals* and Cook's *Voyages*. Two profound influences, inborn and environmental, acted upon the young Jameson and undoubtedly affected his career development. The first, a profound interest in nature, saw Jameson spending his spare time collecting plants and animals. The proximity of the port of Leith formed the second major influence; world-wide trade entered through the port, bringing foreign objects for his collection to augment those acquired along local beaches.

However, Jameson was appointed assistant to John Cheyne, a surgeon in Leith, and aspirations of a maritime life were dispensed with. He struck up an acquaintance with Dr James Anderson (1739-1808) of Leith, also of Aberdeen University,

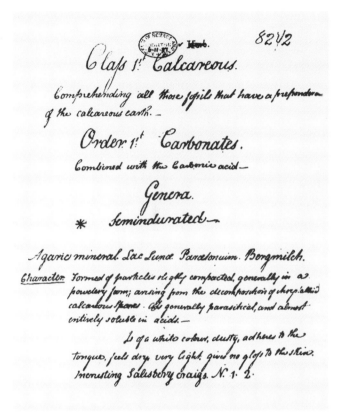

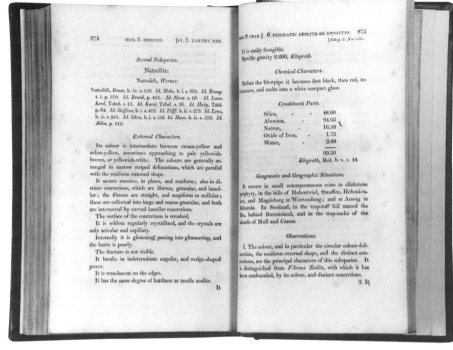

*Above, left:* Jameson's catalogue of fossils (minerals) 1797, entry for first order ie carbonates, written by Jameson. NMS Library archives, catalogue no. 8272.

*Above:* An illustration of text from *A System of Mineralogy* (Jameson, 1820, vol. 2).

an agriculturalist, writer and student of various sciences including mineralogy. Dr Anderson translated *Contributions to a Mineralogical Description of Landeck* (1796) by the eminent and illustrious German geologist and mineralogist L. von Buch (1774-1853) and Abraham Gottlob Werner's (1749-1817) *Theory of the Formation of Veins*, works which engendered a mineralogical bias in Jameson's geological outlook. Between 1789 and 1796 Jameson attended classes in 'Practice of Medicine, Botany and Clinical', and additionally lectures in botany, natural history and medicine (Sweet, 1963). At university he studied chemistry under Professor Joseph Black (Professor from 1766-95/6). During 1792 he studied natural history and enrolled for Professor John Walker's class in spring 1793. In the same year, on 12 August, Jameson sailed from Leith to London to visit relatives, hospitals, zoos, museums, the Linnean Society and eminent scientists, including Sir Joseph Banks, President of the Royal Society 1778-1820 (Sweet, 1963). Numerous natural history and geological specimens were collected, acquired or purchased during this two and a half-months' trip.

In Edinburgh's environs, accompanied by Professor Walker, Jameson embarked on dredging expeditions down the Firth of Forth, and Laurence Jameson (1854) records as 'a few' 72 'zoological treasures' from Jameson's notebook dated 1794.

At this stage Robert Jameson was firmly set on a natural history/geology course, for in 1794 he studied the geology, mineralogy, botany and zoology of the Shetland Islands. The Isle of Arran, geologically unknown at that time, then attracted Jameson's attention. The results of these island visits were published in his first book, dedicated to the Rev. Dr Walker, *An Outline of the Mineralogy of the Shetland Islands, and of the island of Arran, with an Appendix; containing observations on Peat, Kelp, and Coal* (1798). On 28 June 1796 the English mineralogist Charles Hatchett (1766-1847) had arrived in Edinburgh (Raistrick, 1967). On 4 July, Hatchett met Dr Walker, who offered to help with his proposed tour of the western Highlands. Jameson, entrusted with the task, guided Hatchett on a two-week tour of the Highlands. The summer of 1798 saw Jameson exploring the Hebrides and Western Isles.

During 1799 he investigated the Orkneys and revisited Arran. Information gathered on these and other sorties led to Jameson's next major publication, *Outline of the Mineralogy of the Scottish Isles* (Jameson, 1800). The second volume of this work Jameson dedicated to Charles Hatchett.

In the 1790s Jameson also commenced writing a *Catalogue of Fossils* (*ie* minerals; watermark 1794), which is housed within the National Museums of Scotland archives (NMS catalogue no. 8272). Catalogue entries postdate his London and Shetland trips, 1793 and 1794 respectively, and Charles Hatchett's visit in 1796, but precede Jameson's visit to Richard Kirwan (1733-1812) in Ireland, in 1797. Entries do not follow entirely Professor Walker's classification *Classis Fossilium* (1789), printed by Jameson (1820), but appear to be arranged to suit his own collection.

Geological thinking in Edinburgh during Jameson's educationally formative years was dominated by James Hutton until the latter's death in 1797. At this period the teachings of the great German geologist and mineralogist A. G. Werner of Freiberg, Saxony, reached Edinburgh. Jameson avidly followed Werner's teachings; indeed, he expounded Werner's theories during the closing years of the eighteenth century. As a disciple of Neptunistic doctrines, Jameson visited Freiberg to study, at his own cost, under Werner at the Bergakademie. He registered as student number 552 on 19 September 1800 (Hofmann, *pers. comm.*, 1993). For nearly two years he studied geology and mineralogy, and worked in the mines under rules formulated by his master, undergoing the same drudgery and hard work as the miners themselves. Werner entered Jameson in his annual report on 3 June 1801 but not in the following year; in the annual reports of the professors, concerned with the academic work of the completed terms, only progress of the 'Beneficiaten' (students supported by the State) was commented upon. Reports on students studying at their own cost were left to the discretion of the Professor (Hofmann, *pers. comm.*, 1993). Two manuscript volumes of notes on Werner's lectures are housed within the National Museums of Scotland archives. One volume contains notes in English and German, in Jameson's handwriting, the second was compiled by amanuenses and sold to students (Sweet, 1976).

On the death of his father in 1802, Jameson returned to Edinburgh, where he assisted the now very ill and blind Dr Walker with his classes. Hatchett and Kirwan had written to

Sir Joseph Banks in 1801, indicating Jameson's eminent suitability for the Regius Chair of Natural History. Not surprisingly with this calibre of support, and following Walker's death on 31 December 1803, Jameson was elected on 23 January 1804 to the Chair, which he occupied with great distinction for 50 years.

## Jameson, the Mineralogist

At the beginning of the nineteenth century, Jameson's prolific talents resulted in the appearance of numerous mineralogical texts. During 1805 two works were published: *Treatise on the External Characters of Minerals* and *A Mineralogical Description of the County of Dumfries* (Jameson, 1805a, and 1805b), the latter forming the first part of an intended series embracing all Scotland. Subject matter in the first work includes descriptions of colour, streak, fracture, lustre and morphological crystallography, supplemented by two plates detailing 35 crystal drawings. During 1802, after his father's death, and while residing with his younger brother the Rev. Andrew Jameson of St Mungo, Dumfriesshire, Robert Jameson had undertaken fieldwork in that county, the results of which culminated in the appearance of the second work. This publication was dedicated to

> A. G. Werner, The Father of Mineralogy; and Richard Kirwan, whose indefatigable exertions have contributed so greatly to the advancement of mineralogy in the British Empire. This volume is dedicated by their obedient servant. October 10, 1804.

Mines at Wanlockhead in the county, and nearby Leadhills (Lanarkshire) provided Jameson with excellent opportunities to study veins. He describes the Susanna Vein (Leadhills) thus:

> Its usual breadth is about four feet: several years ago it was in one place about fourteen feet wide, but this was owing to a partial enlargement, or what the miners term a belly .... The Belton-grain vein, which was at that time but lately opened .... Its width is from six to eight feet.

Jameson quotes Klaproth's (1795-1802) analyses for cerussite, anglesite and pyromorphite in his descriptions of white lead ore, vitriol and green lead ore from the mines. Werner's teachings were not lost on Jameson, for in this work we see frequent reference to Wernerian doctrines.

A three-volume work (Jameson, 1804-8), *A System of Mineralogy*, appeared during 1804 (vol. 1), 1805 (vol. 2) and 1808 (vol. 3). The first volume, dedicated to Col. Alexander Dirom, Quarter-Master General for Scotland, features on the front cover of *Mineralogical Record* 1990, no. 1. The second volume does not carry a dedication, whereas Sir Joseph Banks, Baronet, is the recipient of the dedication in the third volume. The first two volumes contain, *in toto*, 198 crystal drawings executed mainly by Count de Bournon and Abbé Haüy. So great was the demand for this three-volume series that a second edition appeared in 1816, with a third in 1820. For the third edition Jameson desired a print run of 1250 copies which exceeded his publishers aspirations. The publisher A. Constable, of Archibald Constable and Company Edinburgh, wrote to Jameson on 19 March 1818 stating:

> The former edition was so long in the [sic] that we had carried a considerable expense for paper. Engraving [sic] for years before its publication, but in return I think its right to say that the sale has been more rapid than we had calculated on – the old plates will I presume answer for the new edition? (NLS Constable letter book, fol.790)

For the latter work Jameson largely adopted the arrangements of Mohs:

> It is founded on what are popularly called the External Characters of Minerals, and is totally independent of any aid from Chemistry. This, which may be termed the *Natural History Method*, I have always considered as the only one by which minerals could be scientifically arranged, and the species accurately determined.

In volume 1 of the 1820 edition Jameson, with Mohs's permission, reprinted the latter's hardness scale, which was also published for the first time, in the same year, in Edinburgh (Mohs, 1820). A two-volume work, *Mineralogical Travels through the Hebrides, Orkney and Shetland Islands, and Mainland of Scotland*, had appeared in 1813; a retitled, reissue of his earlier work (Jameson, 1800).

For his publications Jameson drew extensively upon his own material, field observations and authoritative works of the time, of which a brief selection included Saussure (1779-96), Kirwan (1794-96), Haüy (1801, 1809, 1813), Brochant (1803), Ludwig (1803, 1804), Mohs (1804), Leonhard *et al.* (1806),

Hausmann (1809, 1813), Klaproth (1795-1815), Oken (1813) and Aikin (1815). Mineralogy had been rapidly transformed over a very short time span from a 'curiosity' to a respectable subject – the situation aptly epitomised by Jameson (1816):

> Mineralogy, although a science of comparatively modern date, has, within a short period of time, made rapid advances. It was first successfully cultivated in Germany. In Great Britain, so distinguished in all the other sciences and arts of life, it was, until lately, almost entirely neglected. Now, however, it has become with us a subject of general interest and attention, and, like Chemistry, is considered as a necessary branch of education. The establishment of Lectureships and Societies, having Mineralogy as one of their principal objects, is a strong proof of the public feeling of the importance and utility of this science. Within a few years, several of the Universities have founded Professorships of Mineralogy; and that munificent and patriotic association, the Honourable Dublin Society, have lately added to their establishment a Lecturer on this science. This example has been followed by other public bodies, and also by private associations.

Both the first and second editions of *A System of Mineralogy* follow similar subdivisions, beginning with Synonymy then mineral families, *eg* diamond family and zircon family, *etc*. Each family commences with Synonymy, followed by External Characters (physical properties and crystallography), Chemical Characters (blowpipe tests), Constituent Parts (chemical analysis), Geognostic Situation (paragenesis) and Geographic Situation (localities). The two editions contain accurate crystal drawings, showing various morphologies of various species. The foundations of descriptive mineralogy had clearly been laid down; perhaps not surprisingly, modern reference works follow a successful line established some 190 years ago.

Other works by Jameson followed, with the *Treatise on the external, chemical, and physical character of Minerals*, second edition (1816a) and third edition (1817), followed by *A Manual of Mineralogy and Mountain Rocks* in 1821. The 1821 publication was considered 'as the best text-book of its time' (Jameson, 1854). Fifteen hundred copies were sold in the few months immediately after its appearance. In 1837 'Mineralogy' appeared as an article in the seventh edition (1830-42) of *Encyclopaedia Britannica*, and was published separately during the same year in Edinburgh (Jameson, 1837).

Subjects covered by Jameson in his lecture courses included hydrology, botany, zoology, geology and mineralogy. A condensed outline of his lectures in mineralogy is detailed below (Jameson, 1854):

> *Preparative Part of Mineralogy* – Physical Properties of Mineralogy – Morphological Characters of Minerals – Systematic Arrangement of Minerals – A system founded partly on External and partly on Chemical Characters adopted. – In this system there are three Classes; Class 1. *Acrogenous* – On Surface formed Minerals – Gases – Waters – Acids, and Salts. Class 2. *Geogenous*, or Minerals of which the known Solid Part of the Earth is chiefly composed – divided into Haloidal Minerals or Tasteless compounds of Earth and Acids, and Tasteless Compounds of Metals and Acids; Terrigenous or Earthy Minerals – Metalliferous Minerals. Class 3. *Phytogenous Minerals* – Minerals chiefly formed of Mineralized Vegetable matters.
>
> Descriptions of Simple Minerals – Uses of Simple Minerals in the Arts – Medicine – Agriculture – and in the Economy of Nature, Physical and Geographical – Distribution of Simple Minerals.

Robert Jameson was the first to name columbite (1805) and olivenite (1820), and jamesonite ($Pb_4FeSb_6S_{14}$) was named in his honour by Haidinger (1825), for Jameson (1820) had described 'prismatoidal antimony-glance' found at Huel Boys mine, Endellion, Cornwall, as the 'rarest subspecies of grey antimony'. Rose (1827) chemically analysed jamesonite and found iron present; he considered it to be accidental, although nowadays essential to the species.

A fitting tribute to Jameson's outstanding contributions to mineralogy was paid by Mohs in 1820 (who was, incidentally, student number 503 in 1798 at the Bergakademie and succeeded to Werner's position upon the latter's death in 1817):

> In the spring of 1818, I had the pleasure of seeing my much respected friend, the celebrated Professor Jameson, at Edinburgh, to whom Mineralogy has been so much indebted, both by his extending the knowledge of it in Great Britain, and by his exciting a general interest in it, in that country where so much has already been done, and where it may be expected that in a short time so much more will be accomplished. (Mohs, 1820)

Mohs's words rang true, for in 1840 Jameson launched the rare new species greenockite (CdS) from Scotland onto the mineralogical fraternity (Jameson, 1840).

## Jameson and the Museum

Sir Andrew Balfour (*qv*) bequeathed his extensive natural history collection, accumulated over 40 years, to the University of Edinburgh. During Dr Walker's tenure in the university, he related that Balfour's museum was useless, neglected, and sections had decayed. Even after 1750 it still remained a considerable collection, for Balfour's museum was considered one of the finest in Europe at that time. Balfour's collection lay in the College Hall, but soon after 1750 it became dislodged, discarded, and almost virtually demolished. In 1782 Dr Walker retrieved a valuable nucleus of specimens and placed them in the College Museum; the occupant of the Chair in Natural History was *ex officio* Keeper of the Museum (Eyles, 1954). Jameson soon became Walker's favourite pupil after attending his classes in 1793. Shortly afterwards, the College Museum was entrusted to Jameson's charge. Upon Dr Walker's demise, his trustees removed his personal collection; thus for a second time despoliation occurred. Little of teaching value remained when Jameson took over the museum: only a few cases of bird specimens, serpents, dresses, weapons and minerals. However, Jameson deposited his own valuable private collection in the museum and commenced amassing a new collection.

The foundation stone of a New College (today termed the Old College) was laid on 16 November 1789; the new building included space for a new museum. Jameson convinced the college commissioners of the urgent need to finish the museum, and they issued orders for speedy completion of the new galleries. Yet with construction proceeding slowly, it was 1820 before adequate accommodation materialised for the museum.

In 1807 Jameson applied to King George III 'for His Majesty's permission to have circulated from the different public offices printed instructions for collecting objects of natural history for the Museum' (Jameson, 1854). Until 1812, Jameson's salary was £70 per annum, *ie* the same as that of the first Professor in 1767; thereafter it was increased to £100 per annum. An additional sum for expenses, of not more than £100 in any one year, became available on condition that accounts were maintained of additions to the collections, and submitted to the Barons of Exchequer in Scotland. From 1813 Jameson maintained a register of new acquisitions for the museum. To enhance the mineral collection, Jameson actively purchased specimens from local and internationally renowned mineral dealers including Rose (Edinburgh); Heuland, Sowerby and Tennant (London); Bryce-Wright (when the latter conducted business from his Liverpool premises *c*.1843-57); and Krantz (Berlin). Additionally, in 1845-46, entries reveal that Jameson purchased a set of crystal models from A.L.O.L. Des Cloizeaux (1817-97) of Paris, prepared under the latter's superintendence, for the College Museum. As a result of Jameson's request to the King, specimens and collections arrived from many parts of the world, including material from the Arctic expeditions of W. E. Parry, J. Franklin and J. Rae. Apart from the £100 per annum for expenses, monies were granted by the Town Council, and a bank loan of £3000 was obtained to purchase the extensive Bullock and Dufresne natural history collections (Sweet, 1970; 1970a).

The museum was finally opened to the public in 1820, with an admission charge of 2s-6d. Entrance fees went some way towards alleviating interest payments on the loan. Following in Dr Walker's footsteps, Jameson (1817a; in Sweet, 1972) similarly issued 'Instructions to Collectors' (Sweet, 1972). Under Jameson's command, and sometimes at his own expense, the museum continued to expand and gained an international reputation.

Yet inevitably decline set in. On numerous occasions Jameson complained to the authorities about a lack of funds. As Keeper without a salary for that role, he was forced, in 1830, to part with his assistant, 'for the present establishment of the museum is the Professor of Natural History, assisted by a man to keep the door and stove, and a woman to wash and dust the room' (Jameson, 1854). So strong were Jameson's Wernerian principles that he failed to display Hutton's rocks and minerals, plus other collections with a Huttonian bias. Jameson abused his powers (Chitnis, 1970) and in museum business became embroiled with the Royal Commission of Inquiry into the State of the Universities of Scotland, Sir David Brewster and the Royal Society of Edinburgh. (Jameson was elected a Fellow of the latter in 1799 and to the newly formed Geological Society of London, although he never contributed to their publications.)

During 1852 Jameson prepared a detailed statement of the thousands of valuable specimens which could not be exhibited due to lack of accommodation. The statement was laid before the Town Council, who saw the necessity of Government funding for an extension to the west of the College Museum.

The College Museum, University of Edinburgh. Drawn and engraved by W. H. Lizars (1788-1859) c.1822, used as a letterhead by Jameson. As a letterhead it measured 12.5 x 10.0cm on a sheet 25 x 20.5cm. From L. Jameson (1854).

Fortunately the Government of the day was interested in furthering education, including museums, and accepted the offer to transfer the College Museum collections to the nation in 1854. On 23 October 1861, the Prince Consort laid the foundation stone of the new Museum of Science and Art, now the Royal Museum of the National Museums of Scotland.

Throughout Jameson's keepership, a vast natural history collection accrued and a 'Numerical Statement of the Museum Collection' (in 1852), of which by far a greater part could not be

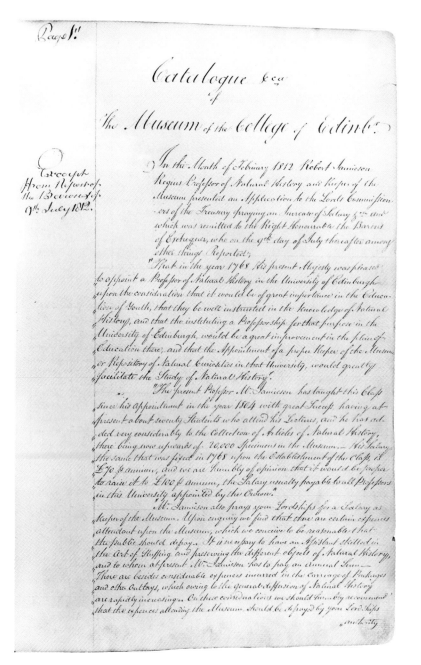

First page of the College Museum catalogue (runs to 358 pages and spans 1812-53), which presents the request for an increase in Robert Jameson's salary to the Lord Commissioners of the Treasury, and for a salary as Keeper of the Museum. NMS Geology and Zoology Department archives.

Labels, rare (note 'J'; there are corresponding numbers on specimens). These red labels were possibly display labels in the Old College Museum. Ink analysis of the numbers on the specimens suggests a date pre-1850's.

exhibited, ran to 74,453 specimens. Included in this total was a collection of 3973 minerals laid out for display and upwards of 3000 minerals in drawers:

> The mineral collection, as every mineralogist knows, is arranged in the highest systematic order. Neither the public nor naturalists have any reason to complain of the apparent want of systematic arrangement of the collections, as Jameson had to restrict the arrangements to suit the accommodation of specimens; but let it be remarked that the natural grouping is always carefully preserved, with a few exceptions, confined not only to specimens exhibited, but also to those unexhibited; therefore, the whole collection may be considered in order either for arranging and cataloguing, or for transferring to the new Museum when completed. (Jameson, 1854)

### Jameson, Journals and Societies

On 12 January 1808, Jameson and a small group of enlightened men congregated in the College Museum and formed the Wernerian Natural History Society, of which Jameson filled the role of permanent President. The society published *Memoirs of the Wernerian Society* at regular intervals, acting as a vehicle for natural history, physical sciences and mineralogical-geological papers. Sweet (1967) published a detailed account of the society which embodied Honorary, Foreign, Non-resident and Ordinary Members. The first volume, as of 12 January 1808, lists Abraham Gottlob Werner, the Right

*Section of page 20 (1813-14) from the College Museum catalogue, showing entries relating to minerals registered on a regional basis (possibly in 'young' Jameson's hand). Later entries became more specific. NMS Geology and Zoology Department archives.*

Honourable Sir Joseph Banks and Richard Kirwan as Honourable Members.

A wide range of scientific, natural history, political and public figures constituted the bulk of Non-residential and Foreign membership. From 1811 and including the 1821 membership, numerous eminent mineralogists, or people with mineralogical attributions or affinities, joined the Wernerian Society. The following is an abridged list from Sweet (1967) and attests to the highest esteem in which Jameson was held by the mineralogical fraternity: J. J. Berzelius (1779-1848, of berzeliite), K. A. Bloede (1773-1820, of blödite/bloedite), A. Boué (1794-1881, geologist and distinguished pupil of Jameson), A. Bruce (1777-1818, of brucite), P. Cleaveland (1780-1858, of cleavelandite), Dr W. Clinton (1769-1828, of clintonite), A. Crichton (1763-1856, of crichtonite), B. Faujas St Fond (1741-1819), C. G. Gismondi (1762-1824, of gismondine), J. W. von Goethe (1749-1832, of goethite), W. K. von Haidinger (1795-1871, of haidingerite), J. F. L. Hausmann

*Right:* Epergene, silver and cut glass (width 42.5cm, height 29cm; made by S. C. Young and Co., Sheffield, 1816-17). Engraved 'To Professor Jameson from the members of the Wernerian Natural History Society, 19 April, 1820. A token of esteem and approbation for his eminent professional ability, ardour and success for his unwearied attention to the interests of the Society, of which he is the founder and President'. NMS Department of History and Applied Art, 1965.877.

*Below:* List of honours bestowed upon Jameson by international learned societies. From L. Jameson (1854).

THE

# EDINBURGH NEW

# PHILOSOPHICAL JOURNAL.

――――

BIOGRAPHICAL MEMOIR OF

## THE LATE PROFESSOR JAMESON,

REGIUS PROFESSOR OF NATURAL HISTORY, LECTURER ON MINERALOGY, AND KEEPER OF THE MUSEUM IN THE UNIVERSITY OF EDINBURGH;

Fellow of the Royal Societies of London and Edinburgh ; Honorary Member of the Royal Irish Academy ; of the Royal Society of Sciences of Denmark ; of the Royal Academy of Sciences of Berlin ; of the Royal Academy of Naples ; of the Geological Society of France ; Honorary Member of the Asiatic Society of Calcutta ; Fellow of the Royal Linnean, and of the Geological Societies of London ; of the Royal Geological Society of Cornwall, and of the Cambridge Philosophical Society ; of the Antiquarian, Wernerian Natural History, Royal Medical, Royal Physical, and Horticultural Societies of Edinburgh ; of the Highland and Agricultural Society of Scotland ; of the Antiquarian and Literary Society of Perth ; of the Statistical Society of Glasgow ; of the Royal Dublin Society ; of the York, Bristol, Cambrian, Whitby, Northern, and Cork Institutions ; of the Natural History Society of Northumberland, Durham, and Newcastle ; of the Imperial Pharmaceutical Society of Petersburgh ; of the Natural History Society of Wetterau ; of the Mineralogical Society of Jena ; of the Royal Mineralogical Society of Dresden ; of the Natural History Society of Paris ; of the Philomathic Society of Paris ; of the Natural History Society of Calvados ; of the Senkenberg Society of Natural History ; of the Society of Natural Sciences and Medicine of Heidelberg ; Honorary Member of the Literary and Philosophical Society of New York ; of the New York Historical Society ; of the American Antiquarian Society ; of the Academy of Natural Sciences of Philadelphia ; of the Lyceum of Natural History of New York ; of the Natural History Society of Montreal ; of the Franklin Institute of the State of Pennsylvania for the Promotion of the Mechanical Arts ; of the Geological Society of Pennsylvania ; of the Boston Society of Natural History of the United States ; of the South African Institution of the Cape of Good Hope ; Honorary Member of the Statistical Society of France ; Member of the Entomological Society of Stettin, and other learned Societies.

(1782-1859, of hausmannite), J. H. Heuland (1778-1856, of heulandite), W. Hisinger (1766-1852, of hisingerite), F. P. N. Laumont (1747-1834, of laumontite), W. Rashleigh (1777-1855), J. Sowerby (1757-1822), L. N. Vauquelin (1763-1829, discoverer of chromium, of vauquelinite), W. H. Wollaston (1766-1828, of wollastonite), and S. Zois (1747-1819, of zoisite).

The society flourished, its heyday ranging from its early inception up till about 1845. On 19 April 1820, society members presented Jameson with a magnificent silver epergne as a token of respect and high esteem. However, by the late 1820s the Wernerian doctrine had lost popularity. Society activities gradually declined from around 1848 and four years after Jameson's death in 1854 the society finally ceased to exist.

In conjunction with Dr Brewster, Jameson produced the *Edinburgh Philosophical Journal* from 1819 to 1 April 1824, whereafter, due to disagreement between the two editors, Jameson continued alone as editor and to use the above title from 1 October 1824 until 1 April 1826. After this period Jameson changed the title to the *Edinburgh New Philosophical Journal*, and remained its editor until his death.

## Epilogue

Although Jameson was primarily a geologist-mineralogist, many branches of natural history benefitted enormously from his labours. Internationally renowned for outstanding and prolific contributions, Jameson was rightfully entitled the 'Father of Modern Natural History'. The esteem with which he was venerated may be gleaned from the prestigious honours bestowed upon him by over 50 international learned societies.

Of slender body, possessed of tremendous drive and limitless energy, Jameson was a careful observer, comprehensive thinker, tireless worker and inspirational character. One of his students, Robert Christison, who later occupied the Chairs of Forensic Medicine and Materia Medica, wrote in 1816:

> The lectures were numerously attended in spite of a dry manner, and although attendance on Natural History was not enforced for any University honour or for any profession, the popularity of his subject, his earnestness as a lecturer, his enthusiasm as an investigator, and the great museum he had collected for illustrating his teaching, were together the cause of his success. (Eyles, 1954)

Charles Darwin attended Jameson's lectures in his second year (1826-27) at Edinburgh University and records that 'they were incredibly dull'.

He died, unmarried, on 19 April 1854 at his residence, 21 Royal Circus, Edinburgh. Venerated by the citizens, who accorded him a public funeral on 28 April, he is interred at Warriston Cemetery, Edinburgh.

## THOMAS ALLAN (1777-1833)

Edinburgh banker and mineralogist

Thomas Allan, the son of Robert Allan (1745-1818), an Edinburgh banker and proprietor of the *Caledonian Mercury*, was born in Edinburgh on 17 July 1777, and educated at Edinburgh High School. From childhood days he took to scientific pursuits and later entered his father's bank. At a very young age he actively purchased mineral specimens and began to develop a collection. After the Treaty of Amiens on 27 March 1802 he visited Paris where he became acquainted with the distinguished French mineralogists Haüy and Brochant. He was also fortunate to be introduced to the German mineralogist A. G. Werner. During a sojourn in the Dauphiné district, Allan enriched his mineral collection, then returned to Scotland in 1803.

During 1806 Charles Lewis Giesecke (1761-1833) sailed for Greenland, where under adverse conditions he collected minerals until 1813 (Sweet, 1974). Material collected then, including cryolite, was sent to Copenhagen on board the Danish ship *Der Freuhlin*, which was captured by an English frigate and brought into Leith (near Edinburgh). Some cargo

*Far left:* Thomas Allan by Watson Gordon (1788-1864). Courtesy City of Edinburgh Museums and Galleries.

*Left:* Thomas Allan's Certificate of Election to the Royal Society London; note Charles Hatchett and Thomas Thomson as two of the signatories. By permission of the Royal Society London. © Royal Society.

I was this morning at Mrs Forster – who has certainly a fine collection of Minerals but I think quite too dear – she does over even Mr. Mawe – she has some fine Corals – but I did not price them. I have little else to add only that I expect to quit this about 8 days hence – If you have any thing farther to say to me pray write in course.

I am Dr Jameson

Yours truly

T Allan

I have re-opened your Box – & got Ato 6 contain the minerals from Bournon – as I doubt not you will be more anxious to see them than any of the rest – I shall forward it as soon as I find time to take it down to the wharf – which I hope to do tomorrow or next day

London 15 May 1804

Allanite (v. of Orthite) 65

Anorthic Melane Ore, Allanite M8

1 y. t. x. s. d. P. r. M. (Isomorphous with Epidote) Cristallised, in the form of a flat prism, engaged in massive allanite, of a dark brown colour nearly black and opaque, with an uneven vitreous fracture. This cristal is unique. I found it among Giesecke's minerals which I purchased at Leith, it occurs only at one spot, namely Alluk, near the southern extremity of East Greenland

2 Portion of a Rhomboidal prism, coated with brown rust like Gadolinite, but internally same as the last – do

3 Massive, and partly cristallised, accompanied with Quartz, felspar and mica, with some minute black brilliant cristals of Zircon do

do 4 another specimen of same, also accompanied with zircon do

do 5 – do intermixed with black mica do

Giesecke 1812 6 do acted on by the weather, which has exposed a peculiar direction in the composition of the substance, do

7 Small fragment of Allanite, with a portion of the brown coating of No 2 sent me by Wollaston in 1812, from Mysore

8 cristallised in dull black prisms not determinable dispersed on massive allanite from Bastnaes, Sweden

9 another mass of the same with indications of cristals do

10 Massive & amorphous, the cerine of Berzelius, do

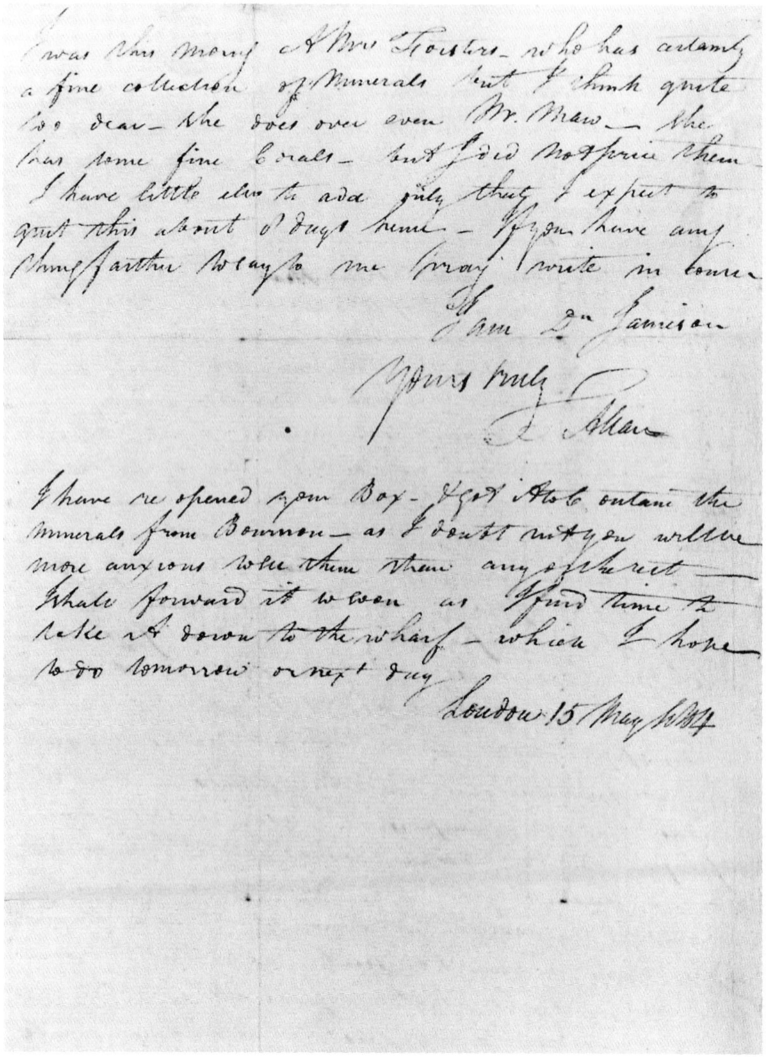

*Above:* Part of a letter from Thomas Allan to Professor Robert Jameson, 15 May 1804, detailing gift of minerals from Count de Bournon to be divided between Allan and Jameson. Edinburgh University, Special Collections Gen. 129/5.

*Above, right:* Allanite entry from the Allan-Greg catalogue. Note the Giesecke – Greenland and Wollaston – Berzelius association. Courtesy of Mineralogy Department, Natural History Museum, London.

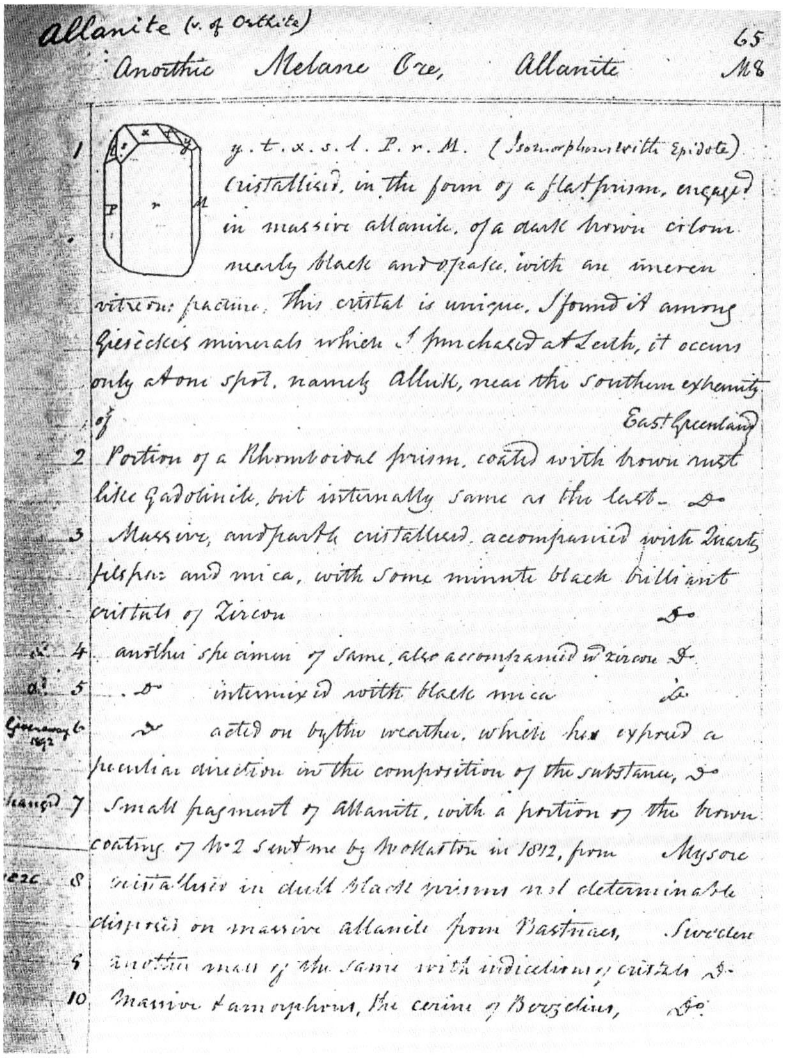

became impounded in a Leith warehouse. By invitation, Robert Jameson examined the natural history specimens and identified the cryolite as 'sulphate of lime'. The mineral collector Lt. Col. Imrie (see Ninian Imrie) also disdained the specimens. Thomas Allan possessed a few grains of cryolite (given to him in Paris) in his fine collection and his friend Imrie took him along to see Giesecke's material. (For more information regarding the extraordinary life of Giesecke, and Greenland minerals, see Petersen and Secher, 1993.) In an autobiographical account (Farrar and Farrar, 1968) Thomas Allan described the state of the mineral cargo:

> I went with Col. Imrie to Leith, and found the minerals which were the contents of 9 or 10 cases and casks thrown down in the middle of a loft, without a single ticket or reference, and all of them released from the packing in which they had been enveloped. They certainly presented as uninviting a spectacle as could be imagined.

Allan correctly identified Giesecke's cryolite and, together with Imrie, bought the sale lot in 1808 for £40; Allan reckoned that the cryolite alone was worth £5000. After cleaning the material, whose provenance at this stage remained unknown, Allan observed a black mineral which he tentatively identified

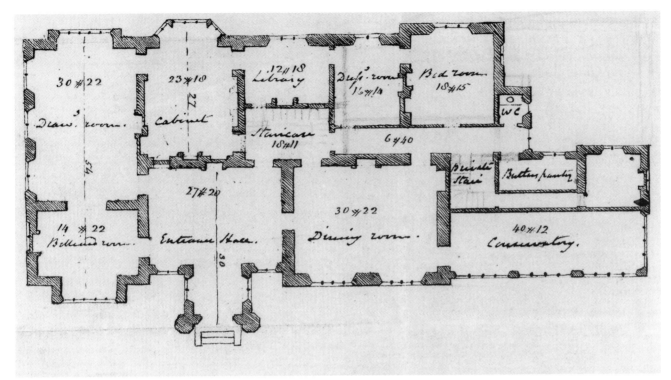

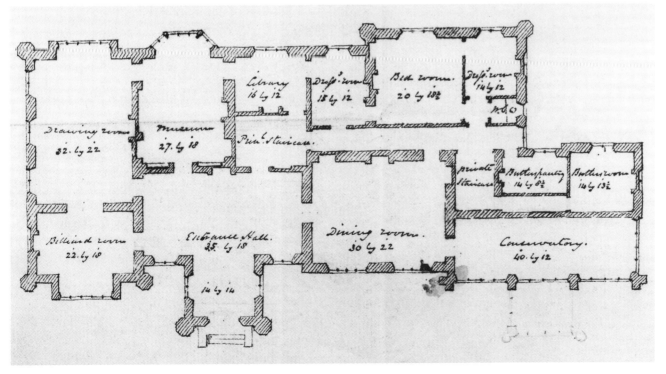

Plans, pre-1827, possibly by the Edinburgh architect William Burn, for a cabinet (upper plan) or museum (lower plan) room, although not constructed, at Thomas Allan's home, Lauriston Castle, Edinburgh. Note the *grande enfilade*-bedroom to cabinet room in the upper plan. The British Architectural Library, RIBA, London.

Lauriston Castle, some 5km from the centre of Edinburgh, is a 16th-century tower house (left-hand side of building) with 19th-century additions. The tower house was built in the 1590s by Archibald Napier whose son John was the inventor of logarithms. Courtesy City of Edinburgh Museums and Galleries.

as gadolinite (Allan, 1808). Not being satisfied with his identification, in 1809 he gave the mineral to Thomas Thomson, who undertook a chemical analysis, recognised it as a new species, and named it allanite in his honour (Thomson, 1810). Currently there are two allanite species: allanite-(Ce) and allanite-(Y), as designated by Fleischer and Mandarino (1995).

Giesecke arrived at Leith in the autumn of 1813 and met Allan, who invited him home to view his fine mineral collection, considerably enhanced by Greenland minerals from the sale. Giesecke presented Allan with additional Greenland specimens, some of which, including cryolite, sodalite, eudialyte and arfvedsonite, are now in the Natural History Museum (London) Allan-Greg Collection. (Allan later became instrumental in placing Giesecke as a candidate for the vacant professorship of mineralogy at the Dublin Society, to which he succeeded.) Farrar and Farrar (1968) give a fascinating detailed account, from a manuscript notebook written by Allan about 1820, of the Giesecke-Greenland-Leith-Dublin association.

In 1822 Thomas Allan became acquainted with the famous Austrian mineralogist W. K. Haidinger who, during his first visit to Britain, visited Edinburgh:

> Mr. Allan, who afterwards invited me to stay at his house like a real home, while I translated and published my translation of Mohs' treatise on mineralogy (1825). And so I came over in 1823 and staied [sic] until 1827, visiting in the mean time Cornwall along with Mr. Allan and his oldest son Robert. (Embrey, 1977)

Allan travelled extensively in pursuit of fine minerals, visiting Cornwall (with Haidinger), Ireland and the Faroe Islands, where he collected excellent zeolite specimens (Allan, 1815).

As an admirer of James Hutton (1726-97) and the Huttonian theory, Allan published supporting papers, together with mineralogical contributions, in the *Transactions of the Royal Society of Edinburgh*. Additionally he wrote an article on 'Diamond' for the *Encyclopaedia Britannica*. Allan's book, *Mineralogical Nomenclature* (1814), passed through three editions up to 1819. For his scientific contributions he was elected to Fellowship of the Royal Society of Edinburgh on 28 March 1805, becoming Treasurer from 26 November 1821 until his death, and curator of the society's museum. He became a Fellow of the Royal Society of London on 6 April 1815, and was also a Fellow of the Linnean Society.

Thomas Allan was a hospitable man who generously permitted serious students to make use of his collection; it became one of the finest in Scotland and Allan also developed an excellent European reputation. He was a highly respected public figure in Edinburgh, being a leading Whig in the capital, a Commissioner of Police, a Commissioner of Improvements and a member of the Merchant Company. In 1823 Thomas Allan purchased Lauriston Castle, a sixteenth-century tower house, while residing at 19 Charlotte Square, Edinburgh. He considerably altered the castle by adding a two-storey extension. Sir Walter Scott notes in his *Journal* that on 3 December 1827, he

> went with Tom Allan to see his building at Lauriston, where he has displayed good taste – supporting instead of tearing down or destroying the old château. (Fairley, 1925)

## TO MINERALOGISTS.

*Sale of the Lauriston Collection of Minerals.*

### NO AUCTION DUTY.

THE splendid and well-known COLLEC-
TION of MINERALS, formed by the
late THOMAS ALLAN, Esquire of Lauriston,
which is not equalled by any in Scotland,
and is surpassed by very few in the Empire,
will be exposed to PUBLIC SALE on 15th
JANUARY, in the OLD SIGNET HALL, Royal
Exchange, at Twelve o'clock noon. Upset
Price £2000.

It consists of 6800 Specimens, contained
in a hundred and fifty-four Drawers, encased
in three handsome Cabinets. The Specimens
rarely exceed five inches in diameter, and
are as remarkable for their characteristic
features as for the singular beauty and ex-
cellency of their arrangement.

The value of the Collection is enhanced
by an ample descriptive Catalogue, which
contains about fifteen hundred correct diagrams of the most interesting crystallized
forms. This valuable work was prepared
with great scientific skill and study by the
late Proprietor, and is considered by judges
to be perfectly unique.

Besides the Scientific Collection above
mentioned, there are about 200 Specimens,
generally of large dimensions, which are at
present contained in a glass case. Some of
these are truly magnificent, and as well as
those of the Scientific Collection, are not
intended to be removed till after the day of
Sale. They will be exposed (including the
Cabinet) along with the above, at an upset
price of £200, but if not sold in this manner,
will be disposed of in Lots, at Mr. Tait's,
11, Hanover Street, on the 17th January, at
One o'clock afternoon. Catalogues are in
preparation, and may be had a fortnight
previous to the Sale.

For further particulars, apply to ROBERT
CHRISTIE, Esq., Accountant, 1, George
Street, Edinburgh.

11th December, 1834.

Allan died on 12 September 1833 at Linden Hall, near Morpeth, Northumberland. His obituary in *The Scotsman* of 18 September 1833 stated:

He was one of the few wealthy men in Edinburgh who zealously expounded the popular cause … the inhabitants of Edinburgh are indebted to him for many able and valuable services … his cabinet of minerals is, we believe, with a single exception, the most complete in North Britain. In private life he was mild, amiable, and unpretending, constant in his friendships, and easy and affable in his intercourse with men of all parties and classes.

He is buried in St Cuthbert's Churchyard, Edinburgh, where a tablet to his memory may be seen (Fairley, 1925).

Two years after Thomas Allan's death his collection was purchased for £1300 by Robert Hyde Greg (1795-1875), a Manchester millowner and merchant. R. H. Greg did not add to the collection but his son, Robert Philips Greg (1826-1906), also a businessman, became an ardent mineral collector. In 1860, on behalf of his father, R. P. Greg negotiated the sale of the collection to the Trustees of the British Museum (Smith, 1907).

## DAVID BREWSTER (1781-1868)
### Journalist, educator, scientist, inventor and university principal

Sir David Brewster was born in the westward room of his father's house in the Canongate of Jedburgh, close to the river and bridge, on 11 December 1781. He was the third child and second son of James Brewster, Rector of the Grammar School of Jedburgh (Gordon, 1869; Mrs Gordon was Brewster's daughter, neé Margaret Maria Brewster). His mother, an accomplished woman, died when he was nine years old, and responsibility for his upbringing fell upon an older sister. An upper window, with a dilapidated pane of glass, in his father's house generated inquiring thoughts which led him in later life to research into the mysteries of refracted light.

As a talented child aged twelve, he commenced studies at Edinburgh University, where in 1800 he obtained an MA degree at the age of 19. One year earlier he had become tutor to the family of Captain Horsbrugh (Horsburgh) of Pirn, Peebles-shire, a post he retained until 1804, residing with the family in Edinburgh during the winter months and at Pirn in the summer. In 1804 Brewster entered employment with the family of General Dirom of Mount Annan, Dumfriesshire, remaining tutor till 1807. Brewster's candidature in 1807 for the Chair of Mathematics at St Andrews University was unsuccessful (although he was appointed Principal of that university in January 1838). During 1807, however, Aberdeen University conferred upon him an LLD, and Cambridge an MA degree. The following year he was elected a non-resident Fellow of the Royal Society of Edinburgh; he subsequently held office with the society as Secretary, Vice President and, from 1864 until his

of Science was organised, Brewster being a founder member, and later in the same year he received a knighthood. On 10 October 1859, he accepted the post of Principal of Edinburgh University – to this day his statue stands outside, and facing, the Department of Chemistry.

Brewster's highly varied and illustrious career spanned many facets, for he became profoundly involved in the institutionalisation of sciences and in the deeply rooted conflict between science and religion; he had initially trained as a minister of religion. Within his sphere of specialisation *ie* optics, optical instruments and the history of science, Brewster published a vast number of papers at a time when an explosive growth in scientific publishing occurred. The *Catalogue of Scientific Papers of the Royal Society* (1868) lists 299 papers written individually by Brewster and five joint contributions to scientific journals. To this list may be added eleven other papers published in the *Transactions of the Royal Society of Edinburgh* and *Proceedings of the Royal Society of Edinburgh* (Gordon, 1869). His most innovative period spanned the years from about 1811 to 1818, although he continued experimenting with light, and minerals, for another 50 years. With Professor Jameson, Brewster edited the *Edinburgh Philosophical Journal* until April 1824.

A now-forgotten, though vitally important, discovery made by Brewster germane to mineralogy is that when convergent light is transmitted by a doubly refracting crystal placed between polariser and analyser, an optical figure is seen. In calcite, Brewster observed interference circles surrounding a maltese cross, whereas topaz displayed two systems of circles. These observations led Brewster to conclude that certain crystals possessed two optic axes:

> In the course of an extensive examination of mineral bodies, in which I was engaged in the years 1816 and 1817, for the purpose of investigating the laws of Polarisation and Double Refraction, I was led to the discovery of two general principles, which connected the optical condition of crystals with their mineralogical structure and their chemical composition. From the number of Axes of Double Refraction which any mineral possessed, I was enabled to determine the Class of Primitive Forms to which it belonged; and while every variation in the position, the intensity, and the character of these axes in similar minerals, was found to be accompanied with a difference of chemical composition, a difference of composition was also found to be accompanied with a difference of optical structure. (Brewster, 1821)

death, President. Seven years later (1815) he was elected a Fellow of the Royal Society of London, which afterwards bestowed upon him the Copley Medal, and three years later the Rumford Medal. Brewster subsequently received six of the Royal Medals, in each instance for new discoveries in light (Gordon, 1869). Throughout 1825 international honours were profusely bestowed upon Brewster, in addition to two Keith Medals from the Royal Society of Edinburgh (1827-29 and 1829-31). In 1831 the British Association for the Advancement

For his mineralogical studies Brewster drew upon the mineral cabinets of Thomas Allan, Thomas Edington, Robert Ferguson, William Nicol, John Murray, Mr Sivright of Meggetland and others including Sir George Mackenzie and the Edinburgh lapidaries Mr Sanderson and Mr Spaden. Mr Thomson of Forth Street, Edinburgh, provided from his cabinet an amethyst which once belonged to the King of Candy (Candy, or Kandy, was a district of the former Ceylon, now Sri Lanka, from which the English exiled the last king in 1815). Additional specimens for optical study were provided by Henry Heuland, and Mr Laing-Meason, who supplied 'various specimens of lead-ores for the purpose of optical analysis … from Wanlockhead' (Brewster, 1820).

A spate of mineralogical papers (in excess of 25) followed, largely between 1818 and 1829 and appearing mainly in the *Edinburgh Philosophical Journal* and the *Transactions of the Royal Society of Edinburgh*, with several published in the *Edinburgh Journal of Science*. Brewster's work also included studies of fluid inclusions and mineral (garnet and diamond) lenses. (Brewster had a microscope lens manufactured from Greenland garnet donated by Sir Charles Giesecke [Gordon, 1869]). Table 1 details many of the minerals worked upon by Brewster who, from optical studies, erected the new species gmelinite, levyne and hopeite, and the withamite variety of epidote (Brewster, 1825, 1825a, 1826 and 1826a).

Hopeite is named after Thomas Charles Hope (1766-1844), Professor of Chemistry at the University of Edinburgh (jointly with Professor Joseph Black from 1795 until Black's death in 1799, then sole Professor until he resigned the Chair in the winter of 1842-43). Brooke (1822) achieved mineralogical immortality for Brewster by erecting the species brewsterite – a strontium, barium zeolite from Strontian.

The Duchess of Gordon possessed a Highland home at Kinrara on Speyside, not many kilometres from Belleville (now Balavil), 5km east-north-east of Kingussie, where Brewster was concerned with factoring property of his sister-in-law. During her sojourn at Kinrara, the Duchess of Gordon befriended him and they became correspondents, chiefly about mineralogy. Lady Gordon's catalogue (1823) of her personal mineral collection refers to 'Dr Brewster' as a donor on several occasions (see Duchess of Gordon).

Brewster's profound interest in the optical properties of minerals would naturally lead to a reference collection of specimens and validated species, although no direct evidence has emerged that he possessed a collection of considerable size. His work has made an inestimable contribution to mineralogy for which today he is little remembered, apart from his name being memorialised in Brewster's Law (*ie nr/ni* = tan *i*, where *ni* symbolises the index of refraction of the medium in which the incident ray travels and *nr* the index of the refracting medium; hence maximum polarisation of the reflected and refracted rays is achieved by adjusting the angle of incidence until its tangent equals *nr/ni*).

David Brewster was an innovator: he invented numerous scientific instruments and the popular kaleidoscope. (The latter was patented on 27 August 1817, although Brewster never received any financial reward, for it was immediately pirated as the patent was defective.) He possessed a nervousness unsuited to public speaking. On occasions he became irritable, impatient, litigious and verbally aggressive, yet he was a man of great personal charm despite strong convictions. (For additional aspects of Brewster's extraordinarily productive life and work, see Gordon, 1869; Harvey, 1972; Morrison-Low and Christie, 1984.)

David Brewster died on 10 February 1868 at his home, Allerly, near Melrose. He is interred in Melrose Abbey grounds on the south side of the abbey close to the graveyard entrance; appropriately, his epitaph reads 'The Lord is my Light'.

| | | | |
|---|---|---|---|
| Alum | Brucite | Gmelinite | Pyromorphite |
| Amethyst | Calcite | Gypsum | Quartz |
| Amphiboles | Cerussite | Halite | Stilbite |
| Analcime | Chabasite | Harmotome | Strontianite |
| Anglesite | Chrysoberyl | Heulandite | Talc |
| Apatite | Diamond | Hopeite | Thomsonite |
| Apophyllite | Emerald | Leadhillite | Topaz |
| Aragonite | Epistilbite | Lepidolite | Withamite |
| Barite | Feldspar | Levyne | Wulfenite |
| Beryl | Fluorite | Micas (various) | Zeolites (excluding denoted) |
| Brewsterite | Glauberite | Peridot | |

Table I. Mineral species worked upon by Brewster, abstracted from his papers.

Alexander Rose, from 'Centenary Celebrations', *Transactions Edinburgh Geological Society* (1934, vol. 13).

## ALEXANDER ROSE (1781-1860)
Mineralogist
and mineral dealer

John Rose, father of Alexander Rose, moved south from the Cromarty area, making a living as an ivory and wood turner. Alexander followed his father's trade early in life, though he later became an eminent mineralogist and mineral dealer. It is rare to encounter boyhood stories (including touches of folklore) of mineralogists from this period. Details of Alexander's boyhood skills are thus worth recounting:

> During the latter years of the eighteenth century there was, at one time, much excitement in the neighbourhood of Blackfriars, or Niddry Street [Edinburgh] caused by the report that a 'Brownie' [fairy or elf] was busy in one of the old houses there. Morning after morning a local turner, on entering his workshop in the upper storey of one of these ancient buildings, was astonished to find that the work he had left unfinished the night before was all beautifully completed and arranged, and the shavings swept up. Curiosity at last got the better of fear, a watch was set, and in the grey light of an early morning the 'Brownie', in the person of a small boy, was discovered climbing a dangerous wall and attic roof till he arrived at the window of the workshop, on entering which he was promptly caught. [Rose, undated; *ie* Mrs Mary Tweedie Stodart Rose, grand daughter by marriage.]

The story ends with the lad being allowed to work in the turner's shop. This lad was Alexander Rose, who later became an expert ivory and wood turner.

In the *Post Office Edinburgh and Leith Directories* Alexander Rose was registered as a Wood and Ivory Turner at two addresses (1805-23). Between 1823 and 1834, in addition to the above profession, he is also mentioned as a 'Dealer in Minerals – Mineralogical Depôt, 63, South Bridge'. The Rose family occupied both 1 and 2 Drummond Street (opposite Edinburgh University Old College, and 50m or so from 63 South Bridge). He was educated at Infirmary Street High School ('around the corner from Drummond Street'). As a meticulous craftsman with turning skills, he became a renowned scientific instrument-maker for the University of Edinburgh.

Sciences formed a major interest in Rose's life. He fraternised with notable scientific men of the day, including Faraday, Daguerre, Sir David Brewster and Hugh Miller. (No. 1 Drummond Street was the last place Hugh Miller visited before his tragic death on Christmas Eve, 1856; Waterston *c.*1960.) Focusing on geology and mineralogy, Rose ultimately attained a professorship at Queen's College, Edinburgh, a lecturing association rather than an institution, which provided lectures at various venues throughout the city. Alexander Rose's mineralogical lectures at the School of Arts, Chambers Street, attracted engineers, architects and general students, who were drawn by his outstanding teaching skills and the superb mineral collection with which he illustrated his talks. As a mark of esteem, students presented him with a silver snuff box bearing the inscription 'Presented to Alexander Rose, Professor of Geology and Mineralogy, Queen's College, Edinburgh, by the Students

## QUEEN'S COLLEGE, EDINBURGH.
### SUMMER SESSION—1841.

The following CLASSES will be opened on TUESDAY the 4th May,—

| CLASSES | | LECTURE-ROOMS | HOURS |
|---|---|---|---|
| BOTANY, (Commencing on the 5th) (Excursions on Saturdays.) | DR GEORGE ATKIN, 10, Nicolson Street. | 24, Brown Square, | 10 A.M. |
| ANATOMICAL DEMONSTRATIONS, | DR KNOX, 4, Newington Place. | 11, Argyle Square, | 11 A.M. |
| PRACTICAL ANATOMY, | DR KNOX, and DR LONSDALE, | 11, Argyle Square, | 9 A.M. to 4 P.M. |
| OPERATIVE SURGERY, | DR KNOX, and DR LONSDALE, | 11, Argyle Square, | 2 P.M. |
| | DR CAMPBELL, 4, Picardy Place. | 11, Argyle Square, | 10 A.M. & 1 P.M. |
| MIDWIFERY, | DR MARR, Dundas Street. | 3, Surgeon Square, | 1 and 4 P.M. |
| | DR MOIR, 33, Abercromby Place. | 4, High School Yards, | 10 A.M. & 1 P.M. |
| FORENSIC MEDICINE, (Commencing on Wednesday 5th May) | MR SKAE, 3, Argyle Square. | 11, Argyle Square, | 2 P.M. |
| PRACTICAL CHEMISTRY, | DR GEORGE WILSON, 28, Gayfield Square. | 24, Brown Square, | 10 A.M. to 4 P.M. |
| NATURAL PHILOSOPHY, | GEORGE LEES, A.M. Clarkburn. | School of Arts, | 1 P.M. |
| PRACTICAL MECHANICS, | MR LEES, assisted by Mr JAMIESON, | School of Arts, | 2 to 4 P.M. |

MATHEMATICS details...

| | MR NICHOL, 32, South Bridge. | 1st Geometry, | 10 A.M. |
| | | 2d Do. | 12 Noon. |
| | | Mechanics, Engineering, Algebra, Practical Mathematics, | at other hours. |
| | MR GALBRAITH, 54, South Bridge. | Practical Mathematics, &c. | 9 & 11 A.M. |
| | | 1st Geometry, | 10 A.M. |
| | | 2d Do. | 12 Noon. |
| | | Mechanics, Engineering, &c. | 1 P.M. |
| | MR MURRAY, House, 7, Pitt Street, CLASS-ROOM, 6, Infirmary Street. | 1st Geometry, | 2 P.M. |
| | | 2d Do. | 1 P.M. |
| | | Algebra, Practical Mathematics, Mechanics, Engineering, &c. | 12 to 1, & 3 to 4. |
| | | Theory and Application of Differential and Integral Calculus, &c. | 4 P.M. |
| MINERALOGY AND GEOLOGY, (with Excursions.) | MR ROSE, 2, Drummond Street. | School of Arts, (Mondays, Wednesdays, and Fridays.) | 3 P.M. |

### MODERN LANGUAGES.

| | | | |
|---|---|---|---|
| GREEK, (Ancient and Modern,) (Class for the Scriptures and Greek Fathers, on Saturdays,) | MR NEGRIS, | 7, Argyle Square, | Junior. 12 Noon. Senior. 1 P.M. Superior. 2 P.M. Junior. 9 A.M. Senior. 10 A.M. |
| ARABIC AND PERSIAN, (Wednesdays and Fridays.) HINDUSTANI, &c. | MR BALLANTYNE, 1, Scotland Street. | 11, Argyle Square, | 1 P.M. |
| FRENCH, (Mondays, Wednesdays, and Fridays.) | MR CHAUMONT, 9, Duncan Street. | Do. Do. | 7 P.M. |
| ITALIAN, (Tuesdays and Thursdays.) | MR RAMPINI, 10, Gloucester Place. | Do. Do. | 7 P.M. |
| GERMAN, (Mondays, Wednesdays, and Fridays.) | MR KOMBST, 6, Great Stuart Street. | Do. Do. | 1st Class.—Grammar, translation, Schiller's Wallenstein, &c. 7 P.M. 2d Do. for Composition, Speaking, Readings in Lessing, Herder, &c. 8 P.M. |

### FEES

FEES to each of the above Courses, £2, 5s., with the following exceptions: Practical Anatomy, Operative Surgery, and Anatomical Demonstrations, £1, 1s. each. Midwifery, £3, 5s. Botany, £3, 5s. Practical Chemistry, £3, 5s. Practical Mechanics, with the Lectures, £4, 4s.

Greek, each Course, £3, 5s. Scriptures, £1, 11s. 6d. Both together, £4, 4s. each Course.

Prizes, and Honorary Certificates, will be given by the College and Lecturers, to the most distinguished Students.

Attendance at Medical Classes in Queen's College qualifies for Graduation at the Universities of London, Oxford, Cambridge, St Andrews, and Aberdeen; and for Examination at the Royal Colleges of Surgeons of London, Edinburgh, and Dublin; the Apothecaries' Hall, the Faculty of Physicians and Surgeons of Glasgow; and the Army and Navy, and other Public Boards.

JOHN ROBERTSON, Solicitor, Secretary to the College.

17, DUBLIN STREET, 20th April 1841.

at 8.30pm. At the December meeting class members formed a society, known as the Geological Society, from which today's Edinburgh Geological Society grew. The first scientific meeting, on 8 December, discussed whether or not Arthur's Seat in Edinburgh was volcanic in origin: its true volcanic nature was recognised. Alexander Rose became the second President of the society in 1835, serving until 1846, the second-longest office term throughout the society's history (Watson, 1934), although he retired from active work in 1856. Mrs Rose, widow of Alexander Rose's grandson, Robert Traill Rose, presented a mineral cabinet, silver cup and the snuff box mentioned previously, to the Edinburgh Geological Society on 17 October 1945 (Campbell and Holmes, 1948).

Entries relating to minerals in the early natural history registers of the Old College Museum reveal that Rose sold minerals to enrich the rapidly expanding collections between 1824 and 1854. His daughter, Miss Agnes Rose (1829-1918), also sold minerals between 1884 and 1885 to the Industrial Museum of Scotland (founded in 1854 and the forerunner of part of the NMS).

Alexander Rose died on 3 July 1860 and is interred in Grange Cemetery, Edinburgh.

## JOHN CAMPBELL (1796-1862)
Wealthy landowner,
aristocrat (second Marquis of Breadalbane)
and public figure

John Campbell was born in the Nethergate of Dundee on 26 October 1796. After an Etonian education, he entered Parliament in 1820 and served until 1826 as the Member for Okehampton. A distinguished public career followed; he succeeded in being elected Member for his own county of Perthshire in 1832, serving for two years. On 5 June 1834 he was elected a Fellow of the Royal Society, being proposed by Rev. Phillip Jennings DD and seconded by John George Children (at that time election certificates were not given to Peers of the Realm; J. Taylor-Reid, *pers. comm.*, 1994). In the same year, upon the death of his father, he became the fifth Earl of Breadalbane. The second Marquis was admitted to membership of the Highland and Agricultural Society of Scotland in 1819 and served as Vice President in 1836 and 1837. During 1838 he was appointed a Knight of the Thistle, and for many

who attended his class during the Session 1839-40'. Additionally, he received numerous testimonials including one from Charles MacLaren, editor of *The Scotsman*, and a Danish knighthood on one of his several visits to Denmark. Long periods were spent away from home either lecturing in England or travelling widely. It is recorded that he visited Iceland to collect zeolites (Rose, undated).

During the early 1830s Alexander Rose had given mineralogy classes in a Drummond Street house adjacent to his residence. On 4 December 1834, eleven class members met in Robertson's Tavern, Milne's Close, and decided to meet at Rose's residence, 2 Drummond Street, every Monday evening

John Campbell, from Gillies (1938).

years served the Queen as Lord Chamberlain. In the following year he took office as Lord Lieutenant of Argyllshire, and became Rector of Glasgow University (1840-42), President of the British Association (1840), Rector of Marischal College, Aberdeen (1843-45), and President of the Royal Society of Antiquaries of Scotland (1852-62).

Throughout the early centuries of Christianity, when the Scots from Ireland settled on the west coast, they called the mountainous range separating them from Pictland, *Druim-alban*, or backbone, and the region beyond *Braghaid-Alban* (Gillies, 1938) – *Alban* being the ancient name of Scotland. The Breadalbane area is generally accepted as extending from the elevated wild tracts of the Grampians in western Perthshire to Lochaber and Atholl in the north and to Strathearn and Menteith in the south. John Campbell possessed the strong belief that great mineral wealth lay concealed in Breadalbane's rocks. His father (1762-1834) had closed down the lead mines near Tyndrum around 1798, but the second Marquis re-opened them in 1838. On the Breadalbane Estates chromite was discovered at Corriecharmaig near Killin, chalcopyrite plus tetrahedrite at Tomnadashan and Corriebuie and galena at various localities. John Campbell roamed the hills, chipping away at the rocks with a geologist's hammer. Workable quartz deposits located on Meall Cruadh, a spur of Ben Lawers, furnished the milky-white picturesque dairy at Taymouth Castle (Gillies, 1938). He befriended eminent geologists including Sir Roderick Impey Murchison (1792-1871) and Sir Archibald Geikie (1835-1924).

The desire to discover gold in Breadalbane provided the motivation behind the Marquis's mining ventures between 1836 and 1862 (Bainbridge, 1980). Indeed gold was found on the Breadalbane Estates around 1840 in nugget form; *eg* the Turreich (Turrick) nugget (see Gold). He employed several prospectors, the first of whom was F. Odernheimer from Hesse-Kassel, Germany, who received a handsome £200 salary for the first year (1838) of a three-year contract, and thereafter £250 per annum for the next two years plus a percentage of profits (Bainbridge, 1980). Surveys conducted by Odernheimer relate to his search for cobalt, barite and sulphur, with the best prospects lying around the southern shores of Loch Tay. Copper ore was discovered in veins at Tomnadashan, on the south side of Loch Tay, during early spring of 1840. The second Marquis of Breadalbane ventured into smelting, acid

Descriptive Catalogue of the Minerals in the Collection of The Marquis of Breadalbane arranged by James Tennant 149 Strand - London September 1845.

1700 Gold Native, Twenty-five isolated crystals, including the cube octoedron tetrahedron and their modifications. a very interesting suit. Brazils
1701 " " a superb group of Cubic crystals. Brazils
1702 " " a large rolled pebble weighing 1oz, 7dwts, 18grs. Wicklow. Ireland.
1703 " " small detached pieces, from the gold washings. Brazils
1704 & 1705 samples of Wash Gold. Do.
1706 & 1707 " ditto with Arenaceous Iron from Brazils and Wicklow.
1708 " " foliated on Quartz. Transylvania
1709 " " ditto with ditto and Blende. Do.
1710 " " ditto and reticulated, on Quartz. rich.
1711 " " minutely cryst. on Quartz. Transylvania
1712 " " pale coloured and reticulated
1713 & 14 " brownish yellow, alloyed with Palladium on Micaceous Iron. Congo Soco. Brazils
1715 " " ditto, detached from the Matrix. Do.
1716 " " ditto in loose grains.
1717 Platina Native, an octoedral crystal. most rare
1718 " " ...
1719 " " ditto, a large porphyta. Siberia

production and chemical fertiliser operations, endeavouring to generate financial gains from his unsuccessful mining venture at Tomnadashan (Bainbridge, 1980). Specimens from the Tomnadashan copper mine were exhibited at the Great Exhibition of 1851 in the Mining and Mineral Products section, the Marquis in the catalogue being listed as a 'producer'. Thost (1860) details chromite, serpentine, silver ore, copper pyrites, grey copper ore, molybdenite, galena and pyrite occurrences in Breadalbane found largely by the Marquis, and an account of the metal mines is given by Wilson and Cadell (1885).

It seems highly probable, considering the Marquis's deep interest in geology, that he personally accumulated a mineral collection. Geikie (1904) comments:

Sir Roderick Murchison, when visiting him [John Campbell] in 1860, after a tour through the western Highlands, remarked to him at dinner that one great difference between the oldest rocks of the north-western and those of the Central Highlands lay in the presence of abundant hornblende in the former and its absence from the latter. 'Stop a bit, Sir Roderick', interrupted the Marquess, 'You come with me to-morrow, and I'll show you plenty of hornblende'. Next day a walk was taken across a tract of moor near the Black Mount, Sir Roderick accompanying some ladies, while the chief marched on in front. At last when the rock in question was reached, the Marquess shouted out in triumph, 'Here's hornblende for you'. And he was right, as Murchison, with a queer non-plused look on his face, had to admit. Nevertheless the geologist's generalisation, though not universally applicable, had in it a certain element of truth.

Title page, and entries, from the Breadalbane collection catalogue. NMS Geology and Zoology Department archives.

Langton House, Duns, Berwickshire, was bequeathed to the sister of the second Marquis of Breadalbane (ninth Earl of Breadalbane, 1958, on NMS files). The house was sold in 1924 by her descendant and a large part of the moveable estate dispersed without evidence of a mineral collection amassed by the fifth Earl coming to light; the latter had instructed F. Odernheimer

> to make a collection of the rocks and minerals of his far-spreading property, and to give also a report regarding them. (Odernheimer, 1841)

The rock collection, including ores, dutifully assembled by Odernheimer formed a component of a later, geographically organised rock collection formulated by R. J. H. Cunningham (1815-42), who presented this collection to the Highland and Agricultural Society of Scotland (established 1784). A new Agricultural Museum of the society was opened in 1843 on the north-west corner of George IV Bridge and facing also Victoria Street, Edinburgh. (At that time the society's soils, ores and rocks exceeded 2200, with arrangements to display them similar to that of the mineralogical exhibits in Edinburgh University and the British Museum.) In 1859 the Highland and Agricultural Society of Scotland donated nearly 3000 specimens to the National Museums of Scotland, where in the Breadalbane section of the Museum's Register specimen numbers and descriptions agree with those in Odernheimer's 1841 paper, and with the 1841 catalogue of the museum of the Highland and Agricultural Society of Scotland.

James Tennant (1808-81), a London mineral dealer and first Professor of Mineralogy at King's College, The Strand, London, in 1845 prepared a mineral cabinet and catalogue of 1907 specimens for the second Marquis which he organised according to the classificatory scheme in *An Elementary Introduction to Mineralogy* (Phillips, 1837). The erstwhile collection, formerly housed in Langton House, Duns, when purchased by Mr Newbigging (also domiciled in Duns) consisted of 1111 specimens which he generously donated to the National Museums of Scotland in 1958 (NMS G 1981-34). Specimens originate from classic British and worldwide localities, are generally fist-size or smaller, and each one is characterised by a distinctive, adhering printed catalogue number.

The second Marquis of Breadalbane was a man of austere cast of countenance and commanding presence (Gillies, 1938). Geikie (1904) remarked:

> Of the great landowners the most striking personality in my time was undoubtedly the second Marquess of Breadalbane. Tall and broad, with a head like that of Jupiter Tonans, having the most commanding presence combined with the most winning graciousness of manner, he was the incarnation of what one imagined that a great Highland chief should be .... A liberal and enlightened landlord, he had done much to improve his vast estates, and was beloved by his tenantry and people.

Some aspects of Geikie's portraiture may well be correct; however, in his own country he was well known for the severity and cruelty with which he evicted tenants from his huge estate during the Clearances, making light of certain Geikie claims.

In failing health he visited Switzerland where he died, *sine prole*, on 8 November 1862 at Lausanne.

## ROBERT ALLAN (1806-63)
### Edinburgh stockbroker and mineralogist

Robert Allan, the eldest son of Thomas Allan (*qv*), was born in Edinburgh on 12 October 1806. After a High School education he proceeded to Edinburgh University, where he attended law classes and lectures delivered by Professor T. C. Hope and R. Jameson (Henry, undated). Even in childhood he had formed a mineral collection of his own, doubtless inspired and guided by his father. From his earliest years, when minerals constituted childhood playthings, Robert Allan acquired an extensive mineralogical vocabulary and knowledge of the external characteristics of minerals. With chemical skills derived from attending Professor Hope's lectures, Robert Allan analysed minerals in Dr Edward Turner's laboratory. The latter's pupillage had been under the German chemist Freidrich Stromeyer (1776-1835; who was the first to analyse a silver, copper sulphide subsequently named stromeyerite). During Haidinger's residency with the Allan family (see Thomas Allan), Robert Allan received advanced mineralogical schooling as Haidinger classified their minerals according to the Mohs system. From Haidinger Robert Allan acquired his crystallographic expertise.

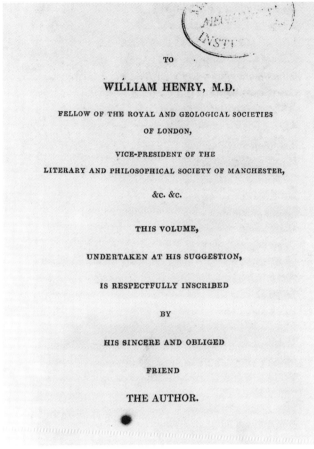

*Far left:* Robert Allan, frontispiece to W. C. Henry (undated), Manchester Central Library, Local Studies Unit.

*Left:* Dedication to W. Henry, M.D., in *A Manual of Mineralogy* by R. Allan (1834). William Henry (1744-1836), a manufacturing chemist and widely respected Manchester figure who had attended chemistry lectures at Edinburgh University, was the father of William Charles Henry FRS (1804-92). The latter married Margaret, daughter of Thomas Allan.

In the summer of 1824, at the age of 17, Robert Allan accompanied his father and Haidinger on their visit to Cornish mines. During 1825 and 1826 he visited, again with Haidinger, Norway, Sweden, Denmark and Germany. When in Stockholm they met the illustrious Swedish chemist J. J. Berzelius (1779-1848), who demonstrated his analytical blowpipe techniques. Robert Allan commented:

1825 – August 17, from 8 to 11, Berzelius showed us in his laboratory several elegant processes for discovering the presence of bodies in minerals by means of the blow-pipe. He works with great facility, neatness, and accuracy, always mixing the body under examination with the proper re-agents on the back of his thumb and using a very small portion for each experiment. (Henry, undated)

Robert Allan observed numerous rare minerals in Berzelius's collection, which had been systematically organised. (Berzelius developed a mineral classification based upon the electronegativity of elements which gave rise to oxide, halide, phosphate, sulphate and silicate classes, the foundation of today's scheme.)

Knowing little of crystallography, he has his minerals arranged quite according to his own principles of chemistry, and instead of the form, always gave us the contents. The Professor supplied us with several dozen good specimens from his duplicate drawers. (Henry, undated)

Haidinger and Robert Allan departed from Stockholm on 19 August for Uppsala, furnished by Berzelius with letters of introduction plus a complete itinerary for Sweden. On 22 September they reached Copenhagen and devoted a week to the examination of various mineral cabinets and porcelain manufactory under the guidance of professors Forchhammer and Oersted:

September 24, Dr. Forchhammer took us to the King's Museum, which contains the finest collection of Kongsberg silvers in the world. The immense masses of pure metal are quite astonishing, and the fantastic shapes of the fibres and beauty of the crystals must be matter of admiration to everyone. Many of the specimens of pure metal are three or four pounds weight. (Henry, undated)

Via Lübeck, Hamburg, Hannover and Brunswick, Haidinger and Robert Allan arrived at the mining town of Clausthal in the Harz district, where again they examined mineral cabinets. After thoroughly exploring the Harz district, including Andreasberg, they eventually arrived in Berlin on 25 October. Robert Allan spent four months in Berlin, where he attended chemistry lectures delivered by Professor Mitscherlich and devoted some time daily to chemical analysis in the Professor's laboratory. From Berlin the duo travelled to Freiberg where they remained for ten days. Under Professor Mohs's guidance, they visited mines and examined the mineral cabinets of Werner. After Freiberg, mineral-collecting trips were executed in the south of Germany, Styria (Austria) and Idria (former Yugoslavia). Towards the end of October 1826, Haidinger and Robert Allan returned home via Paris, where they examined mineral collections housed at the Ecole des Mines and Jardin des Plantes (Henry, undated).

Even at the age of 19 Robert Allan was a meticulous diarist, daily recording mineralogical observations or information gathered during conversation. In 1828 he became a Fellow of the Geological Society, and he was admitted Advocate in 1829, although he never practised at the Bar. Elected a Fellow of the Royal Society of Edinburgh on 6 February 1832, he was for many years curator of its museum and library. Three geological papers were written by Robert Allan (1834a, 1850 and 1855); however, his most celebrated scholarly work, *A Manual of Mineralogy* (1834), became a reference text for later works, *eg* Greg and Lettsom (1858) and Dana (1837; Palache *et al.*, 1966), as had his father's books before him. Amongst other mineralogical authorities of the time, Robert Allan drew upon contemporary works of Mohs, Haidinger, Phillips, Jameson, Leonhard, Monticelli, Lévy, Cleaveland, Brooke and Berzelius. He additionally capitalised upon his father's collection for data:

Prepared so far under the superintendence of a late lamented parent, whose skill in discriminating the objects of this science,

and zeal in its pursuit, had enabled him to form a collection of highly characteristic minerals, with which the descriptions hereafter given have been diligently collated. (Allan, 1834)

Information gleaned during the European tour with Haidinger was also incorporated in his work, for Robert Allan (1834, p. viii) states under 'Observations' that he

noticed the principal localities and modes of occurrence of the different varieties; which the author, having had opportunities of visiting many of the European mining districts, has been enabled to assign with greater accuracy than is generally found in mineralogical books.

The wide reputation generated by Robert Allan's *A Manual of Mineralogy* resulted in the publishers of *An Elementary Introduction to the knowledge of Mineralogy* (Phillips, 1823) inviting him to prepare a fourth edition of that work, which appeared in 1837. Robert Allan retained the chemical classification of Phillips, but added details of 150 new species and nearly 60 new figures in the new edition.

Robert Allan was of a calm, equable, thoughtful temperament, and of modest nature. He possessed dextrous, delicate fingers, great manual expertise in the use of tools, and wielded the mineralogical hammer with singular adroitness (Henry, undated). Mineralogy for Robert Allan occupied leisure hours, for he was engrossed in business with his father. The *Post Office Edinburgh and Leith Directory* for 1848-49 lists one R. Allan as a Stock and Share Broker at 11 St Andrews Square, Edinburgh, the Stock Exchange then being located at 23 St Andrews Square. He resided at the family home, Lauriston Castle, Edinburgh. Although robust, even youthful in mind and body, he died on 6 June 1863 at 4 Hillside Crescent, Edinburgh after a simple fall in the garden opposite his house, and is buried in Dean Cemetery, Edinburgh.

## THOMAS JAMESON TORRIE (1808-58)
### Advocate, geologist and mineral collector

Thomas Jameson Torrie was a reticent, private character who became interested in geology and mineralogy undoubtedly through the profound influence of his uncle, Professor Robert Jameson. Torrie was of independent means which enabled him to travel widely, accumulate sizeable collections, and become well acquainted with the geology of Europe. By profession Torrie was an advocate, though in early life he showed a fondness for natural science. In 1824 he was admitted to membership of the Plinian Society, being elected President in 1827. Between 1826 and 1831 Torrie was elected to the Wernerian Natural History Society. Professor Jameson wrote letters of introduction to Capellan, Humbolt and Brochant in July 1830 for Torrie to study geology and, in particular, mineralogy, during the winter of 1830-31 (EDU Special Collections Letters 9/13, 9/14 and 9/15). On 15 December 1834 Jameson Torrie was elected a Fellow of the Royal Society of Edinburgh, and he also served on the museum committee of the Highland and Agricultural Society of Scotland in 1841.

Heddle and Greg (1855) published a paper 'On British Pectolites' and Heddle (1880) relates that several years after the pectolite paper he 'became the purchaser of the collection of minerals amassed by Mr Jameson Torrie'. Within this collection one of Torrie's specimens was marked 'Strontianite Leadhills' (Heddle, 1882); this radiating greenish specimen, a plumboaragonite (see Aragonite), now resides within the National Museums of Scotland Heddle Collection.

Torrie died on 7 August 1858 at Roslin, just south of Edinburgh, and is interred in Warriston Cemetery, Edinburgh.

## PATRICK DUDGEON (1817-95)
### Public figure and mineral collector

Little is known of Patrick Dudgeon's early life. He was born at Marionville House (which still stands), Restalrig in Edinburgh, on 27 September 1817 (J. Williams, *pers. comm.*), and was educated at Edinburgh Academy. His grandfather, Peter Dudgeon, was an officer in the Royal Navy and his father, Robert Dudgeon, followed a mercantile career in Liverpool and became one of the founders of the Royal Insurance Company. In his late teens Patrick Dudgeon went to China where he spent 16 years, returning about 1849, for in 1850 he married his second cousin Cecilia. Some of his earliest mineralogical specimens were acquired in China and Japan, chiefly rock crystal, which are now accessioned into the National Museums of Scotland General Mineral Collection (especially NMS G 1890.114.497-

Patrick Dudgeon, by Marshall Wane of Edinburgh. Oil on canvas, 74 x 62cm. A hospital minute book records that the Board decided to have the portrait painted posthumously in April 1895 at a cost of 30 guineas, the sum being paid in November 1895 on completion of the portrait. Presumably his son Col. Robert Francis Dudgeon C.B., who succeeded his father on the Board, loaned a family portrait for the purpose. Courtesy Crichton Royal Hospital, Crichton Royal Museum, Dumfries.

506). Possibly during this period he became competent in mineralogy, geology and analytical chemistry. Apparently he did not attend Edinburgh University, for although four people named 'Patrick Dudgeon' are recorded as matriculating, none was born in 1817 (J. Currie, *pers. comm.*, 1997).

Patrick and Cecilia lived for a short time at Kailzie, Peeblesshire, then in 1853 purchased the Cargen Estate near Dumfries, from William Stothert; Dudgeon lived here until his dying day. For a description and history of Cargen House (now virtually demolished), see *The Scottish Field*, September 1922:

> Within the grounds also is an octagonal building which formerly housed the mineralogical collection of Mr. Patrick Dudgeon. In the more rigid days of a generation or two ago this was the smoking den, for the use of tobacco was strictly forbidden within Cargen House.

Dudgeon's birthplace, Marionville House, Restalrig, Edinburgh. From Grant (1882).

For many years, along with Professor Heddle and Sir Archibald Geikie, an annual geological excursion was made to the North of Scotland, the Western Islands, and even to the Orkney and Shetlands. For many summers these excursions were carried through, and were the means of enriching the store of accurate geological data, and of adding as well to the splendid private Scottish collection of specimens, which has now found a permanent home in the Royal Scottish Museum in Chambers Street, Edinburgh. (*The Gallovidian*, 1910)

In the reprint of 'The County Geognosy and Mineralogy of Scotland', by M. Forster Heddle (1878) from *Mineralogical Magazine*, Heddle (1878a) dedicates the work thus:

To
Patrick Dudgeon Esq.,
of Cargen.

Without your aid, and without your companionship, my dear Friend, this book would never have been written.

It is a record of our scientific work in several departments, during many years; and I desire, in attaching your name to it, that it should stand as a record of the enduring friendship which has subsisted between us for all that time.

M. Forster Heddle.

A man of great compassion and kindness, his servants spoke of him in terms of warmest affection; some served him for 25 to 35 years. Throughout the Dumfries district it is claimed that Dudgeon pioneered technical education, giving lectures in popular sciences and chemistry in the servants' hall at Cargen. He also lectured on chemistry at the Dumfries Mechanics' Institute. Dudgeon, a major figure in public life, was a Commissioner of Supply, Magistrate, and Deputy Lieutenant of the Stewartry. For a lengthy period, he was also a Director of the Highland and Agricultural Society, being admitted in 1851 (S. Rose, *pers. comm.*, 1997), and at his hands agricultural affairs received considerable attention.

Dudgeon enjoyed a wide scientific reputation far beyond Dumfriesshire, in recognition of which he was elected a Fellow of the Royal Society of Edinburgh on 2 April 1860 (proposed by Sir William Jardine). He developed a lifelong friendship with the mineralogist Professor Heddle. Together (and with others) they became founder members of the Mineralogical Society of Great Britain and Ireland in 1876; Dudgeon also served as a Trustee. Pure geology did not escape Dudgeon's attention, for he struck up an association with the distinguished geologist Sir Archibald Geikie:

Dudgeon made mineral exchanges with the Swedish Natural History Museum, Stockholm (100 specimens in 1856 and 13 in 1857), to which he supplied, in addition to north of England specimens, several good vanadinites from Wanlockhead (M. Cooper, *pers. comm.*, 1995). A number of Dudgeon specimens from Dumfries Museum were exhibited in 1995 in the museum of the Crichton Royal Hospital, Dumfries, to commemorate the centenary of his death. (The hospital pioneered mental health care in south-west Scotland for over 200 years. Dudgeon served as Director of the Board of Crichton Royal Hospital 1855-85 and Trustee 1885-95; M. Williams, *pers. comm.*, 1999). Minerals were also despatched to Sydney Museum, Australia (now renamed the Australian Museum). The Scroll Register of Minerals and Exchange Schedules in the Australian Museum, Sydney, reveals that Professor Anderson-Stuart at the University of Sydney participated in exchange arrangements with the Observatory Museum, Maxwelltown, Dumfries. Most of the 52 specimens sent from the latter originate from Leadhills-Wanlockhead; Tongue, Sutherland (amazonite); South Harris (moonstone); and

Shetland (hornblende) (R. Pogson, *pers. comm.*, 1997), and are clearly from Heddle-Dudgeon localities.

Dudgeon organised a series of exhibits, including one on the mineralogy of Dumfriesshire and Galloway, plus a loan collection of jade which helped to popularise the Dumfries and Maxwelltown Observatory. Passionately fond of meteorology, Patrick Dudgeon commenced observations on rainfall in 1860 at Cargen. He religiously maintained daily records for 35 years, the work being continued by his eldest son (R. F. Dudgeon), who published a paper in 1910 on 'Fifty years rainfall at Cargen, near Dumfries'.

Between 1865 and 1895 Patrick Dudgeon published numerous notes, many mineralogical in nature, and papers in the *Transactions of Dumfriesshire and Galloway Natural History and Antiquarian Society* detailing well over 30 minerals from different localities within Dumfriesshire and Galloway (see Dudgeon, 1867, 1871 and 1890). Other works included an extensive paper (1877) on gold in southern Scotland, a description of linarite in slag (1884) and a description of arsenopyrite from a vein at Palnure Burn, Talnotry, Kirkcudbrightshire (read 1894, published 1897).

At a Mineralogical Society meeting held within Glasgow University on 6 September 1876, Dudgeon exhibited a large series of minerals including 20 species new to Britain at that time. Six hundred Dudgeon specimens, mainly Scottish, were donated to the Edinburgh Museum of Science and Art in 1878 (NMS G 1878.49.1–594). A second major donation (2938 specimens), registered in 1890 (NMS G 1890.114.1–2122), contained 390 species from numerous world localities. As a measure of Dudgeon's dedication and expertise, it is worth noting that this latter collection contained about half of the then-known total number of mineral species, approximately 800, in 1890 (Skinner and Skinner, 1980).

Patrick Dudgeon died suddenly at his home on 9 February 1895. He is buried in the north-east section of the burial ground in Troqueer Churchyard, Dumfries (J. Williams, *pers. comm.*). As a mark of respect, in the small village of Islesteps near Cargen, the Patrick Dudgeon Memorial Hall (1922) was named. Understandably Heddle wrote Dudgeon's obituary (Heddle, 1897) and noted:

Mr Dudgeon ultimately presented the whole of his Scotch collection to the Nation, to form the nucleus of a collection of Scottish minerals.

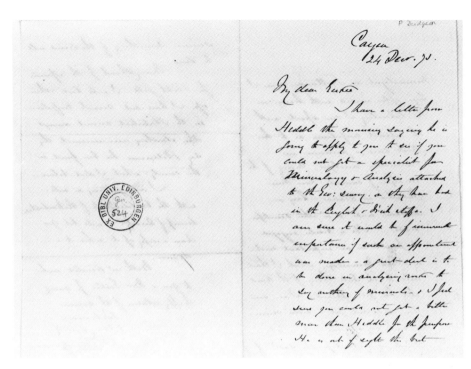

Letter to Sir Archibald Geikie, 24 December 1878, requesting that a mineralogical-analytical post be created at the Geological Survey. Edinburgh University, Special Collections, Gen. 524.

… his sole interest lay in the enriching of the Scottish collection; and his delight was unbounded when he found that the larger collection of his life-long associate was to rest beside and be incorporated with his own. The chief palliative of that associate's grief at his irreparable loss lies in the thought that each day's labour brings him nearer the completion of that which is to stand as an abiding monument to the *originator of the collection of Scottish minerals.*

## MATTHEW FORSTER HEDDLE (1828-97)

Medical doctor,
Professor of Chemistry and mineralogist

Matthew Forster Heddle was born at Melsetter, Hoy, Orkney, the second son of Robert Heddle, on 25 April 1828. His ancestors, on both sides, were of Scandinavian descent (Goodchild, 1897). Orphaned at 14, he was brought up by three relatives. From Orkney the young Heddle attended school at Edinburgh Academy, after which he went to Merchiston between 1842 and 1844. With characteristic vigour, he soon founded a natural history society at Merchiston and thus began a life's obsession with collecting. He built up a splendid herbarium which, it is said (Goodchild, 1897), he loaned to a friend who unfortunately ruined it. From that point onwards, Heddle resolved to collect indestructible objects – rocks and minerals: a lifelong interest had begun.

In the late 1840s Heddle entered Edinburgh University to study medicine. During 1851 he graduated MD after studying chemistry and mineralogy at Clausthal and Freiberg in Germany. That same year Heddle became President of the Geological Society of Edinburgh. His MD thesis of 1851, *The Ores of the Metals* (NMS archives MS 553.3), was prefaced by the following statement:

A feeling that the systematic works on the Chemistry were deficient in instruction as to the substances or Mineral species from which the rarer Metals were to be obtained, induced me to collect information on the subject.

A knowledge of their Ores would certainly place the Metals themselves more within our reach.

We might indulge the hope that out of the number whose properties are as yet investigated but superficially, some might have a Physiological or Therapeutic action on the Human frame, – nay some might even act specifically on some disease.

For the next five years he practised medicine in the then-unsavoury Grassmarket region of Edinburgh, although never cared for the profession. Having only a few patients provided Heddle with ample time and opportunity to advance major interests. He directed his attention towards chemistry, geology and mineralogy. In 1856 Heddle hired a boat to visit the Faroe Isles, where he succeeded in acquiring an extensive zeolite suite. Numerous duplicates enabled him to exchange advantageously with other mineralogists. These early Heddle specimens, and exchanges, created a nucleus around which he accumulated his large mineral collection.

During 1856 Dr Heddle was appointed Assistant to Professor Connell, Professor of Chemistry in St Andrews University, whom bad health prevented from lecturing. The appointment was made on the understanding that Heddle would eventually succeed to the Chair, which he did in 1862, holding the post until 1884.

Academic life brought long periods in the summer which saw active engagement in fieldwork, whereas winter months were devoted to the study of collected material. He produced thin-sections, polished rock slabs and agates by the thousands, and amassed a superb collection of Scottish minerals. (Heddle also compiled a catalogue of James Blackwood's [1823-93] 1200 mainly geological slides after the latter's death. The slide collection is held in the Dick Institute, Kilmarnock.) A profusion of high-quality scientific papers followed, mainly between 1876 and 1884, in which he detailed the geology of various areas, described minerals and erected 'new species', for example totaigite, tobermorite, craigtonite and rubislite. Today, only tobermorite stands as a valid species (Heddle, 1880). Heddle produced several papers detailing minerals new to Britain, while in Spencer's lists of 1898 and 1931 (see Embrey, facsimile reprint [1977] of Greg and Lettsom's 1858 *Manual of the Mineralogy of Great Britain and Ireland*), Heddle was responsible for 24% of British species entries (Livingstone, 1990).

In his relentless pursuit of minerals, Heddle expended considerable amounts of time and large sums of money. He travelled throughout Scotland long before detailed Ordnance Survey maps and railways to remote parts appeared. (Correspondence in Elgin Museum, Moray, reveals that Heddle stayed in the Station Hotel, Portsoy, on 12 September 1877, and with Professor Nicol and Patrick Dudgeon was on his way northwards.) Fortunately Heddle traced the course of his collecting routes on maps bequeathed to the Scottish Mountaineering Club. (Within NMS Geology and Zoology Department archives are a number of Heddle signed maps, dated 6 November 1897; other maps on which routes are marked are possibly Heddle's. The maps cover Sutherland, Perthshire, Aberdeenshire, Inverness-shire, Oban and Loch Awe districts and Skye):

He diligently explored nearly every mountain and glen, and almost every part of the coast of Scotland, in search of minerals. With Mr Patrick Dudgeon of Cargen, Mr Harvie-Brown [naturalist] of Dunipace, Mr James Currie, junr., and a few other chosen friends of similar tastes, he visited every locality for minerals which previous observers had recorded, and furthermore, himself added a very large number of new localities to the list of those previously known. (Goodchild, 1897)

During vacations Heddle and Dudgeon sailed together along extensive stretches of the Scottish coastline. Many of Heddle's specimens originate from secluded coastal localities of difficult access, a feature which now serves to enhance their scientific and historical values. One such locality, Sgurr nam Boc, on the south-west shore of Skye, is aptly summed up by Heddle (1893a):

This shore is, except under the most exceptional circumstances, altogether unapproachable. With a south-west wind there lies no breakwater whatever between it and the Atlantic; with a south wind only the narrow width of Canna, distant 10 miles. The special danger of the coast is, however, due to the extraordinary distances to which great masses of rock have rebounded seawards in their fall from cliffs, which are at most points over 700, and in several attain to almost 1000 feet in height.

A maze of hidden danger has to be threaded; the altogether unexpected swell and break of a surf all around the boat has to be encountered; only upon two occasions has a landing been successfully accomplished at certain spots by the present writer. The almost immediate find of laumontite, altogether unrivalled, and of stilbite and heulandite, little inferior to that of Iceland, led him to conclude that a like success, as regards landing and remaining on shore, could not have been attained to for a very considerable period. It was indeed with longing eyes that he had upon one occasion to hurry to the boat, leaving undetached a specimen of analcime with a surface of about one square foot, which was covered with pale blue crystals, nearly all of 2 inches diameter.

Only two landings must have frustrated Heddle somewhat, for in boyhood he had been 'accustomed to trust himself alone in a small boat, in which he often traversed the wild seas of the Orkneys' (Goodchild, 1897).

Heddle wrote over 50 papers, published principally in *Mineralogical Magazine* and *Transactions of the Royal Society of Edinburgh*. In the latter journal he wrote under the general title 'Chapters on the Mineralogy of Scotland' (a selected listing of these papers being detailed by Livingstone, 1990). Heddle was equally well known for his contributions to geology. Mineral validatory methods were primarily chemical, for Heddle was a prolific analyser, as attested by the very large number of mineral analyses presented in *The Mineralogy of Scotland* (Heddle, 1901). A considerable number of these analyses contain seven to ten reported oxides, especially in the

|  | £ | s | d |
|---|---|---|---|
| Laminated Specular Iron from Norway | " | 3 | " |
| White Iron Pyrites from Alva Stirlingshire | " | 2 | " |
| s Spearkies from Freyburg | " | 3 | " |
| Spearkies from Freyburg | " | " | 3 |
| W. Thomson — Iron Pyrites | " | " | " |
| Crystal of Iron Pyrites | " | " | 6 |
| Cubical Crystals of Iron Pyrites | " | 1 | " |
| Silicate of Iron in Octohedral crystals from the Mournes Ireland | " | 2 | 6 |
| Mr Rose — Earthy Phosphate of Iron from Salisbury Crags Edinburg | " | " | " |
| Vivianite with Sordawallite from Bodenmais in Bavaria | " | 2 | " |
| Pharmacolite from Cornwall | " | 4 | " |
| Arseniate of Iron from Cornwall | " | 5 | " |
| s Arseniate of Iron Cumberland | " | 1 | " |

|  | £ | s | d |
|---|---|---|---|
| Childrenite from Cornwall | " | 1 | 6 |
| Crystallized Carbonate of Iron with Crystallized Quartz from Cumberland | " | 2 | " |
| s Crystallized Carbonate of Iron from the Fichtelgeberge | " | 2 | " |
| s Carbonate of Iron from Cornwall | " | 1 | 6 |
| s Carbonate of Iron from Cornwall | " | 1 | 6 |
| Dr Knapp — Stahlactitic Carbonate of Iron Pseudomorph after Selenite from Cornwall | " | " | " |
| Dr Knapp — Carbonate of Iron Pseudomorph after Selenite Cornwall | " | " | " |
| Dr Knapp — Carbonate of Iron in hollow pseudomorphic cristals after Selenite from a mine called Virtuous Lady Cornwall | " | " | " |

Entries from a 'Cabinet of Minerals, 1850', Heddle's specimen purchase lists. Note earthy vivianite from Salisbury Crags, Edinburgh, purchased from Mr Rose. Courtesy Hunterian Museum, University of Glasgow.

silicate section. It must be recalled, in order to place Heddle's analytical prowess into perspective, that when utilising classical methods (largely gravimetric techniques) of silicate analysis, only approximately one to one and a half full analyses of eleven oxides could be performed per week even as late as the mid-1950s. In Heddle's day, alkali determinations were particularly laborious. Some of his analyses are excellent, *eg* saponite and pyrope, whereas others are difficult to interpret. Very little of Heddle's analytical techniques, apparatus or notes are known, as few documents relating to this field survive.

Three manuscript volumes, *Scottish Mineral Analyses* by Heddle (NMS Geology and Zoology Department archives, NMS G 555.411), contain mineralogical notes, chemical analyses of minerals, and occasional details of analytical techniques. For his researches into rhombohedral carbonates, and the feldspars, he received the Keith Gold Medal from the Royal Society of Edinburgh in 1876, the same year in which he became a Fellow.

Heddle accumulated a large collection of mainly Scottish minerals. Although field collecting with a 28lb hammer and black powder explosive was his forté, he purchased numerous specimens, both Scottish and foreign, with which to enrich his collection. A manuscript notebook entitled 'Cabinet of Minerals 1850', once the property of Alexander Thoms although written by Heddle, details mineral purchases. The most expensive acquisitions listed are 'Vanadiate of Lead in hexagonal prisms plus other Leadhills/Wanlockhead specimens all at £3-0-0 (from Dr. Brown), Columbite from Haddam, Connecticut £4-0-0, and Meteoric Iron, fell at Newcastle, County Down Ireland £5-5-0' (Hunterian Museum archives).

Heddle contributed greatly to the *Manual of the Mineralogy of Great Britain and Ireland* (Greg and Lettsom, 1858) and the authors, who consulted Heddle's collection, comment:

> Dr. Heddle has kindly undertaken the general, and especially the chemical revision of this work, preparatory to its going through the press; and the Authors take this opportunity of acknowledging the great obligations they are under to that gentleman.

*Above:* Heddle's manuscript notes on breunnerite, with details of chemical methods employed (Note 'xx' means crystals). Data contained in three volumes of notes formed the foundation to *The Mineralogy of Scotland* (Heddle, 1901). NMS Geology and Zoology Department archives 555.411.

*Left:* Obverse and reverse of the Royal Society of Edinburgh's Keith Gold Medal (NMS G 1982.2.1). Actual size 4.5cm diameter, weight 93.13gm, 9ct gold. Purchased from Mrs Littlejohn (née Heddle). The medal was first established in 1820 (Keith *et al.*, 1820).

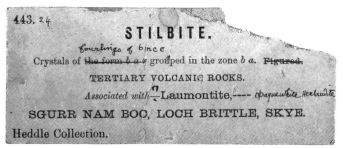

He also wrote the article 'Mineralogy' for the ninth edition of *Encyclopaedia Britannica*, published in 1883.

Heddle was a co-founder of the Mineralogical Society of Great Britain and Ireland (instituted 3 February 1876) and its second President (1879-81). During the first three years of its existence, Henry Clifton Sorby, as President, gave a single address to the society, although of Professor Heddle no formal presidential address is recorded (Fletcher, 1889). Heddle played an active role endeavouring to recruit members; correspondence in Elgin Museum (letter 80-29), written by Heddle on 23 July 1880 to one of Scotland's leading natural historians, Rev. Dr George Gordon (1801-93), states:

> I write to recommend to you the Mineralogical Society of Gt. Britain with a view or in the hope that you will become a member thereof.
> The Society, instituted for the notance of Geology, Mineralogy and Petrology, has been in existence for four years and has published 3 vols of transactions ....

(In 1882 the society had a little over 100 subscribing members.)

In 1889 Heddle was invited to act as a consulting mineralogist to some gold mines in South Africa (Goodchild, 1897). He left St Andrews and went to South Africa, where events turned out contrary to his expectations and he returned to Britain.

During 1892 Ramsay H. Traquair (1840-1912; Keeper of Natural History 1873-1906 in the then-named Edinburgh Museum of Science and Art, now the Royal Museum within the National Museums of Scotland) invited Heddle to critically appraise the mineral display, which included Dudgeon's minerals, within the museum. Heddle was offered an Honorarium of £1 per day, for four hours work each day. However, in March 1893, with a reduced Professional Assistance budget of only £30 available due to Treasury cuts, Heddle was offered only 14 days employment (Traquair file, NMS Geology and Zoology Department archives).

Some 7000 Scottish mineral specimens comprise the Heddle Collection, which was partially donated and partially purchased by the Industrial Museum of Scotland in 1894 (NMS G 1894.212). This truly heritage collection went on display in the gallery of Scottish Geology and Mineralogy during 1895. Heddle arranged the display, which J. G. Goodchild, a member of the Geological Survey billeted in the museum, completed.

In retirement Heddle commenced writing his most famous work, *The Mineralogy of Scotland* (a two-volume octavo publication), in which he drew extensively upon his field data and chemical analyses. He meticulously drew over 600 crystal diagrams and completed the greater part of the manuscript. With failing health, he formed a small committee consisting of Wilbert Goodchild, Heddle's daughters, and Alexander

# MUSEUM OF SCIENCE AND ART.

## REOPENING OF THE SCOTTISH MINERAL GALLERY.

### THE HEDDLE-DUDGEON COLLECTION.

——:o:——

[*From* "THE SCOTSMAN" *of September 6, 1895.*]

THE hall of the Scottish Mineral Collection on the upper floor of the west wing of the Museum of Science and Art has now been opened to visitors. For the greater part of the nine months during which this hall has been closed, Professor Heddle, of St Andrews, has been engaged in laying out and labelling the collection of Scottish minerals lately acquired from him, and which it has been a great part of his lifework to gather together. This collection, partly by gift, and partly by purchase, has now become national property, and finds a permanent resting-place in the Edinburgh Museum.

The collection at one time seemed likely to find another destination, but at a fortunate moment it was brought under the notice of Sir Robert Murdoch Smith, and he secured it for Edinburgh. By the arrangement he then made, Dr Heddle agreed to accept £1000 for 1000 selected specimens, and to present the remainder of his collection, consisting of between six and seven thousand specimens and valued at £2600, to the Edinburgh Museum. The funds for the purchased portion were obtained chiefly by subscription, one personal friend and former student of Dr Heddle's generously giving £500 ; so that this, the largest collection ever acquired by the Museum at one time, has come to it almost entirely as a gift. The Museum was already in possession of a small but fine collection of Scottish minerals, given for the most part by the late Patrick Dudgeon of Cargen, which at Mr Dudgeon's request has been incorporated with that of Dr Heddle, and the joint collections, as now laid out and labelled, bring together in scientific arrangement specimens of all the Scottish minerals known to exist.

The extent of the collection may be gathered from the fact that it contains specimens from the Shetland Islands and from the Cheviot Hills, and ranges over the breadth of Scotland from St Kilda and the Flannan Islands to its easternmost point at Peterhead. Within these four corners of Scotland and its islands certain districts have been especially fertile in their yield of minerals, and the localities from which the greatest number of specimens have been obtained may be set down as Shetland, Sutherland, the Western Islands, the Central Highlands, Kilpatrick and Dumbarton Hills, the Lead Hills and Wanlockhead, and the Sidlaw and Ochil Hills. The arrangement of the collection is not, however, in geographical order. The locality of each specimen is clearly given on its label, but the grouping is determined by chemical composition, following for the most part the plan adopted by Dana in the sixth edition of his "Mineralogy." Dana's arrangement is adhered to in the systematic collection, which fills the wall cases and flat cases on the floor of the hall, and is complete in itself. A number of large specimens which are of individual interest are, however, set out in nine upright cases specially designed and devoted to their display.

The labelling, which has not yet been completed, formed an important part of Dr Heddle's work. Each label gives the name, geologic formation, locality, and facts of special interest relating to the specimen ; and in many cases is accompanied by an analysis, this being of the very specimen to which it is attached, and by drawings showing its crystallographic form. The collection is thus fitted to be of the most complete use to the student of mineralogy. But it cannot fail to interest even the most casual visitor, who will be astonished to find that things of such great beauty have lain hid, and still lie hid, among the rocks or on the hillsides upon which he may have often trod.

Amongst the oxides, to quote the chemical phraseology, the quartzes are specially attractive. The cairngorms and amethysts, the fine series of agates and jaspers, as well as the curious form assumed by quartz in the "graphic granites," carry their own claims to attention. The cairngorms are shown both in their natural state and also after being boiled, cut, and polished. It will be noticed that the cairngorm in its natural state is of a dark brown colour, dull and almost opaque. To discharge this colour the stone is boiled in oil, and its future fate depends on the success of this operation. It is only if it comes out of the oil clear, lustrous, and of fine colour, either yellow or brown, that it will pass into the hands of the lapidary. But a characteristic of the Scottish cairngorm will be observed in the specimens in this collection. The Scottish stone never quite loses all trace of a banded clouding, which discloses an interrupted growth. Similar stones found under tropical skies will boil out clear and cloudless, but the genuine cairngorm of our northern mountains always retains a trace of this zonal clouding in the heart of its brightness. But of value far surpassing the well-known cairngorm, Scotland can boast of other gems. The Heddle collection contains, for instance, cut specimens of pyrope from Elie—far more truly a ruby than that passed off as such from Africa ; blue topaz, from Ben a Bourd ; beryl, from Braemar ; amethyst, from Morar ; moonstone, from Harris ; flèches d'amour, from Campbelton ; and huge crystals of amazonstone from Sutherland.

Two cases are filled with agates and jaspers, sliced and polished to show their beautiful markings ; and in the adjoining cases are specimens of showy purple and green fluorspars, chiefly obtained from steam cavities in the Carboniferous lavas of Gourock. Perhaps less striking at first glance, but quite as beautiful on a closer examination, is the extensive series of crystalline calcites. These represent every known Scottish locality where this mineral is found, and show a larger number of combinations of various faces than has yet been recorded for any other single country.

Almost as important as the calcites is the unrivalled series of zeolites — hydrous non-magnesian silicates, obtained chiefly from the Tertiary volcanic rocks of the Inner Hebrides and the Carboniferous volcanic rocks of the basins of the Forth and Clyde ; and among the larger specimens special notice should be made of the fine series of serpentines from Portsoy, Ballantrae, and other Scottish localities, exhibiting a wide range of pattern and colouration. In connection with the decorative use of these serpentines it is interesting to note that the pillars of the Grand Hall at Versailles came from the quarries at Portsoy.

From a geological point of view, the most instructive feature of the collection are the felspars, of which there are fully three hundred specimens, representing all the recognised

2

species. These are accompanied by drawings illustrating their crystalline forms, and by nearly seventy analyses of the specimens exhibited. Hardly less important than the felspars is the fine collection of amphiboles and pyroxenes, which extend to nearly four hundred specimens. Here, as in the case devoted to felspars, every known spécies is represented. The chemical composition of about sixty specimens is also given.

In these minerals there is always a definite connection between the chemical composition and the natural crystalline form. An exception to this rule is, however, illustrated in the series of specimens arranged in the railing cases on the south side of the hall. The minerals there shown are commonly called Pseudomorphs, and, as the name implies, appear under a form of crystallisation which does not belong to the species. Many of these specimens show that an imbedded crystal has been dissolved out by water percolating through the soil and rocks, the water at the same time, or subsequently, depositing the elements of another mineral whose chemical composition and natural crystalline form are entirely different from those of the mineral which it replaces. The new-comer perforce adapts itself to the new conditions. While retaining its composition it surrenders its accustomed form, and as it fills up the matrix of its predecessor, becomes moulded to a shape which is not its own—it becomes a pseudomorph. Other specimens of this class illustrate stages of progress in gain, or loss, or substitution, with alternations in these processes, and also provide important examples of the molecular changes brought about by the same influence of percolating water or still more powerful solvents. The changes here illustrated are general and widespread as the globe itself. To quote from Dr Heddle's article "Mineralogy," in the *Encyclopædia Britannica*, the interest and importance of these pseudomorphs lie in the fact that they are "the types and evidences of vast metamorphic transformations—processes either of decay or of reformation, which have modified wide-spread rock masses, and which are at the present time altering the structure of the crust of the earth."

It may be noticed in passing that the pseudomorphs have been placed at the meeting of the mineral and geological collections. In the Mineral Hall the elements which go to the formation of the rocks of Scotland are shown, and the Survey collection in the adjoining gallery exhibits specimens of the rocks themselves. With both these collections, the pseudomorphs, so to speak, claim connection, and thus afford a natural transition from the one collection to the other. This relationship, in fact, was one of the main considerations which determined the emplacement of the collection in the Museum in juxtaposition to the Gallery of Scottish Geology, the officer in immediate charge of which, Mr Goodchild, of the Geological Survey, zealously aided Professor Heddle in the arrangement of the minerals.

All the minerals already referred to—principally silicates—are those which enter most largely into the formation of rocks. There is yet another section of the collection which remains to be noticed, viz., the minerals which are themselves ores of metals, or which arise from the decomposition of ores of metals. These represent not only the more commonly known metals, using the word metal in its popular acceptation, but also some whose names are scarcely known outside the laboratory. There is, of course, gold from Kildonan, from the Lead Hills, and several specimens from Wanlockhead. Among the latter is the famous Gemmell Nugget, found in 1872, and then unfortunately fractured. At first the property of the Duke of Buccleuch and others, the pieces were eventually reunited, and the nugget passed into the charge of Mr Dudgeon, who arranged that it should find a place with the rest of his collection in the Museum. Silver, copper, antimony, cadmium, and nickel are all represented by the various minerals of which they form the bases, and also in the cases on the north wall will be found the fine collection of galena, an ore of lead, and blende, an ore of zinc. The mines at Lead Hills have yielded a large number of specimens. Some of these, arising from the decomposition of ores of lead and copper, are of great beauty, and others are remarkable for the perfection of their crystalline forms. The compounds of such rarer metals as vanadium in vanadiate of lead, and titanium in rutile, may chance to be passed as minerals of interest only to the scientific inquirer. But rare and little known as these metals are, theyhave already been turned to practical account in the arts ; for even at its present high price vanadium in the form of vanadic acid is used commercially to give a deeper dye to aniline black, and the dentist values titanium for the delicate trace of pink with which it tints the cold white of the artificial tooth.

The case of hydro-carbons, coals, and bituminous shales stands somewhat apart from the rest of the collection. That these minerals are of the greatest economic value goes without saying. It is only within the last fifty years that the shales have been really exploited. The result, however, has been not only to give us from this unpromising looking material a burning oil as clear as water, and a paraffin wax as white as alabaster, but to call into existence a vast industry unknown in the first half of our century. The late Sir James Gowans used to tell of his first acquaintance with shale. It was in the forties, while he was busy as a contractor laying the Edinburgh and Glasgow railway line, that he came across many outcrops of the mineral. Specimens from these he handed to his friend Professor George Wilson, who obtained from them by distillation a tarry substance which eventually yielded a black wax. This wax was made into black candles, which were sent to the Exhibition of 1851, and one of them is shown in the hydrocarbon case in this collection. It was said that Mr Gowans was offered the rights of the whole of the shale on the estate for a trifle, but he was busy with his railway work and took no further interest in the matter. It is scarcely likely that any one then dreamed of the future development of the oil industry, or of the immense sums which would be paid in shale royalties within a few decades.

Many will yet remember the interest excited by the famous trial, Gillespie v. Russell, in which the issue was whether a certain mineral was or was not coal. The fact that Dr Heddle was engaged in this case as chemical expert has enabled him to give a fuller interest to his collection of hydrocarbons, for he has preserved the specimens used in evidence at the trial, and they are now shown with their analyses. It is fortunate that these have been preserved, as the Torbanehill mineral is now completely worked out, and such specimens, with their historical associations, are practically irreplaceable. All through the collection there is evidence of the fortunate circumstances which have aided Dr Heddle in gathering these specimens. He was on the ground when the great railway cuttings were made, and the navvy's pick turned up many an interesting specimen. While the quarries of the metamorphic limestone and the mines in the Lead Hills were still open, he was on the watch for everything of interest they might yield, and some of the finest and rarest specimens in the collection were thus obtained. The greater number of these mines are now finally closed, and such an opportunity canscarcelyrecur. These were happy chances, but it was no doubt to Dr Heddle's untiring energy, which carried him, hammer on shoulder, to the top of almost every mountain in Scotland, and to every spot by sea or land where a specimen was likely to be found, that the collection owes its completeness and value. The size of many of the specimens, still more than their number, vouches for the personal vigour and indomitable endurance necessary for the securing of such prizes, and the result shows that the labour has been well spent, for the Heddle-Dudgeon Collection is not only the most representative gathering of Scottish minerals that exists, but is also, it is safe to say, the finest national collection of the minerals of any one country in the world.

An account of the Heddle-Dudgeon specimens, from their donated/purchased collections, on display in the Museum of Science and Art (NMS Royal Museum) in *The Scotsman*, 6 September 1895.

Note 'It is safe to say, the finest national collection of the minerals of any one country in the world'. NMS Library archives.

Stellate actinolite, with titanite from Ord Ban, Loch an Eilean, near Aviemore, Inverness-shire: the specimen featured in *The Mineralogy of Scotland* (Heddle, 1901, vol. 2, opposite p. 35) is now housed in the Hunterian Museum, University of Glasgow; maximum dimension 35cm. There is a residual Thoms number attached; no. 4923. Courtesy Hunterian Museum, University of Glasgow.

## MUSEUMS, PRIVATELY OWNED AND PUBLIC

Mineral collections reside in numerous provincial museums and institutions run by local authorities or trustees, or in privately owned ventures throughout Scotland. The demise of collections is an all-too-familiar picture and even today some collections are under threat from diminishing resources. In recent years the urgent need to gather information regarding past and present botanical, geological and zoological collections in Scotland has been addressed by Stace *et al.* (1987). In their publication *Natural Science Collections in Scotland*, 632 institutions throughout Scotland were contacted, of which 286 held natural science collections (as did 180 private individuals). A considerable proportion of the 286 collections itemised geological or mineralogical collections of 100 or more specimens whether as stand-alone, or as components of larger collections. Stace *et al.*'s (1987) data disclose that numerous old geological/mineralogical collections, or parts thereof, are extant albeit somewhat dispersed. For example, the catalogue of Dr Grierson's (1818-89) museum at Thornhill, Dumfriesshire (Black and Bisset, 1894) reveals approximately 400 accessioned mineral specimens. Dispersal of Dr Grierson's collection (Truckell, 1966) occurred when the trustees donated material in 1965-66 to the National Museums of Scotland, Edinburgh, and Aberdeen, Glasgow and Dumfries museums. In a different vein, an educational collection formulated during 1859 by Dr W. Chambers, and displayed during that year in the Chambers Institution, Peebles, remains within the institution although it is currently inaccessible to public viewing. A catalogue (Turner, 1927) divulges that the collection, with a tripartite division into petrology, mineralogy and palaeontology, was housed in 17 cases. Of the 206 pages in the catalogue, 99 relate to accessioned mineral specimens.

Current thematic mineral exhibitions, in popularist settings, continue Weir's museum tradition. The Wanlockhead Lead Mining Museum features mineral displays and humanised

Thoms, James Currie and J. G. Goodchild 'for the purpose of bringing out my works on the Mineralogy of Scotland'. (Heddle's Will, SRO SC 20-50-78).

Matthew Forster Heddle was a giant of a man with great physical strength and powers of endurance. Of a genial and kind nature, he possessed high and honourable principles. Although somewhat of quick temper he was devoid of malice. Fond of mimicry, with a strong sense of the ludicrous, he became a great storyteller. He never sought honours (nor were any conferred upon him), nor appropriated other people's ideas (Goodchild, 1897; Goodchild also wrote Heddle's obituary – Goodchild, 1898).

Scotland's foremost mineralogist, whose principal objective was to collect the largest and finest specimens of Scottish minerals, passed away at St Andrews early on the morning of 19 November 1897. In the grounds of St Leonard's at St Andrews a wall-mounted memorial commemorates the Heddle family and 'M. Forster Heddle MD Professor of Chemistry in the University of St Andrews'.

scenes depicting arduous and dangerous mining conditions of the nineteenth century. Two privately owned enterprises, Treasures of the Earth, at Corpach near Fort William, and Creetown Gem and Rock Museum, close to Newton Stewart, successfully project minerals in an attractive simplistic manner. Innovative and imaginative displays endear these ventures to the general public, and they have become popular visitor attractions. Both establishments contain high-quality large aesthetic crystals, including Europe's largest uncut emerald (in matrix) at 12kg, a 1.45kg idiomorphic topaz, and a wide variety of minerals and cut gems including diamond, aquamarine, garnet, gold and silver, and large quartz crystals.

## GLASGOW MUSEUMS
### Art Gallery and Museum, Kelvingrove

Glasgow Museums was founded in 1870. Acquisition of minerals began at this time and the bulk of the collection dates from the late nineteenth and early twentieth centuries. The large geological collection contains an estimated 8000 mineral specimens with no particular areas of specialisation, either mineralogical or geographical. Scottish material is not strongly represented, but includes small collections of specimens from Leadhills, Strontian and the Clyde Plateau lavas (zeolites and others). Much of the collection, particularly the older material, has little or no data.

By far the most significant collection is that of David Corse Glen (*qv*). This forms a major part of the collection. Also of interest is the collection of Professor John Fleming (*qv*), one of the leading early figures in Scottish geology. The collection, which includes over 800 specimens from both Scottish and other localities, dates from the first half of the nineteenth century and was donated to the museum in 1902 by Major J. A. Fleming. It contains specimens from Shetland, whose minerals Fleming described in his first published paper in 1807. Of local interest, the museum has a number of specimens of 'pseudogaylussite' (see Ikaite) dredged from the River Clyde near Cardross, including figured specimens (Macnair, 1904).

## NATIONAL MUSEUMS OF SCOTLAND

The National Museums of Scotland came into being on 1 October 1985 as a result of the National Heritage (Scotland) Act 1985, which combined two of Scotland's long-established national museums. The first began in 1780 as the Museum of the Society of Antiquities, the second in 1854 as the Industrial Museum of Scotland, becoming successively named the Edinburgh Museum of Science and Art, the Royal Scottish Museum, and the Royal Museum of Scotland. With the opening of a new 'Museum of Scotland' on 28 November 1997, the Royal Museum of Scotland shortened its name to the 'Royal Museum'.

Within the Geology and Zoology Department at the National Museums of Scotland are major, public mineral displays which draw upon the Scottish and General Mineral Collections. In the collections specimens are housed in transparent-lidded, specially sized boxes secured in air-conditioned, custom-built storage. The Scottish Mineral Collection, *c.*15,000 specimens, has core Heddle (7000) and Dudgeon (600 and 2938) collections and is systematically organised. Dr J. Wilson and R. Brown (*qv*), both of Wanlockhead, and A. C. Christie, of Morningside, Edinburgh, formed collections, which by purchase and donation also became assimilated into the Scottish collection. Wilson and Brown accumulated some outstanding and rare Leadhills-Wanlockhead minerals, and the former in 1878 and 1880 presented Heddle with 'several small specimens … which carry a good deal of the second mineral [hydrocerussite] … but no larger quantity apparently being procurable' (Heddle, Scottish Mineral Analyses, Geology and Zoology Department archives NMS G 555.411). The Scottish Mineral Collection contains some 315 of the 552 or so species known from Scotland. Named collections are not retained as entities, but are disseminated throughout in a systematic manner. Specialities include Leadhills-Wanlockhead primary and secondary minerals, and zeolites from Skye, Mull, the Glasgow environs and Strontian. Exquisite suites of lead-hillite, lanarkite, susannite, caledonite, and calcite on sphalerite grace the collection.

The General Mineral Collection, *c.*45,000 specimens, contains material from worldwide localities including superb bournonite from Wheal Boys, St Endellion, Cornwall; matlockite from Derbyshire; and a rare platinum nugget (350gm) from Tagilsk, Russia. From simple scroll register

67

records commencing 1813, P. Davidson (*pers. comm.*) studied the development and growth patterns of the collections *sensu lato* (*ie* rocks, fossils and minerals). His findings reveal that the rates of donated to purchased material remained fairly constant overall, being approximately 5:4. In the early years of the donors, academics and aristocrats predominated, whereas after 1918 the middle classes became more important, until post 1945 where a considerably greater spread throughout all classes emerged.

A disquieting feature of both the Scottish and General Mineral Collections is a virtual total absence of archival records, thus complex collection and personality histories have vanished with time.

## UNIVERSITY COLLECTIONS

### UNIVERSITY OF ABERDEEN

Showcase specimens, gems, plus a systematic collection characterise the mineral holdings within the Department of Geology and Petroleum Geology. Departmental visitors are corridor-coerced to view some 130 exquisite mineral specimens in well-illuminated, floor to head-height showcases. Adjacent, a small desktop gem exposition reveals numerous faceted stones including benitoite, spodumene, chrysoberyl, cassiterite and a deep indigo cordierite (1.5 x 0.5cm). A second series of display cases contains over 100 large specimens ranging from 15 to 40cm. Noteworthy in this collection is a triangular-shaped (with 14cm sides) Cumbrian specimen endowed in euhedral alstonite (0.8cm) and deep, honey-coloured Frizington barite. A Russian, classic, slender aquamarine crystal measures 22cm. Teaching collections of minerals are also held in the department.

The systematic collection occupies eleven oak cabinets, each containing 30 drawers 9cm deep. Based on Dana's system the collection commences with native elements, simple sulphides, sulphosalts and oxides, all of which occupy four cabinets. Uncommon species are disseminated throughout this section *eg* cooperite, patronite, altaite, coloradoite and joseite, with many originating from world-famous localities. Rock-forming minerals are housed in four cabinets; the remainder retain carbonates, sulphates, hydrocarbons, phosphates, vanadates

and tungstates. Based upon approximately 30 specimens per drawer, the estimated collection size is *c.*10,000 specimens (housed in open trays), although some register number allocations are in the 11,000 to 12,000 range. Two manuscript catalogues, which run to 12,372 entries, indicate gradual collection enhancement, for during 1969 Professor R. A. Howie donated howieite (register no. 10,827) from Langvale, California, to the department. A purchase acquisition policy enriches the collection for acquisitions emanate from Richard Tayler Minerals, Surrey; Minerals Unlimited, California; David New, Washington, U.S.A.; the late Richard W. Barstow, Cornwall; and other dealers.

In common with many institutional collections current holdings represent an amalgamation, two significant incorporations include those donated by T. C. Phemister, mainly crystals (1902?-82, Professor of Geology at the University 1937-71), and the Rev. J. M. Gordon (1849-1922; Gordon *qv*).

### UNIVERSITY OF EDINBURGH
#### Grant Institute of Geology

The mineral collection is principally a conglomeration of three collections amassed by Thomas Brown of Lanfine (*qv*), James Davidson (*qv*) and James Currie (*qv*). Small glazed cabinets house regional and/or thematic displays. Part of Brown's late eighteenth-century early nineteenth-century collection of minerals (*c.*600 specimens) was donated by his daughter Miss Brown and received in the Grant Institute during 1874-75.

### UNIVERSITY OF GLASGOW
#### The Hunterian Museum

The Hunterian Museum, named after its founder William Hunter (*qv*), has major mineral displays and is the oldest surviving museum in Scotland and one of the oldest in the United Kingdom (Rolfe, 1985). Although opened to the Glasgow public in 1807, its contents had been available for scholastic purposes since 1770 at William Hunter's purpose-built museum in Windmill Street, London. Nowadays the museum houses in excess of 500,000 specimens of which approximately 25,000 are minerals. Historic mineral collections include those of Thomas Brown (*qv*), Frederick A. Eck (d.1884?, minerals from Mourne Mountains, Ireland, Europe,

Mexico and South America donated in 1884), James Currie, Sir George Steuart Mackenzie (*qv*), William Nicol (*qv*, including some of the earliest thin-sections ever made), Frank Rutley (1842-1904, of eponymous mineralogy fame; a collection of originally *c.*1500 minerals with manuscript catalogues by Rutley; the worldwide collection was purchased in 1907) and Adolarius Jacob Forster (mineral collector and dealer, who kept his large personal collection in Gerrard Street, London; William Hunter purchased minerals from Forster *c.*1771; the latter's collection formed the basis of his nephew John Henry Heuland's cabinet; Forster catalogues are in the Hunterian Library).

## UNIVERSITY OF ST ANDREWS

The mineral collections within the Department of Geography and Geosciences originated in the museum of the Literary and Philosophical Society of St Andrews. Founded in 1838 by Sir David Brewster, the museum was a joint venture between the university and the society and its collections reflected that bygone era of exploration and discovery. A small catalogue, 14.5 x 8.5cm, dated 1849, lists 135 rock and 398 mineral specimens, of which 400 are extant. All are hand-specimen size being housed in open trays within metal drawer cabinets. Although few specimens state identifying catalogue numbers, a small number possess adhering F. H. Butler labels (1849-1935; a London mineral dealer) and one label indicates Imrie's name. Scattered throughout the collection are Heddle-written labels, and possibly a few Heddle specimens. Additional mineral collections include teaching and reference collections. The former is housed in glass-topped desk cases whereas the latter, totalling 1216 specimens, of which 800 are non-silicates, is currently being computer catalogued. This systematic collection, of small specimens assembled from worldwide localities, comprises native elements, arsenates, carbonates, halides, phosphates, oxides, sulphides, sulphates and vanadates. The mineral collection of C. F. Davidson (1911-67; Professor of Geology at St Andrews 1955-67) reflects his authoritative specialisation in radioactive minerals and ore genesis, especially strata-bound ores. His collection embraces ore minerals with related rocks from Hungary, Czechoslovakia, banded sulphide ores from the Rammelsberg district, Germany, and Canadian and South African gold mines. A collection of 54 ore

specimens, including argentite from Altenberg and Freiberg, was presented to the department during April 1969.

## OTHER INSTITUTIONS

### THE SCOTTISH MINERAL AND LAPIDARY CLUB

Founded in 1958 in Edinburgh; houses a *c.*3000-specimen collection, predominantly of minerals, and successfully accommodates popular lapidary and mineral interests. One of the club's highlights is the annual Heddle Lecture delivered by a guest speaker.

### COLLEGE OF THE FREE CHURCH OF SCOTLAND

Situated in Edinburgh, the college trained ministers and missionaries for the Free Church of Scotland, and possessed a natural history museum to assist in teaching students the wonders of divine Creation. People, and especially missionaries, forwarded specimens from all over the world. Numerous mineral specimens were acquired and presented to the college between 1849 and 1851. In 1966 the collection was presented by the Senatus of the college to the National Museums of Scotland (NMS G 1966.40). Transferred minerals included a large group of amethystine quartz, calcite; radiating tremolite from Glen Tilt, Perthshire; prehnite; a fluorite-smoky quartz geode; and fossil wood from Egypt and the West Indies. Also relocated was a geological map of Scotland by Dr John MacCulloch, dated 23 March 1840, third issue.

### HISTORIC MINERAL COLLECTION AT NAIRN LITERARY INSTITUTE

The geological collections within the museum of the Nairn Literary Institute (founded 1858) include many undocumented rocks, fossils and minerals. Very little original documentation remains associated with the specimens, which were stored for a long period in a dusty, randomly jumbled fashion.

From published records it is known that internationally important vertebrate fossil and mineral collections were presented to the museum. The historic Brodie Cabinet of

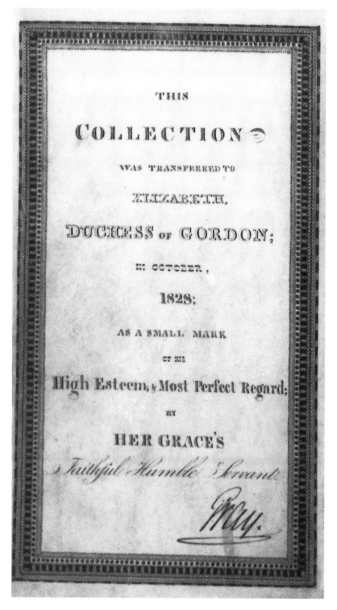

THIS

COLLECTION

WAS TRANSFERRED TO

ELIZABETH,

DUCHESS OF GORDON;

IN OCTOBER,

1828:

AS A SMALL MARK

OF HIS

High Esteem, & Most Perfect Regard;

BY

HER GRACE'S

Faithful Humble Servant

Gray.

Minerals was donated to the museum 'some few years' before 1884, its nucleus being Sir Charles Lewis Giesecke's collection of rare Greenland minerals. Prior to that period, the cabinet had passed through the hands of several noted mineral collectors including Ninian Imrie (*qv*); see also Thomas Allan for an account of the arrival of Giesecke's minerals in Leith, Edinburgh, and their purchase in 1808; Lord Gray (*qv*), the Duchess of Gordon (*qv*) and Brodie of Brodie, each of whom made significant additions. In 1884 it numbered some 2000 specimens made up largely of minerals from Norway, the British Isles and Australian minerals, largely ores from New South Wales, plus gems and semi-precious stones. Its marketable value ranged from £700 to £800, the collection being reckoned the museum's most valuable possession.

Writing about the museum in the *Guide to Nairn*, Bain (?1875) related that it

> possesses a Cabinet of Minerals of great value, presented by the late Brodie of Brodie. It belonged originally to the late Earl Gray, who received 120 valuable specimens from Lt. Col. Imrie. In 1820 his Lordship presented it to the late Duchess of Gordon, in whose possession the collection was largely augmented, so that it now illustrates with great completeness the geology of this country and also of Norway and Greenland, besides embracing specimens of the produce of almost every part of the world.

The author, together with S. Moran of Inverness Museum, in 1990 sorted several thousand dusty rock and mineral specimens and retrieved 77 high-quality mineral specimens, including one of cryolite bearing the number 120. Throughout the search 15 labels were retrieved, of which eleven could be matched to specimens. The labels (one watermarked 1794) clearly demonstrate that material came from both Greenland (schorl) and Norway (labradorite, Fridricharven; and sphene, Arandal). The 77 minerals have been cleaned and identified (including sodalite, cryolite and eudialyte) and are currently housed in the Nairn Literary Institute (Livingstone, 1993b).

## Catalogue

### Cryolite.

**Specimen N°. 120.**

The Cryolite of Jameson Vol. 3. page 299.
Alumine fluatée alkaline of Hauy Vol. 2? page 398.
The Cryolithe of the Germans.

**Synonymes.**

**Its Gen.¹ Colours.**

Are pale Greyish white, Snow white and yellowish brown

**Gen.¹ Forms.**

It occurs Massive, and Disseminated in straight and thick lamellar concretions. It is shining, and the lustre is somewhat Vitreous inclining to Pearly.

**Transparency.**

It is Translucent in the air, and is rendered much more so when put into Water, but it does not dissolve in Water.

**Sp.¹ Gravity.**

Is 2.949 by Hauy, and 2.953 by Karsten.

**Hardness.**

It is softer than Fluor spar.

**Its Chemical Characters.**

It may be melted by a less degree of heat than that which is denominated Red heat, and it may also be melted even by the flame of a candle. Before the Blow pipe it immediately runs into a very liquid fusion, then hardens, and at last assumes the appearance of a Slag from which it is difficult to udret it.

## Catalogue.

### Cryolite continued.

**Specimen N°. 120.**

**Constituent Parts.**

|  | By Klaproth. | By Vauquelin. |
|---|---|---|
| Alumina. | 24.0 | 21.0 |
| Soda. | 36.0 | 32.0 |
| Fluoric acid & Water. | 40.0 | 47.0 |
|  | 100.0 | 100.0 |

**Observations.**

This curious Mineral was discovered by M.ʳ Giesecke in West Greenland among the Rocks of the Fiord or arm of the Sea called Arksut in Latitude 60° about 30 Leagues from the colony of Juliana Hope. M.ʳ Giesecke found it in two thin layers in Gneiss. One of these layers consists of the Greyish and Snow white Cryolite, which is pure, and unmixed with any other Minerals. The other layer is composed intirely of the Yellowish brown Cryolite mixed with Galena, Iron-pyrites, Sparry Iron ore, Quartz and Felspar. These two layers are situated near to each other. The first, or pure layer, is washed by the sea at high water, and a considerable portion of it is laid open to the eye by the decay and removal of the superincumbent Gneiss in which it is situated.

This layer varies in thickness from one foot to two feet and a half. Such is M.ʳ Giesecke's account of the Geographic and Geognostic situation of Cryolite which was so denominated from its similarity to Ice in its fusion, from xρυος Ice, and λιθος Stone, or Icestone.

Entry for, and specimen number 120 – cryolite; one of the earliest, classic Greenland mineral specimens to reach Europe. No. 107 is a eudialyte arfvedsonite syenite, also collected by Giesecke.

## MISCELLANEOUS COLLECTORS

ROBERT BROWN (7 July 1864-29 June 1941) – a native of Wanlockhead, resided at George's Square in the village and left the local school when twelve years old to take up employment in a nearby lead-washing plant. A period in the smelting works followed, after which carpentry became his profession. Although collecting minerals for 40 years constituted his hobby he amassed extensive collections of fossils, china and books, many on geology and mineralogy. Additionally he authored numerous articles and as a keen horticulturist won medals and badges at shows. His mineral collection, which extended to several thousand specimens (Brown, 1919), was largely acquired from Scotland and housed in artistically designed, self-built cases and cabinets. Foreign specimen highlights included gold on quartz from Africa and Australia, azurite from Zambia, Sri Lankan spinel rubies, Canadian garnets and an exquisite pyrargyrite from Vichyne, Hungary. Leadhills-Wanlockhead minerals occupied pride of place in Robert Brown's collection which included 'beautiful specimens of every mineral found in Wanlockhead and Leadhills districts' and 'a fine collection of hemimorphites' (*Hamilton Advertiser*, 24 July 1937). He published two papers in the *Transactions of the Dumfriesshire and Galloway Natural History and Antiquarian Society* detailing the minerals and mining history of the area (Brown, 1919, 1925). Robert Brown is credited with being the first person to record witherite from Scotland (Brown, 1919), whereas Spencer (1931) lists crocoite (Brown, 1925) for the first time from Great Britain. Between 1919 and 1927 Robert Brown sold 69 Leadhills-Wanlockhead minerals (predominantly aragonite, calcite, cerussite, hemimorphite, hydrocerussite, pyromorphite, witherite and vanadinite) to the National Museums of Scotland. An additional sale of 43 minerals from the orefield was made in 1951 by his son, also named Robert. Robert Brown senior died on 29 June 1941 and is buried at Wanlockhead.

ROBERT JAMES HAY CUNNINGHAM (4 April 1815-15 May 1842) – born at George Square, Edinburgh, was a man of exceptional geological talent and great promise, considering that his working life lasted less than ten years. In his biography (Waterston, 1959), there is great acclaim for Cunningham's regional geological surveys of the Lothians, the Inner Hebrides, Banffshire and Sutherland. As a field mineralogist he excelled, recorded numerous localities for various minerals, and collected comprehensively. The following minerals are denoted by Waterston (1959): fluorite, stibnite, kyanite, actinolite, epidote, apatite, magnetite, psilomelane, sphene, staurolite and kaolinite. Cunningham specimens were displayed in the Highland and Agricultural Society of Scotland Museum (he became a member of that society in 1839 and also of the Wernerian Natural History Society sometime between 1832 and 1837) in George IV Bridge, Edinburgh. In a reorganisation of the exhibition after a fire in 1851, the geological collections were transferred to the Industrial Museum, and lately the Royal Museum within the National Museums of Scotland, and numerous Cunningham specimens are traceable. James Hay Cunningham died at his home, 47 Minto Street, Edinburgh (Waterston, 1959).

JAMES CURRIE FRSE (13 April 1863-? November 1930) – born at Leith, was educated at Edinburgh University and at Cambridge, where he gained a BA in the Mathematical Tripos (Jehu, 1932). His scientific interests embraced numerous disciplines including botany and archaeology, although he specialised in crystallographic mineralogy, an extreme subject for an amateur, for his principal business was that of a shipowner. As a highly active member of the Edinburgh Geological Society (President 1904-6) he published numerous papers in the *Transactions* of that society including his Presidential Address 'The Mineralogy of the Faroes, arranged Topographically' (Currie, 1905). He was an ardent mineral collector and travelled widely in search of rare specimens. For a number of years he was closely associated with Professor Heddle, and penned the section on pseudomorphs in Heddle's *The Mineralogy of Scotland* (1901). James Currie gathered together a large, valuable collection which he housed in a museum at his Edinburgh residence, Larkfield. The Currie Collection is characterised by Scottish and Faroes zeolites, a worldwide locality representation and good platinum-group minerals. Following his death, the collection of 4275 specimens was donated to the Grant Institute, Department of Geology, University of Edinburgh, by his widow in 1932.

JAMES DAVIDSON (17 January 1856-? December 1929) – a chemist, County Analyst for Dumfriesshire, and friend of Patrick Dudgeon, commenced collecting in the 1870s. His collecting areas included Leadhills-Wanlockhead (he collected from the working mines), the Lake District and Northern England, and his collection (932 specimens, including *c.*100 from Leadhills-Wanlockhead) was presented to the Grant Institute, Department of Geology, University of Edinburgh, by his son during 1947.

DAVID DEUCHAR (1743-1808) – born at Boscham (Beauchamp) Kinnell, Angus, became an artist of considerable repute and also became well known for discovering the world-famous portrait painter Sir Henry Raeburn. His father Alexander Deuchar (1711-73), a farmer, suffered hardship during the 1745 Jacobite Rebellion and moved to Edinburgh in the same year. He was interested in 'Scotch Stones' (agates), garnets, cairngorms and minerals, which he kept in barrels (to conceal them from raiders) and transported to Edinburgh. Young David Deuchar entered his father's lapidary seal-engraving business, which produced fashionable pebble buttons, and eventually set up his own business and worked from several Edinburgh premises. He became highly proficient as a seal engraver, trained as a goldsmith, and shortly after commencing his own business obtained the appointment of seal engraver to HRH The Prince of Wales.

David Deuchar developed an interest in mineralogy, for he acquired two fine specimens of iridescent labradorite from Mr Shaw (see John Jeans). *The Bee* (1793, vol. 15) refers to David Deuchar's mineral cabinet as 'the elegant collection in Morningside': in 1795 he purchased Morningside House, Edinburgh, set in some 33 acres of beautifully laid gardens, from its probable builder Lord Gardenstone. Following Deuchar's death at Morningside House, his vast collection of prints, paintings and *objets d'art* was auctioned from 25 February 1810 for three weeks, at Mr Ross's saleroom in Drummond Street. An advertisement placed in the *Edinburgh Evening Courant*, 30 October 1815, by Alexander Deuchar, successor to the latter and son mentioned, states 'a Grand Display of Scots Pebbles, Minerals and Heraldry, at 10 Hunter's Square, South Bridge … Open for inspection, … N.B. Cabinets of minerals systematically arranged.' [Further

research may throw light upon David Deuchar's mineral collection; much of his work and possessions remain with his present-day descendants (Smith, 1986). However, a letter from Col. Alexander Deuchar (David Deuchar's great-great-great-grandson) of 28 November 1995 to the author states: 'I have come across no mention of the onward passage of any 'large' collection of minerals during my researches into my family history and the generations I have mentioned are quite well documented.'] The Laing Papers in Edinburgh University Library contain the last two pages of a manuscript mineral catalogue belonging to a collection conjectured to have been located in a natural history collection in Morningside, Edinburgh (Carswell, 1950). A specific reference to an independent natural history museum in Morningside has not been discovered by the author: the collection was conceivably owned by David Deuchar in Morningside prior to his purchase of Morningside House.

The systematically organised catalogue in French (EDU La III 352/1/98) extends to 1152 entries with the collection totalling 120 Livre (approximately equal to £120). This collection was examined by Rudolf Erich Raspe (1737-94) who reinstated it according to the catalogue descriptions. (Raspe was Professor at the Collegium Carolinum, Curator at Kassel, author of the Baron Munchausen's *Travels* – written in a Cornish copper mine; assay master, geologist, mineralogist, collector, prospector, scientific adventurer, an expelled member of the Royal Society, an associate of Joseph Black and James Hutton, and a swindler; Carswell, 1950). He also dealt in minerals, selling specimens to Philip Rashleigh (1729-1811) and Charles Hatchett (Russell, 1952a) and his mineral collection, which included numerous Scottish specimens, was noted by Jameson (1800).

The catalogue pages, headed 'CL IV Métaux', encompass antimony, manganese, molybdenum and tungsten entries with specimens mainly emanating from Freiberg, Schneeberg, Altenberg, Kapnik and Keswick (in the north of England). Following the catalogue entries are three pages bearing extensive notes in Raspe's hand detailing results of his curatorial activities on the collection:

This well chosen Collection I have revised and rectified according to this Catalogue; and to give to its pretty scientifical Arrangement its due permanency numbers have been fixed to each article, which correspond with those of the Catalogue and the boxes of the Cabinet.

Many pieces had been misplaced into wrong boxes nay over into wrong drawers; and I have revised and rectified the whole to the best of my Knowledge and with the closest attention to the pretty accurate descriptions of Dr. Titius.

The report is signed and dated by Raspe 'Morning Side. Febr. 11. 1790'.

THOMAS EDINGTON – of Glasgow, who owned the Phoenix Iron Works at Garscube Road, acquired a fine mineral collection with an excellent representative suite from the Glasgow-Dumbarton area. Between 1812 and 1817 Thomas Edington was elected to the Wernerian Natural History Society of Edinburgh. From his cabinet Haidinger (1825a) received the only specimen bearing edingtonite. Significantly, close to the present-day Phoenix Recreational Ground is Edington Street. Thomas Edington died in the Royal Asylum, Perth, on 26 July 1859. Neither the Deed of Settlement executed by him on 27 June 1846, nor an Edington Inventory (SRO 36/51/40 and 36/48/45 respectively), mention minerals.

HARRY EDMOND (1898-1994) – born at Wanlockhead and started his working life at the age of 15 with a lead-zinc mine, desirous of becoming a mining engineer. He served his time as a fitter and turner, finishing his apprenticeship with the Wanlockhead Lead Mining Company. He went to Africa in 1926 and spent 37 years there on numerous engineering projects, many associated with mines in Tanzania and Zimbabwe, including a period at Broken Hill. Work took him to the Shabani Asbestos Mines and other famous mines in the Copper Belt.

As a keen mineral collector, by 1963 he had accumulated over 1100 ore specimens and semi-precious stones; part of this collection he donated to a museum in Vancouver, Canada, and to the National Museums of Scotland (*Roan Antelope*, 16 February 1963; a newspaper for mine employees published by the Rhodesian Selection Trust Ltd). In his Scottish home Harry Edmond constructed a mineral museum of considerable size, with purpose-built glazed desk and wall cases, drawers and display shelving. His fine collection, in excess of *c*.5000 specimens, strongly featured African minerals, Leadhills-Wanlockhead specimens, faceted gems, cut and polished stones

and agates. Between 1962 and 1991, 18 lots totalling some 300 mineral specimens were received in the National Museums of Scotland from Harry Edmond. Of the African minerals presented, lepidolite, eucryptite, bikitaite, pollucite, dioptase, azurite, smithsonite, malachite and a 3cm wulfenite crystal are noteworthy. Many of his specimens came from world-famous localities: Bikita, Tsumeb, Shinkolobwe and, nearer to his home, Leadhills and Wanlockhead.

JOHN FLEMING (10 January 1785-18 November 1857) – born near Bathgate some 29km south-west of Edinburgh, studied for the ministry yet possessed an intense passion for natural science. He occupied the Chair of Natural History at King's College, Aberdeen, 1834-45, then became Professor of Natural History at the Free Church College, Edinburgh. As a prolific writer he authored more than 100 books and papers on natural history, including geology and mineralogy. In his early twenties he collected data for the *Economical Mineralogy of the Orkney and Zetland Islands* (1807), of which his biographer relates:

> There are evidences in it of great descriptive power, readiness in the application of the nomenclature of the science, correctness of eye, and such a quick appreciation of the economical value of the rocks described, as would not discredit the ablest mineralogist at the present time.

John Fleming's mineral collection, *c*.800 specimens, was donated in 1902 to the Kelvingrove Museum, Glasgow, by Major J. A. Fleming. The collection, from worldwide localities, contains specimens mainly from Scotland, England, Cornwall, Europe, Iceland, the United States, Mexico and Australia.

LORD GARDENSTONE (24 June 1721-22 July 1793) – Francis Garden, born in Edinburgh, educated at Edinburgh University and admitted an Advocate on 14 July 1744. He succeeded to the family estates, and a large fortune, which enabled him to construct a museum at Laurencekirk, Kincardineshire. As an itinerant with a love for natural history and pigs, he came into contact with John Jeans and Alexander Weir and acquired a mineral collection which Hosack (1822), on his tour through Scotland while in residence at the University of Edinburgh during 1792-93, mentions:

I afterwards became more enamoured with this science ['natural knowledge'] by an opportunity which was afforded me of examining at Lawrence Kirk, the small but beautiful collection made by Lord Gardenston.

DAVID CORSE GLEN (12 January 1824-92) – born at Hawkhead, near Paisley, lived in Glasgow, became an engineer in shipbuilding and around 1854 set up business in partnership. He actively pursued scientific interests, taking membership of the Glasgow Geological Society in 1866 and becoming a life member in 1881, also joining the Natural History and Archaeology Societies in Glasgow, and in 1876 the Geological Society of London. Mineralogy formed his specialist interest, and he accumulated a collection of many thousands of specimens from all parts of the world. At the East-End Exhibition (1890-91), held at Dennistoun, Glasgow, a large portion of his collection was exhibited. He wrote several papers, including, with John Young, 'List of Minerals and Rock Specimens found in the Central, Southern, and Western Districts of Scotland' (Glen and Young, 1876). In 1886, he acquired by purchase a collection which had previously been put up for sale by auction in 1805, and allegedly contained minerals from the cabinet of the Earl of Bute (1713-92, whose collection reputedly contained 100,000 mineral specimens); the Museum Geversianum (which contained the 1787 collection of Abraham Gevers; see Wilson, 1994, p. 212); Professor Gaubius (1712-80); Baron Calmeti; a Mr Voigt, and William Babington (1757-1833, who purchased a very considerable portion of Bute's collection). According to the sale catalogue of 1805, which runs to 2701 entries with relatively few Scottish entries, the collection belonged to 'a Gentleman deceased'! In the catalogue it states:

> The whole is comprised in three Mahogany Cabinets, each about 6 Feet high, 2 Feet 7 Inches wide, 1 Foot 6 to 1 Foot 8 Inches deep, each Cabinet containing Forty-two Mahogany Drawers.

The collection was purchased in 1805 by one of the Frasers of Lovat, where it remained in the Fraser family until bought by Glen in 1886 from the executors of Archibald Thomas Frederick Fraser (A. Gunning, *pers. comm.*, 1995).

Glen's collection, acquired by the Corporation of Glasgow in 1896, is now held by Glasgow Museums.

DUCHESS OF GORDON (20 June 1794-31 January 1864) – Elizabeth Brodie, whose father was a younger son of Brodie of Brodie in the north of Scotland, was born in London. Carefully educated and heiress to great wealth, she married George Gordon, the Marquis of Huntly, in 1813. Unblushing vice in high society distressed her considerably, so she turned

Duchess of Gordon, portrait, 68.5 x 58.5cm. In the collections of the National Trust for Scotland.

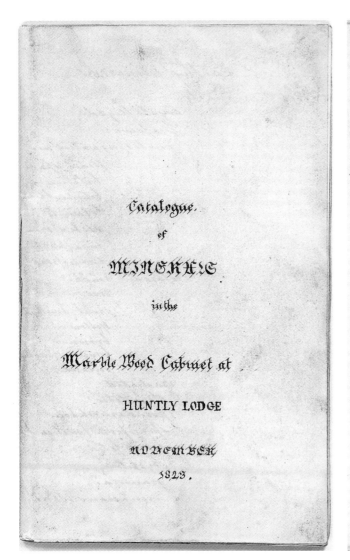

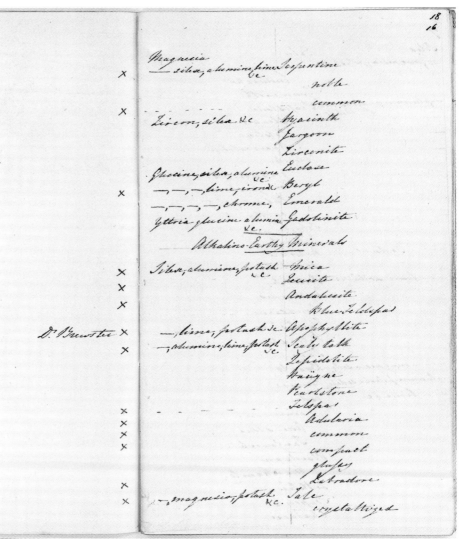

to Bible studies for solace, and to make a complete renunciation of the world. She became Duchess of Gordon in 1827 and began a life of earnest devotion.

Elizabeth Brodie, along with her uncle James Brodie, became interested in science; her financial position encouraged indulgence in her passion for collecting. Sir David Brewster (*qv*) and the Duchess corresponded regularly:

> chiefly upon mineralogy, in which Lady Huntly was most intelligently interested, and she was able also to understand and appreciate Dr Brewster's optical experiments. (Gordon, 1869)

Lady Gordon's mineral collection appears to have been quite extensive, for Elizabeth Grant (Strachey, 1911, p. 381) in *Memoirs of a Highland Lady* writes of 'sorting of what became afterwards a very fine collection of shells and minerals, left by Lady Huntly … to her first cousin Brodie of Brodie'. A catalogue in Nairn Literary Institute of Lady Gordon's mineral collection, in her own handwriting, reveals a deep appreciation of minerals. Her 'catalogue', loosely bound with thread (*c.*60 pages) reveals that her collection was systematically organised: the long-hand work is a direct copy, with annotations, of the classificatory scheme published in the second (1819) edition of Phillips's *An Elementary Introduction to Mineralogy* (1816-23). Crosses in the catalogue possibly allude to representational material in her collection, some species being from more than one locality. A few possible remnants of the Duchess of Gordon's collection were rescued from the admixed collections of the museum at the Nairn Literary Institute (Livingstone, 1993b).

JOHN MORE GORDON (1849-1922) – the eldest son of George More Gordon of Edinburgh, as a student had a distinguished career in classics at Balliol College, Oxford University. Ordained during 1875, he became Vicar of St Johns, Redhill, Surrey (1882-1913) where he began to take an interest in minerals, finding the study a relaxation from pastoral duties. He spent his latter years at Charlton, Montrose. As a keen mountaineer, and member of the Swiss Alpine Club, his mineral collection became considerably enriched in Swiss minerals. Shortly after the new Marischal College of Aberdeen University was opened in 1906, Rev. Gordon visited the new geology department in the college. Forever on the lookout for new specimens, especially from Scottish localities, he fervently believed that Scottish minerals should be fully represented in a Scottish university.

At the time of his death his collection totalled over 7000 specimens. His typed catalogue runs to 3235 entries which encompass worldwide sources. Part of his collection was gifted during 1922, in accordance with his will, to the university. He also bequeathed a fine collection of microscopes, goniometers, refractometers, spectroscopes and petrological apparatus. In the systematic collection of the Department of Geology and Petroleum Geology at the University of Aberdeen, 2997 Gordon specimens, including numerous gemstones, became incorporated, retaining the original Gordon numbers, prefixed by 'G'. Specimen label data in the phosphate section reveals that Gordon pursued a purchase acquisition policy, since a Leadhills pyromorphite is accompanied by a label reading 'James R. Gregory and Co., 139 Fulham Road, London' (G624). Similarly a pyromorphite from Wheal Alfred, Phillock, Cornwall, bears a label reading 'Russell and Shaw, Mineralogists, 11 John Street, Theobalds Road, London WC'. (The London firm of mineral dealers founded by J. R. Gregory, 1832-99, commenced business around 1850 as J. R. Gregory and Co., and eventually as Gregory Bottley and Lloyd, who absorbed several other dealers including Russell and Shaw, founded in 1848 by Thomas Russell; Embrey and Symes, 1987.)

Between the years 1916 and 1921 J. M. Gordon presented a collection of 151 rocks and 54 minerals to Montrose Museum and his widow donated a small rock and mineral collection to the National Museums of Scotland during 1924.

DUKE OF HAMILTON (tenth Duke, 5 October 1767-18 August 1852) – the Duke's earlier years were spent in Italy where he acquired a taste for fine arts. On the accession of the Whigs to power in 1806, he was despatched as Ambassador to St Petersburg. His mineral collection, *c.* 500 specimens and all of which are numbered, but devoid of accompanying catalogue or list, is housed in Lennoxlove House, Haddington, East Lothian (Stace *et al.*, 1987). The tenth Duke, a contemporary of fellow Whig Robert Ferguson, died at his home in Portman Square, London.

THOMAS CHARLES HOPE (21 July 1766-13 June 1844) – Professor of Chemistry at Edinburgh University, is known for

1) his discovery that water attains a maximum density at a few degrees (4°C) above freezing point, and

2) his work on strontium salts extracted from strontianite.

Jameson (1800) refers to Hope's mineral specimens, and Svedenstierna (Dellow, 1973) in 1802-3 related that 'Dr Hope also possesses a nice collection of minerals, particularly of those graduated steps which serve to illustrate the above-mentioned subject' (*ie* geology). According to the NMS Geology and Zoology Donor's Index, Mr James Hope gifted 'a few specimens of Strontites [to Jameson's collections]' (NMS G 1847.23), although currently they can no longer be specifically identified in the Scottish Mineral Collection. Mr James Hope additionally gifted a large collection of geological specimens (NMS G 1847.21), property of the late Dr Hope, Professor of Chemistry.

*Right:* James Hutton *c.*1782, by
Sir Henry Raeburn, oil on canvas,
127 x 101.6cm. Scottish National
Portrait Gallery.

JAMES HUTTON (3 June 1726-26 March 1797) – the 'Father
of Modern Geology' (of uniformitarianism), was born in
Edinburgh, attended Edinburgh High School and entered
Edinburgh University in November 1740 at fourteen and a
half years old. From 1744-47 he studied medicine there, and
graduated MD in September 1749 from Leyden. Hutton's
interest in geology may well have commenced as early as 1748,
when he studied in Paris. Passionately fond of chemistry, he
may have attended G. F. Rouelle's lectures on science which
included geology and mineralogy (Dean, 1992). Hutton relin-
quished his Norfolk agricultural work and moved to
Edinburgh in 1768, whereupon his personal scientific studies
advanced in the company of Joseph Black. Hutton derived
important stimuli from Cronstedt's chemical analysis of
minerals, which led to his own analysis of zeolite, prior to 1772,
in which he discovered the presence of soda. In Hutton's
(1788) second edition of the *Theory of the Earth* he devotes 22
out of 25 pages to minerals which he maintained formed
(fused) underground, for example ores, mineral veins, rock
salt, septarian nodules, agates and marble. He discusses
gold, galena, blende, pyrite, fluor, calcareous spar and quartz
crystals.

Upon Hutton's death, his surviving unmarried sister,
Isabella Hutton, gave his rock and mineral collection to
Joseph Black, who presented it to the Royal Society of
Edinburgh as an illustration of the 'Huttonian theory of the
Earth'. The society accepted the collection in December
1799 (Jones, 1984) and appointed Professor John Playfair and
Sir James Hall as trustees. Under its charter the Royal Society
of Edinburgh had no power to retain its own collections, but
was duty-bound to deposit them where scholars and the
public could study the specimens – in the University
Museum. Hutton left his collection in little semblance of
order (Jones, 1984) and the trustees delayed work on it for
several years before handing it over; the collection appears to
have arrived in the University Museum in 1808. (Possibly
during early rearrangement, some of Hutton's minerals were
sold to Robert Ferguson of Raith, *qv.*) Hutton purchased
specimens from dealers and actively exchanged, one exchange
being a box of minerals which he forwarded to Russia for
Matthew Guthrie (1743-1807), who resided in that country at
St Petersburg from 1769 until his death on 30 August 1807
(Sweet, 1964). An account of Hutton's collection was pieced

together from fragmentary evidence by Jones (1984). He was
interested in mineralogy *per se*, and even on the day of his
death Hutton wrote down some remarks on new attempts at
the classification of minerals (Dean, 1992). Professor Robert
Jameson was very unwilling to display Hutton's specimens,
since the latter's geological principles clashed with Jameson's
staunch Wernerian discipline, and Hutton's collection
disappeared.

Hutton is buried in Greyfriars Churchyard, Edinburgh,
where his epitaph promulgated by Geikie reads: 'Founder of
Modern Geology.'

JOHN JEANS (*c.*1724-1804) – an enthusiast for mineralogy,
he lived in Aberdeen, although he travelled extensively on foot
over the greater part of mainland Scotland, including the
Highlands, and visited London annually. On his tours he

collected extensively until he became eminent as a dealer. Jeans was a man of some note: in London in 1769 he resided at Dr Johnson's house with James Ferguson the astronomer (*Notes and Queries*, 1904), and was probably a man of considerable means.

The first specimens of gold-blue labradorite from Labrador were shown

> in the year 1790, in a ware room on the South Bridge, [Edinburgh] by one Shaw, from London, a native of Aberdeenshire, who I think keeps a shop of Natural History in the Strand; and was the same person who sold that wonder of nature, the Elastic Stone, to the honourable Lord Gardenstone …. Mr Shaw again paid us a visit as late as November 1792, when he exhibited some most brilliant specimens of Labrador spar; particularly one of fine extremely bright and variegated colours; one pretty large, of the scarce fine colour with the purple tinge, and one with gold, blue and green shades; the first was sold to the celebrated Dr. Black: the two last are in the elegant collection at Morningside. (*The Bee*, 1793, vol. 15, p. 99)

John Jeans (the Scottish fossilist) discovered a similar labradorite in Aberdeenshire, although it exhibited only golden colours. He supplied numerous collectors with minerals, including William Hunter between 1774 and 1777. Jeans supplied, on 21 April 1774, 16 specimens to Hunter costing £2-3s, which included minerals from Banff, Leadhills and Wanlockhead and an old silver mine in West Lothian. Other specimens, serpentine and asbestos with copper ore from Ross-shire, were supplied in March 1777 for £2-6d, and seven specimens in April of that year for £2-3s-6d (Hunterian Museum manuscript M553.554). A friend of Professor Black, J. S. Hermann, when visiting London on his way back to Germany, corresponded with Black and noted:

> A certain Jeans from Aberdeen is here with Scotch fossils and asks great prices for them. He has indeed very fine specimens of white and green asbestos, exceeding fine schorl, amienth, mica and serpentine stone …. I enclose here a Frank catalogue of the Collection with the price fixed to each specimen. If you or Dr Hutton should be pleased to make such a bargain, you may give your orders to Mr. Jeans … he is here till the 24th June. (EDU. Library MS Gen. 873/1/89)

ISAAC LAWSON (d.1747) – born in Scotland, studied medicine and botany at Leyden University, became a friend of the celebrated Linnæus, and Physician General to the British Army. The Rev. William Borlase (1696-1772), a Cornishman and mineral collector of immense generosity, who readily donated specimens to all enquirers, dispatched three shipments of minerals to Leyden for John Andrew, a medical student at the university. Andrew divided the specimens between Boerhaave, Gronovius and Linnæus, who was there on an extended visit (Wilson, 1994). Isaac Lawson was the first patron of Linnæus, enabling him to emerge from obscurity to notoriety. Mendes da Costa (1717-91), a London mineralogist of Portuguese-Jewish descent, purchased Isaac Lawson's collection:

> Dr Lawson was extremely well skilled in fossils [minerals], had an extensive correspondence, especially in Germany, and had made large and elegant collections of fossils. In 1767 I bought a large parcel of his collection, and numbers of specimens his brother gave away and were lost. (Mendes da Costa, 1812)

(In November 1747 Mendes da Costa, a conchologist and mineralogist, was elected a Fellow of the Royal Society, becoming Clerk on 3 February 1763. From that date he also served as Librarian and Keeper of the Museum (Hall, 1992) until dismissal in December 1767 for dishonesty. After arrest his library and collections were seized and sold by auction during May 1768.)

SIR GEORGE STEUART MACKENZIE (22 June 1780-? October 1848) – became known to the scientific community in 1800 after obtaining 'decisive proof of the identity of diamond with carbon' by a series of experiments on steel formation from iron plus carbon (Mackenzie, 1800). All the diamonds used in the experiments were cut and polished. In these experiments he is said to have made free use of his mother's jewels and by means of diamond powder converted iron into steel (Gordon, 1869). He was elected as an ordinary member of the Royal Society of Edinburgh in 1799. As a pupil and friend of Professor Jameson, Sir George Mackenzie developed an intense interest in geology and mineralogy. During 1810 he undertook a pioneering six-month geological tour of Iceland, where he collected 300 rock specimens (Peacock, 1927) and zeolites. The rock collection was presented to the Royal Society of Edinburgh which, in 1909, donated it to the Hunterian Museum, Glasgow. Sir Charles Lyell (1881) referred to his Icelandic minerals as 'the magnificent collection of mineralogical treasures'. To support the conclusions he had formed with regard to Icelandic geology, Mackenzie visited the Faroe Islands in 1812 accompanied by Thomas Allan. Mackenzie (1815) described the geology of the Faroes, whereas Allan (1815) published an account of the minerals. In 1813 Sir George Mackenzie donated his minerals to the Hunterian Museum, Glasgow. As a keen collector he developed a particular interest in chalcedony and its formation. Of chalcedony he wrote:

> although my cabinet contains a variety of forms, perhaps unequalled in any other collection, I have, for many years, refrained from describing them, because the oftener I contemplated my specimens, the less able I found myself to apply any single agent, so that its operations alone could account for their forms. (Mackenzie, 1826)

DAVID MAYNE (fl. 1700s) – a Scottish field collector who built up a large collection of minerals, ores, petrefactions and gems, all self-collected, chiefly from Scotland during the course of 22 years travelling through Great Britain and Ireland. His collection, all catalogued, was appraised and valued at 200 guineas by Mendes da Costa who occupied the position of Librarian and Keeper of the Museum at the Royal Society. The collection was advertised for sale on several occasions in different papers including the *London Chronicle* 21 December 1765, although eventually the collection was sold by public auction during 1766 (Mendes da Costa, 1812).

GILBERT LAING MEASON – an Edinburgh collector active in the late eighteenth and early nineteenth centuries. Gilbert Laing Meason is not listed in the *Edinburgh Post Office Street Directories* for 1795-1806; however, a Gilbert Laing, merchant, of 26 St Andrews Square, is mentioned. In 1806-7 a person with the same name is listed as Advocate, whereas in 1807-9, at the same address, resided a Gilbert Meason, and Gilbert Laing. The 1809-10 listings mention only Gilbert Meason at the same address. However, Gilbert Laing Meason correspondence (1802) in the National Library of Scotland bears the address

'St Andrews Square, Edinburgh' (NLS MS 3418, fol.151). In 1815 he was elected an ordinary member of the Royal Society of Edinburgh. Gilbert Laing Meason, a long-standing friend of James Sowerby (1757-1822), provided numerous specimens, especially Scottish, for illustration and description in Sowerby's five-volume work *British Mineralogy* published between 1804 and 1817. In 1820 minerals were also donated to Brewster for optical study:

> In the month of February 1820, Mr Laing Meason, to whose kindness I had been indebted for various specimens of lead-ores for the purpose of optical analysis, sent me some crystalline fragments of a yellowish-green substance from Wanlockhead. (Brewster, 1820)

In his capacity as manager of a mining company at Wanlockhead, Gilbert Laing Meason possessed an excellent opportunity to acquire and donate mineral specimens, for he was interested in science. Two years after donating secondary lead minerals to Brewster, in a letter to Thomas Traill, he stated on 2 March 1822, 'I have given up my lead business to the manager by proper agents and sold my house in Edinburgh' (NLS MS 19339, fol.56).

There is no index entry for Gilbert Laing Meason in Jameson's Old College Register.

REV. JAMES MORGAN (d.1869) – of Barclay Street, Stonehaven, Kincardineshire, and an honorary member of the Montrose Natural History and Antiquarian Society, collected Scottish and European minerals during the mid-nineteenth century. He amassed a collection of approximately 1500 minerals which he catalogued in no particular order. A species or chemical name is detailed for each numbered specimen, although locality information is rarely recorded. His mineralogically diverse collection was bequeathed to Montrose Museum in his Will dated 1857. Two almost identical manuscript catalogues, by Morgan, accompany the collection (M. H. King, *pers. comm.*, 1995).

JOHN MURRAY (d.1820) – a lecturer on chemistry at the University of Edinburgh who critically evaluated the Huttonian and Neptunian explanations for rocks and the distortion of strata (Murray, 1802). Occasionally delivered lectures on mineralogy and wrote a lengthy paper (Murray, 1815) on diffusion of heat at the earth's surface. Svedenstierna (Dellow, 1973), on his 1802-3 tour of Scotland, commented that 'Dr Murray … is an active and knowledgeable man, and has a nice little collection of minerals'.

## ALEXANDER THOMS (1837-1925)

ALEXANDER THOMS (1837-1925) – the son-in-law of M. Forster Heddle, who also resided in St Andrews, continued the Heddle reputation for mineral collecting. During the late nineteenth and early twentieth centuries, Thoms accumulated over 5000 mineral specimens (his undated manuscript catalogue runs to approximately 5200 entries) and in excess of 1200 agates. His considerably depleted collection went to the Hunterian Museum via the geology department of Dundee University, where it was formerly housed. Highlights include a 20cm Japanese stibnite specimen, tetrahedrite after bournonite, a 15cm Cornish specimen enriched in bournonite, and a good representative Leadhills-Wanlockhead suite. From Bryce-Wright, Thoms acquired mendipite from the type area. Matlockite, gold in matrix from Landlord's Brae, Wanlockhead, and a large euhedral schorl are also notable.

Thom's collection contains specimens accompanied by early F. H. Butler labels produced shortly after the death of R. Talling (1820-83, a renowned Cornish dealer whose business was taken over by Butler). During 1915 Thoms presented to the National Museums of Scotland a set of 888 wooden crystal models (NMS G 1915.11. 1-888) illustrating Haüy's (1801, 1822) *Traité de Minéralogie*. Alexander Thoms was a member of a small committee consisting of Heddle's daughters, W. and J. G. Goodchild, and James Currie, formed 'for the purpose of bringing out my works on the Minerals of Scotland' (Heddle's Will, SRO SC 20-50-78). He also wrote *Memoir of Dr Heddle* (in Heddle, 1901). Accompanying Thoms's collection are three small notebooks inscribed 'Heddle' and 'Notes from Mr Rose's lectures', apparently written by Rose for Mr Nisbet, one of his class.

## WILLIAM THOMSON (1761-1806)

WILLIAM THOMSON (1761-1806) – of English birth, matriculated at Queen's College, Oxford in 1776 and then moved to Edinburgh to matriculate in medicine during 1780-81 and 1781-82 (Waterston, 1965). Thomson's interest in mineralogy had already nucleated during his Edinburgh years, yet he must have been influenced by the principles of the Rev. Dr John Walker. In 1782 Thomson returned to Oxford, where four years later he was elected physician to the Radcliffe Infirmary, and a Fellow of the Royal Society. He suddenly left England and eventually settled in Palermo, Sicily, where he studied volcanic rocks. Thomson's extensive collection of 'ten thousand geological specimens' arrived in Edinburgh in May 1808. The arrival was publicised by Professor Jameson:

> It gives us the greatest pleasure to inform the public that the magnificent and extensive collection of minerals bequeathed by that celebrated mineralogist the late Dr Thomson of Naples to the University of Edinburgh has arrived in this country in a state of perfect preservation and is now deposited in the Museum of the University. (Waterston, 1965)

Thomson made numerous contributions to mineralogy, including the discovery of natural oil inclusions in quartz and the original sarcolite specimens (Waterston, 1965; Livingstone, 1984). He also discovered etching phenomena on polished acid-treated meteoritic iron, and in 1808 published the earliest revelation of the Widmanstätten structure (Paneth, 1960).

Very few Thomsonian specimens can now be recognised in the National Museums of Scotland's collections. (For a more detailed history of the Thomson Collection up until 1866, see Waterston, 1965.)

## THOMAS STEWART TRAILL (29 October 1781-1862)

THOMAS STEWART TRAILL (29 October 1781-1862) – born at Kirkwall, Orkney, graduated in medicine from Edinburgh University and in 1832 was appointed to the Chair of Medical Jurisprudence. He contributed mineralogical papers to the *Transactions* and *Proceedings* of the Royal Society of Edinburgh, to which he was elected in 1819. Around 1842-43, he wished to dispose of his extensive mineral collection, which he kept in handsome cabinets, of 3000 specimens (plus 1500 rocks and 500 fossils including excellent fossil fish from Orkney). The collection was noted for its fluorspars, barite, lead and copper minerals, zeolites, feldspars, apophyllites, scapolites and ores of silver and tellurium. It also included an outstanding transparent euclase (2 x 1cm), and gold in matrix from Leadhills (Jameson, 1843). Traill toured the Shetland

Islands in 1803 (Traill, 1806), making extensive mineralogical observations where during the tour he discovered strontianite in the lead mine near Stromness (Traill, 1823). He also compiled a manuscript catalogue of the collection at the Royal Institution, Liverpool, which later became incorporated into the Department of Earth Sciences at the University of Liverpool (Rushton, 1972) and more recently Liverpool Museum. Traill's handwritten catalogue extends to 3098 entries of which 410 are Scottish. Seven hundred and twenty Traill specimens are extant, including 72 minerals from wide-ranging Scottish localities (W. Simkiss, *pers. comm.*, 1997). Of the extant Scottish material, 16 specimens are from Wanlockhead (witherite 1, barite 1, pyromorphite 6, cerussite 6 and anglesite 2), and four from Leadhills (calcite 2 and anglesite 2). Traill may possibly have accepted an invitation to collect at Wanlockhead, for a letter from Gilbert Laing Meason, dated 12 November 1809, states:

> I go to Wanlockhead mines about 28 or 29 May till 4 or 5 December. If you could be there at that time you would have an opportunity of getting additions of the collection there to your minerals. (NLS MS 19337; fol.182)

Traill arranged the sale of Thomas Allan's collection to R.H. Greg (Embrey, 1977).

REV. DAVID URE (1749-98) – born in Glasgow the son of a weaver, went to the city grammar school and afterwards attended the University of Glasgow. He turned his attention to the undeveloped science of geology. While assistant to the minister at East Kilbride, he studied the local history, natural history and geology. The results of his labours were published as *The History of Rutherglen and East-Kilbride* (Ure, 1793). The work is worthy of special mention for it represents one of the earliest attempts to detail geological features of a specific area. Ure's work revealed the presence of a variety of minerals, including calcite, quartz, pyrite, gypsum, zeolite, galena, mica and barite. His collections of ores, rocks and other natural history specimens were advertised for sale by auction in Edinburgh on 26 and 27 December 1798, and his fossil collection went to the museum of the Royal Society of Edinburgh (Murray, 1904) and ultimately to the Hunterian Museum, Glasgow.

HENRY WITHAM (1779-1844) – born Henry Thomas Silvertop, changed his name by royal licence to Witham when in 1800 he married Eliza Witham of Headlam, Durham, heiress to the estates of William Witham of Cliffe, near Piercebridge, County Durham (Rackham, 1983). He was a kind and generous man with a passion for geology, and gambling. When overwhelmed by financial difficulties in 1826, he moved from the family residence at Lartington Hall, County Durham, and spent the following six years in Edinburgh (*Edinburgh Post Office Directories* reveal that throughout that time, he resided at 14 Great King Street). During this period he actively pursued his geological interests: Jameson (1829) records Witham's geological and mineralogical collection as being in Edinburgh. Witham undertook a mineralogical excursion to the Scottish Highlands in August 1824, during which he collected a deep carmine-red mineral filling vesicles in Glen Coe lavas. Brewster (1825) published results of an optical and morphological study on this mineral which he named withamite – a manganese-bearing variety of epidote. Henry Witham became a member of the Wernerian Society in 1822-23 and is well known for being the founder of palaeobotanical structural studies; he learned thin-section preparation techniques from William Nicol.

## CONTEMPORARY COLLECTORS

THOMAS JAMES KEMP MEIKLE – a retired civil engineer, was born on 23 March 1918 at Glassford near Strathaven, Lanarkshire. Called up in 1939, he served in the Royal Engineers and attained the rank of Major before demobilisation in 1945. Although involved professionally with rock characteristics, he became keenly interested in lapidary and mineralogy as a hobby during the mid-1950s. His work brought him into contact with the late Dr J. E. Richey FRS, FRSE (1886-1968) of the Geological Survey, who presented him with a comprehensive rock collection. Civil engineering construction works afforded opportunities for *in situ* mineral studies and collecting throughout Scotland. Extensive collecting trips, targeted at mineralogically renowned areas, were undertaken to Cumberland, Northumberland and Derbyshire (1968-74, and 1983-84); Somerset, Devon and Cornwall (1977 and 1988); Wales (1987); and Norway (1975).

Thomas Meikle's collection, largely self-collected, is in excess of 1100 hand and cabinet-size specimens displayed in cases and on shelving. Additionally, 1500 (plus) micromount to thumbnail-size specimens exhibiting well-developed crystal forms are housed within individual plastic display boxes. Displayed specimens are labelled, although only micromount/ thumbnail minerals are fully catalogued. Some curatorial data, which includes a specimen description, locality and area, can be computer manipulated to generate species and regional listings. Eleven hundred micromount/thumbnail specimens are of Scottish origin, of which 300, collected from Boyleston and Loanhead quarries and Hartfield Moss, are representative of Clyde Plateau lavas mineralisation. The Leadhills-Wanlockhead orefield mineralogy, and minerals developed in slags produced from smelter processes, are represented by 230 specimens covering most of the 70 or so species known from the area. Other suites include material from Strontian (30), Mull (40), Alva (30), Kirkcudbrightshire (40), Solway Firth (20) and Glen Gairn (105). An additional 300 specimens were collected from as far afield as Shetland, the Inner and Outer Hebrides, and the Southern Uplands, with 160 from Caldbeck Fells and 100 from Devon and Cornwall. Classic foreign localities have contributed via exchange – 50 from Mont St Hilaire, Canada, and 108 from Arizona, New Mexico, Nevada and Utah.

As a founder member (1947) and President (1952) of the Strathaven Camera Club, Thomas Meikle's keen interest in photography naturally turned later to mineral specimens. Consequently, an extensive photograph library (750 slides, 500 prints) of macro, micro and thin-sections (self-prepared) was accumulated. Running parallel with photographic interests, lapidary activities, including faceting, flourished, for Thomas Meikle is a founder member, and past Chairman, of the West of Scotland Mineral and Lapidary Society (Glasgow) in addition to being a member of the Scottish Mineral and Lapidary Club (Edinburgh). Thomas Meikle is a member of the British Micromount Society (winning the Founder's Cup 1992-93) and of the Russell Society.

In 1968 Thomas Meikle wrote a 15-page printed booklet *Lapidary Notes – Basics for Beginners* and he then published two articles in *Gems* (Meikle, 1970, 1970a). *Gems* magazine, in the summer of 1971, published an issue devoted mainly to mineral and gemstone sites in the British Isles. Owing to popular demand, the issue rapidly went out of print; however, the relevant papers, including those by Mr Meikle, were reprinted by Lapidary Publications (*c*.1971) in *Finding Britain's Gems*. Other published works include papers on minerals from the Clyde Plateau lavas (Meikle, 1989, 1989a; 1990), acanthite (Meikle, 1992), greenockite (Meikle, 1992a) and native silver from the Hilderston Mine, West Lothian (Meikle, 1994).

Thomas Meikle has donated many specimens from Scottish localities to the National Museums of Scotland, including veszelyite, anatase, a specimen bearing diaspore crystals, djurleite and elyite.

GORDON SUTHERLAND – born on Stronsay, Orkney Islands, during 1918, combined a farming life with 30 years of mineral collecting. His collection (*c*.5000+ specimens), virtually self-acquired and characterised by great diversity in specimen size, species and localities, is displayed on open shelving or within glass-fronted cabinets and desk cases in characteristic Sutherlandian fashion. Specimens in profusion rest in juxtaposition on loose, hand-written labels which detail locality information.

As an avid collector, he has roamed difficult and remote terrains in his quest for Scottish minerals. Consequently, displayed material contains rare species intermixed with excellent examples of rock-forming minerals. Outstanding in the former respect is a 1cm phenakite crystal group from Ben Macdhui, Aberdeenshire (Scotland's second highest mountain), while an example of the latter is a 45cm muscovite crystal extracted from the Loch Nevis mica prospect, Knoydart, Inverness-shire. Regional collections feature prominently, with remarkably fine representative minerals from Skye, Mull, Aberdeenshire, Strontian, the Shetland Islands and Leadhills-Wanlockhead. Skye apophyllite (1.5cm) originates from the famous Heddle locality at Sgurr nam Boc, while Mull zeolites include Ben More scolecite displaying fans of 12cm radii. An aesthetic radiating aragonite specimen from the Laverock Hall Vein area, Leadhills, typifies high-quality specimens collected many years ago, a quality that is immensely difficult to repeat today.

Gordon Sutherland's collecting forte, *ie* large, numerous, or unusual, is mirrored throughout his collection. A 9cm beryl crystal from Pitcaple, Aberdeenshire, a group of 30 x 20 x 15cm calcite rhomb-bearing specimens extracted at Bleaton Quarry, Kirkmichael, Perthshire, and a deep blue-green amazonite boulder (*c*.40 x 32 x 32cm) from Ben Loyal, Sutherland, are constellated with 'monomineralic' displays. Included in the latter are 60 rutile specimens found at Loch na Lairige, Killin, Perthshire, and around 130 smoky quartz crystals (<20cm) from Beinn a Bhuird, Aberdeenshire. The combination 'large and unusual' is exemplified by a corundum-spinel specimen (40 x 20 x 20cm) bearing numerous corundum plates (6.5 x 5 x 0.5cm) in part sapphire, found in drift at Glenbuchat, Aberdeenshire.

Previously housed at Kirkmichael, the collection was then described by Cooper (1976) as

> the most wonderful collection in Scotland, if not in the whole British Isles, of 'rock' – no museum has a collection like it – all collected by Gordon Sutherland – cut – polished – labelled – stones from all over Scotland – Cairngorm bigger than your fist – lumps of amethyst weighing over two cwt, agates by the hundred – I simply could not take it all in – stone from Shetland – Norway – hundreds of them.

A significant specimen in the National Museums of Scotland, the 'Beinn Bhreac Boulder', which measures 76 x 61 x 46cm and comprises fist-sized euhedral amazonite studding a vein-cutting syenite, formed the subject of a description by Heddle (1883). Renowned for finding large and unusual specimens

and relentless in his pursuit of minerals, Gordon Sutherland located the precise spot from where the Beinn Bhreac Boulder was obtained and an article relating to its re-discovery appeared in the *Highland News* of 10 November 1984. (Of the Beinn Bhreac Boulder, Heddle reported in 1883, 'I regard this as mineralogically the most wonderful and interesting stone in Scotland. It was not only in itself a perfect mineral casket, containing in its small bulk [it is a fragment from a block estimated to weigh 100 tons] as many minerals as do some counties, but it contained these in numerous crystalline forms, and also in crystals of large dimensions'.)

Heddle presented the specimen to the Duke of Sutherland, and subsequently the Countess donated it to the National Museums of Scotland during 1966 (NMS G 1966.41).

For some part of 1988-89 (at the age of 70) Gordon Sutherland undertook, for Bergen University, a botanical study of the Lofoten Islands and the arduous Svartisen glacier area (covering some 1000 square kilometres of ice) in northern central Norway on the Arctic Circle. Consequently, Norwegian minerals feature prominently in the collection, amongst which is an excellent hexagonal blue beryl crystal (3.8cm face to face), from Bjornfost. In the largely unmapped terrain, he named nunataks after conspicuous geological or mineralogical features, *eg* black-banded nunatak, rose quartz nunatak – names retained on today's maps.

Gordon Sutherland generously presented material to be fashioned into friendship tokens for international celebrities or accompanied them on mineral collecting trips to the Cairngorms. For over 25 years he has donated Scottish mineral specimens to the National Museums of Scotland, including paratacamite from Dererach, Loch Scridain, Mull, and mullite from Stron nam Boc, Carsaig Bay, Mull.


ROBERT TALBOT GARDNER – born on 24 September 1920 at Bridge-of-Weir, Renfrewshire, worked for 34 years as a rice miller in Glasgow. Between 1939 and 1946 he served in the Royal Naval Volunteer Reserve. After joining the West of Scotland Mineral and Lapidary Society during 1967, his hobby activities leaned particularly towards collecting and studying minerals. His collection, of which 70% is personally collected with the remainder accumulated via gifts, exchanges and purchases, totals some 800 specimens, 200 from Scotland.

Specimen sizes range between 1cm and 25cm overall, averaging 5cm. An exceptional specimen from Loanhead Quarry, Beith, Ayrshire, presents a white and salmon-pink thomsonite spray extending over a 20 x 20cm face. Rarities include a 5mm greenockite crystal found in prehnite near Bishopton, Renfrewshire, and chalconatronite from Anglesey. One section is devoted exclusively to calcite exhibiting various morphologies; others feature quartz and lapidary materials.

The collection is housed in a large, modified, glass-fronted oak bookcase containing concealed strip lighting. An inscribed label bearing the main species name, ideal chemical formula and locality information accompanies each specimen. Arranged on a systematic basis, the collection contains silicates, phosphates, sulphates, sulphides, carbonates, oxides, halides, tungstates, molybdates, vanadates, arsenates, borates and native elements. Self-taught qualitative chemical analysis, plus crushed grain optical identifications utilising a polarising microscope, are Robert Gardner's species validatory methods. Within his chemical armoury are flame, closed and open tube tests, plus reagents for qualitative cation determinations. Little emphasis is placed upon regional representations, although collecting trips concentrated upon Skye, Mull, Tiree, Leadhills-Wanlockhead, the Shetland Islands, Caldbeck Fells, Cornwall and Wales, all at various intervals between 1968 and 1990.

Curatorial details for each specimen, ie number, main species name and locality, are entered into a leather-bound catalogue under the appropriate systematic group. Additionally there is a cross-referencing numerical index.

Robert Gardner donated to the National Museums of Scotland a sample of chalconatronite from Trearddur Bay, Holy Island, Anglesey, and a specimen of ilmenite (7 x 0.5cm plates) in quartz, found at the lochside of Loch na Lairige, by Ben Lawers, Perthshire.

JAMES WILLIAM MILLER – born 10 February 1926 at Bannockburn, Stirling, served 26 years in the police force then pursued a career spanning 17 years in local government. His interest in geology stemmed from schooldays. Mineralogy enthralled and captivated him after his mid-1950s discovery of a galena-bearing cherty rock exposed in Hopeman Harbour, north-west of Elgin. During 1960 he met and received encouragement from the late Ernest Stollery, a Portsoy fossil and mineral collector and dealer. A visit to Oxford University Museum proved fruitful, for he struck up an acquaintance with the late Arthur Kingsbury (1906-68). With Kingsbury's characteristic enthusiasm and encouragement, mineral collecting developed into James Miller's main hobby.

His collection, in excess of 2500 specimens, which is displayed on shelving, contains material from both United Kingdom and foreign localities. Forty-five per cent is self-collected, the remainder accruing from personal exchange or international connections. Specimens range from micro-mounts through hand-size to 30cm. Latterly exchange activities have diminished; collecting endeavours previously concentrated upon localities in Aberdeenshire, Banffshire, Moray, Nairn, Inverness-shire, Skye and Leadhills-Wanlockhead. Each specimen is numbered and catalogued, with 70% of the collection additionally detailed in a card-index system. Highlights include xonotlite specimens from the Bin Quarry near Huntly, Aberdeenshire, and specimens exhibiting rare, bright-green octahedral fluorite associated with laumontite from Limehillock Quarry, Grange, Banffshire. Another noteworthy specimen manifests excellent 1cm-long stibnite crystals in calcite with colourless, purple-banded fluorite from Blackhillock Quarry, Keith.

Around 1986 the Moray District Council museum curator requested assistance from James Miller in acquiring a collection of local minerals for the Falconer Museum at Forres. The J. W. Miller catalogued collection of minerals, including species donated prior to 1986, at the Falconer Museum, runs to over 340 specimens. In 1987 and 1988 during early quests, unbeknown he followed in the footsteps of Dr Keith Nicholson, who had previously studied manganese mineralisation within the Dalroy environs (Nicholson, 1990). A specimen collected from Dalroy by James Miller, and subsequently presented to the National Museums of Scotland, contains black, lustrous crystals (<2mm) of hausmannite – the only known occurrence in Scotland (Livingstone, 1993a). Additional donations include fluorite, stibnite, braunite, laumontite, marcasite (cockscombe), rhodochrosite and hydrozincite.

A description of his mineral collecting activities at Leadhills-Wanlockhead, Newton Stewart and Cairnsmore appeared in print during 1973 (Miller, 1973).

ALAISDAIR JOHN MACKAY – a civil engineer and project management consultant, born on 27 August 1941 in the hamlet of Bettyhill, Sutherland. At Robert Gordon University and Aberdeen University he read civil engineering, and allied geological studies provided opportunities to commence a rock collection. Interest in rocks stemmed from boyhood; aged nine, he accompanied Geological Survey geologists mapping around his Sutherland home. During this early indoctrination he learned about schist and gneiss, and discovered how to recognise hornblende and other minerals – an interest perpetuated to this day.

Upon graduating and commencing employment, Alaisdair Mackay discovered that minerals could be purchased, thus a window of opportunity opened and a lifetime's passion to acquire selected, quality species began. His first purchased minerals were an exquisite smoky quartz, and barite (each cost £1) from a £7 weekly wage. Towards the late 1930s, Bottley of Old Church Street, London, had purchased a collection of Pala (California) tourmalines to construct a crystal grotto, although the war intervened and the grotto was never commenced. From this consignment, Alaisdair Mackay acquired a tri-coloured tourmaline for 10/- (50 pence). Foreign business trips now afford the opportunity for mineral acquisition and recently a fossil bivalve containing vivianite crystals was purchased in Moscow.

Part of Alaisdair Mackay's purpose-built, private mineral museum.

The collection is housed in a purpose-built room, with roof spotlights and continuous shelving around the walls; some minerals are displayed on units, whereas others are retained in drawer systems. A truly international flavour characterises the collection, with some 3500 specimens, from approximately 70 countries, presented. Specimens range from a 54.4kg (82cm long and 67cm girth) rock crystal, to cuboid quartz; plus other species on microscope slides. In general, specimens average 3.8cm across. Few spectacular cabinet-size specimens have been acquired primarily due to paucity of display space. Although the collection range is highly varied, from a Pacific Ocean-bed manganese nodule to Libyan glass, specific suites concentrate upon quartz, feldspar, meteorites, pegmatite minerals, and crystals. One section is allotted to the species collection (approximately 800 species).

Each mineral is catalogued, with a unique number entered into a ledger and computer. The specimen reference number, *eg* 74-10-35, is painted onto the rear as a black number against a white background. This tripartite numerical code breaks down into 74 = year of registration, 10 = silicate, and 35 = the consecutive number of that chemical system catalogued for that year.

In its entirety the chemical classification adopted is:

    1  =  elements
    2  =  sulphides and sulphosalts
    3  =  oxides and hydroxides
    4  =  halides
    5  =  carbonates
    6  =  nitrates and borates
    7  =  sulphates
    8  =  phosphates, vanadates and arsenates
    9  =  tungstates, molybdates and uranates
   10  =  silicates

Specimen labels are printed by computer on 220g/m² paper cut to 4.4 x 1.6cm and are then encased in adhesive tape and sealed by a hard roller. Under these conditions the labels are durable for 15 to 20 years.

A long-standing interest remains in rock identification and use of the petrological microscope. The seemingly endless morphological and colour variations expressed by quartz and feldspars generates an enduring fascination for the collector.

JAMES WILLIAMS – born on 27 February 1944 at Stourport-on-Severn, near Kidderminster, Hereford and Worcester, although his mother's family stem from Dumfries where he has resided since boyhood. He attended St Joseph's College (founded during 1875 by the French Marists teaching order) in Dumfries and his early mineralogical interests developed around the age of twelve. A mineral collection housed within the college's chemistry department kindled a lifelong interest. Eventually the collection residue passed into his keeping (82 largely unlocalitated common minerals). Early interest in minerals and chemistry developed simultaneously, an inter-relationship that has remained throughout his career as an industrial chemist to the present day.

His mineral collection, all stored in boxes, currently totals 1254 specimens. Specimen sizes range through micromounts and thumbnails to 15-18cm, plus a supplementary microscope slide collection of minerals. James Williams specialises in collecting minerals from south-west Scotland; in particular Dumfriesshire, Galloway and Kirkcudbrightshire; with note-worthy representations from Wanlockhead, Dumfriesshire, and Hare Hill, New Cumnock, Ayrshire. Uncommon Scottish species in his collection include bismuthinite, cervantite(?), ferrimolybdite, greenockite, lanarkite, litharge, mixite, scheelite, skutterudite represented by smaltite, tetrahedrite and witherite. Mineral validatory methods are primarily chemical analyses, an approach which naturally leads to interest in the overall chemical signature of an assemblage rather than to mineral aesthetics. Subsequently many specimens do not achieve show-case standards, although mineral quality is fully appreciated. Petrological studies are undertaken to assist verification. The whole collection is computer catalogued thus enabling numerous interrogative approaches. Foreign minerals from over 40 countries, largely acquired via museum exchanges, form a minor section to the collection. A limited purchase acquisition policy enables species enhancement, whereas a small number of Scottish specimens have been sold to mineral dealers.

During the formative years an interest in Dumfries Museum emerged which fortuitously lead to exposure to a mineral collection. He re-catalogued the whole collection and during 1965-67 became actively involved in the museum's new exhibition scheme. At this period James Williams additionally provided mineralogical cataloguing expertise for the Grierson Museum at Thornhill prior to dispersal of the collection. This stimulating phase opened new avenues to explore including various branches of natural history and archaeological and antiquarian topics. A progressive gravitation towards the Dumfriesshire and Galloway Natural History and Antiquarian Society (founded 1862) resulted in membership during 1963 and Council membership in 1965. He served on the Editorial Committee eventually achieving joint editorship from 1975-76 continuously to the present day. Current duties involve indexing the society's *Transactions* from 1862 to date, thus creating a master index. He has published over 60 notes and papers (largely single authorship) including several mineralogical papers (Williams, 1964, 1965, 1970 and 1976) on a wide variety of local history topics in the *Transactions of the Dumfriesshire and Galloway Natural History and Antiquarian Society*.

Combined mineralogical, chemical and microscopy interests created a curiosity into how people achieved some-thing for the first time or initially approached a particular problem and this avenue nurtured antiquarian book collecting. Within this sphere part of his collection includes the second edition of *Encyclopaedia Britannica* (1777-84), volume 15 of the *Encyclopaedia Perthensis* (*c*.1800), which contains a lengthy article on mineralogy and mineral identifications, and *A Mineralogical Description of the county of Dumfries* (Jameson, 1805b). Further works include *Manual of Mineralogy* (Jameson, 1821) and *Memoirs of the Wernerian Natural History Society* (vols. 2-6; 1811-31).

Donations to the Royal Museum include the first find and recording (Williams, 1970) of scheelite in Scotland from Craignair Quarry, Dalbeattie, bismuthinite from Beeswing, Kirkcudbrightshire and from the same county ferrimolybdite (Macpherson and Livingstone, 1982). A suite of minerals from Needle's Eye, Southwick, Kirkcudbrightshire and a rare find of lanarkite at Pibble Mine, Creetown were also donated.


JOHN and CATRIONA McINNES – John was born in Glasgow in 1948. After graduating as a geologist at Glasgow University in 1973, he undertook research towards a PhD, first at Bristol, then at Glasgow. He met his wife, Catriona, at Glasgow University through her interest in geology. Members of the Glasgow Geological Society since the early 1970s, both had independently collected minerals and rocks. In 1977 they moved to Edinburgh, where John took up a post in the Institute of Geological Sciences, now the British Geological Survey, where he still works. Catriona, who gained an MA in economics and politics and a Postgraduate Diploma in computing, worked in 1976 and 1977 cataloguing mineral specimens for the Hunterian Museum. After moving to Edinburgh, she taught mathematics before becoming Principal Teacher of Computing at an Edinburgh school from 1985-98. Involved for many years with the Scottish Examination Board in the development of computing in Scottish schools, Catriona was awarded an MBE for services to education in 1996.

In the 1980s collecting mainly revolved round family holidays on the west coast of Scotland, especially in Mull or Skye, which offered great opportunities to collect zeolites and associated minerals.

As the family became independent, John and Catriona increased their collecting activities and widened their perspectives. Inspired by the book *Gem and Crystal Treasures*, by Peter Bancroft (published in 1984 by Western Enterprises, Mineralogical Record), they focused holidays round visits to famous mineral localities in many parts of the world. Both panned for sapphires at Anakie and Inverell in Australia and in Montana, collected red beryl at the former Harris Mine in Utah on two separate occasions and ventured underground to 3658m in the Sweet Home rhodochrosite mine, Colorado. They studied sufficient German to allow them to read German mineral journals and to communicate at international Mineral Shows, particularly the Munich Show, which both regularly attended since 1990.

An impulsive decision to attend an evening class in gemmology led to another extension of interests. John and Catriona are now qualified gemmologists and Fellows of the Gemmological Association of Great Britain. They were actively involved in re-establishing the Scottish Branch of the Gemmological Association, of which Catriona is currently Secretary and John a committee member. Both are highly committed to teaching gemmology through classes at Edinburgh University, Open Days, and Gem and Mineral Shows. A website has been established by John and Catriona to promote interest in gemmology, where new and informed articles by Scottish gemmologists are published.

Their mineral collection consists of around 800 catalogued specimens. Three glass-fronted cabinets display some of the best specimens, one being reserved for Scottish minerals. The rest are housed in drawers in mineral cabinets, arranged by accession number, except for larger specimens, which are free-standing in separate areas. Thumbnails and fragile specimens are stored in clear plastic boxes. The Scottish specimens, which constitute about half the collection, are mainly self-collected and around 120 species are represented. The collection is strong in zeolites and associated minerals, from the Clyde Plateau lavas. In particular there are excellent specimens of prehnite, thomsonite, analcime, grossular garnet and other more unusual minerals from Loanhead Quarry, Beith. Zeolites from Skye include rare cowlesite. From Mull are zeolites and good specimens of lustrous schorl tourmaline, plus other metamorphic minerals. Many classic localities are represented: antigorite serpentine, sea-green talc and chromian clinochlore (kämmererite) from Shetland, pink and blue tourmaline from Glenbuchat, strontianite and harmotome from Strontian, flourite from Craigmushet, red heulandite and stilbite from Langcraigs, hübnerite, beryl and topaz from Glen Gairn.

The remainder of the collection comes from famous localities throughout the world and was mainly acquired by purchase at international mineral shows, exchange and donation, although some specimens were self-collected during foreign visits. Highlights include museum-quality Indian zeolites and apophyllite crystals, colourless and various shades of green, up to 8cm. Among the Colorado minerals is a comprehensive collection from the Sweet Home Mine, including rhodochrosite, fluorite, hübnerite, tetrahedrite and goethite. From Colorado also are fine specimens of topaz (a self-collected, euhedral crystal 2.5cm), zircon, astrophyllite and blue barite. Mexican specimens figure largely in the collection with several good gem-quality crystals of danburite up to 10cm. There are both purchased and self-collected, partially transparent, euhedral red beryl crystals to 2cm excavated out of matrix and sherry coloured topaz, both from the Harris Mine, Utah. Additionally are good representative, self-collected specimens from the pegmatites of Maine, including a 3cm bluish-green tourmaline, garnet and purple apatite. Another speciality is British fluorite, with a large range of north of England specimens. Fluorites from various localities range in colour from yellow (Hilton), through green (Heights and Rogerley), pink (Cambokeels) to colourless, purplish-blue, purple and electric blue (small crystals from Bekermett). Numerous specimens are of gem quality and cubes normally attain 8mm to 2cm, although the largest is 6cm. Many fine specimens were collected from Heights Quarry in 1992, and also from the Fraser's Hush Mine (including matrix specimens up to 36 x 20cm covered in transparent to translucent blue/purple crystals to 2cm from the find in 1989).

Details of all catalogued specimens have been entered into a computer database, which may be searched on any field thus enabling easy specimen retrieval and catalogue production with different data arrangements.

John and Catriona also possess a catalogued collection of cut stones, among which are some excellent examples of Scottish material; most of the stones are used for teaching purposes. The collection includes unusual cut stones, the latest synthetics, simulants and treated materials.

As a background to their collecting, John and Catriona have always maintained a profound interest in the history of geology, mineralogy and gemmology and, consequently, acquired an extensive library of modern and antiquarian books on these subjects. For them part of the enjoyment of collecting in Scotland has been the research into old literature and retracing the steps of mineralogists, such as Heddle. The book collection includes *The Geognosy of Scotland, Shetland to Sutherland* (Heddle, 1878a, a reprint of the *Mineralogical Magazine* papers by Lake and Lake, of Truro), *The Mineralogy of Scotland* (Heddle, 1901), *The Mineralogy of the Scottish Isles; with Mineralogical Observations Made on a Tour through Different Parts of the Mainland of Scotland and Dissertations upon Peat and Kelp*, in two volumes (Jameson, 1800), *A System of Mineralogy* (Jameson, 1820), and several original editions of *Memoirs of the Geological Survey of Scotland*.

Specimens of talc, antigorite serpentine and dypingite from Unst have been donated to the National Museums of Scotland. The Hunterian Museum will also receive specimens from Unst and a suite of Scottish minerals has been donated to a new museum in Sri Lanka.

JAMES GORDON TODD – son of a Church of Scotland medical missionary, was born on 28 June 1951 in Poona, Bombay, India. Currently he practices as a consultant anaesthetist in a Glasgow University Teaching Hospital. At the age of five, while residing in India, he acquired his first specimen, an amethyst crystal group, which kindled childhood interest in mineralogy. After the family returned to Scotland during 1957, and encouraged by his father Dr William Todd (who also shared an interest in minerals), many family holidays were spent collecting on the Ayrshire coast, Arran and around Montrose.

From 1966 Gordon Todd dwelled for a year in Chingola, Zambia, where his father worked as a medical officer with the Anglo-American Mining Corporation. Excellent opportunities prevailed to collect copper, lead and zinc minerals, especially the rarer secondary phosphates: tarbuttite, hopeite, parahopeite and libethenite. For the remainder of his secondary school and university education he returned to Scotland, although holiday visits to Zambia and other South African countries provided opportunities to amass a fine mineral collection.

Following in his father's footsteps, Gordon Todd graduated from the Faculty of Medicine, University of Glasgow, in 1975. Apart from a year in New Zealand, he has pursued his medical career in the west of Scotland, with leisure time over the past 20 years devoted to Scottish topographical mineralogy.

His Scottish mineral collection, comprising approximately 500 catalogued specimens representing 110 species, is displayed in glass-topped cabinets and wall-mounted units. Reference and duplicate specimens are stored in 18 drawer-fronted units within the display area. The collection, organised on a geographical basis, follows areas covered by the six Scottish regional geology guides published by the British Geological Survey. Specimens range from miniature to large cabinet size with excellent displayed suites of zeolites and associated minerals from lavas of the Clyde Plateau, Mull and Skye. Mining area minerals are well represented, especially from Strontian, Muirshiels, Mannoch Hill (near Muirkirk) and the Leadhills-Wanlockhead area. Approximately 60% of the specimens are self-collected with the remainder acquired via exchange, purchase and donation.

The world collection, comprising around 600 catalogued specimens (200 species), is housed separately from the Scottish material. Some 250 specimens are displayed in glass-topped and wall-mounted cabinets with the residue stored in drawered units. This collection, which contains specimens from the rest of the British Isles and every continent, is arranged systematically according to major chemical groups. Apart from silicates, data for all groups are currently on computer disc, the ultimate aim being to catalogue the entire collection in this manner.

A small collection of 200 micromount and thumbnail specimens is stored in clear plastic boxes. Diversification into micromineralogy primarily resulted from contact with Thomas Meikle, who extolled the virtues of, and persuaded Gordon Todd to purchase, a stereomicroscope.

Gordon Todd regards his collections as working entities, for specimens have been exhibited in the National Museums of Scotland 'Fine Minerals – Special Exhibit' (April 1988-March 1991) and in Paisley Museum's display 'Mineral and Fossil Collecting' (May 1990-December 1993). Minerals have also been exhibited at numerous 'Members' Nights' of the Glasgow Geological Society. Additionally he has executed slide presentations on aspects of Scottish mineralogy, palaeontology and geology to the Glasgow Geological Society, the Earth Sciences Group at Paisley University and to local natural history societies. During 1991, for eminence at mineral collecting and curation, the Geological Curators Group in conjunction with the Geologists' Association and the Russell Society presented Gordon Todd with an award and certificate of merit for a collection considered to 'exhibit a high standard of curation, preparation, documentation and sense of purpose'. Six published articles on Scottish topographical mineralogy appeared between 1989 and 1993 (Todd, 1989; Todd and Laurence, 1989; McMullen and Todd, 1990; Todd and McMullen, 1991; and Ingram *et al.*, 1992, 1993). Other works include accounts of Devensian marine fossils (Todd, 1992), boltwoodite on stamps (Todd, 1992a), and the mineralisation at Trearne Quarry, Beith (see Burton and Todd, 1992). Other papers on topographical mineralogy also followed.

Gordon Todd's passion for collecting extends outwith the mineral kingdom to include books, fossils and postage stamps depicting minerals and mining locations from 77 countries. His library embraces a wide range of present-day and antiquarian mineralogical, palaeontological and geological books and journals. Works include original editions of *The Mineralogy of Scotland* (Heddle, 1901), *The Ganoid Fishes of the British Carboniferous Formations* (Traquair, 1877) and a second edition of *The Old Red Sandstone* (Miller, 1842). Since December 1991 Gordon Todd has served on the Council of the Glasgow Geological Society and is also a member of the Russell Society. Mineral donations were received from Gordon Todd by the Hunterian Museum, Glasgow, between 1991 and 1994, and the National Museums of Scotland, Edinburgh during 1989-90.

DAVID GAVIN ANDERSON – born on 24 February 1955 in the Wirral, Cheshire, returned five years later with his family to his father's native Scotland, to reside in the Annan area, Dumfriesshire. Following his graduation from Dundee University Medical School in 1979, David Anderson sought employment in the west of Scotland, working initially as a hospital doctor and later in general practice in Ayrshire and Wigtownshire.

His first find, at the age of seven, a crystalline nodule retrieved from a ploughed field behind the family home, remains in his collection. Fossil collecting from nearby road-works soon followed. Interest in rocks and minerals had kindled, becoming increasingly stimulated by museum visits, frequent trips to Leadhills-Wanlockhead and the Criffel area, plus opportunistic collecting on family holidays.

Focused mineralogical interest intensified to embrace the mineral kingdom whether by direct purchase or personal collecting. Outstanding purchased specimens include Namibian azurite and dioptase, and an exquisite pink tourmaline crystal (8 x 3cm) from the renowned Himalaya Mine, Mesa Grande, San Diego County, California. A trip to Iceland presented an opportunity to collect zeolites; however, David Anderson's overriding interest is self-collected Scottish minerals. The latter theme has dominated his collecting activities throughout the past decade. His Scottish collection now stands at 1200 specimens, of which approximately 80% is self-collected with the remainder obtained via exchange or donation.

The collection is housed in drawers, glass-fronted cabinets and on shelves. Mineral validatory techniques include use of a geiger-counter, long and short-wave ultraviolet light and a binocular microscope for morphological assessment and assemblage determination. Each specimen is numbered, labelled and cross-referenced with details entered into a computerised database. Specimen data are filed by geographic region along with appropriate literature references, associated minerals and clues gathered leading to the find. A locality description supplemented by an eight-figure map reference plus details of surrounding geology are also listed. The database may be interrogated in several ways.

Collection highlights are pink intergrown pectolite fans from Lendalfoot, chromite in serpentinised dunite found at Colmonell, and fine mineral suites discovered in Dumfriesshire and Galloway, and Ayrshire. Recent enthralling finds include large ferruginous quartz crystal groups up to 15cm (individual crystals 3 x 3cm) from Cairnsmore-of-Fleet, and impressive interpenetrant cubic pyrite crystals (4cm) occurring in mudstones at Goat Quarry, near Aberdour, Fife, which resulted in a co-authored paper by the collector (Ingram *et al.*, 1992). Rarities are Dalbeattie radioactive suites and powellite won at Glen Creran, Argyllshire, only the second known occurrence of this mineral in Scotland. Molybdenite associated with the powellite formed the subject of a second paper to which David Anderson contributed (Ingram *et al.*, 1993).

The powellite discovery stemmed from meticulously planned field trips, based upon Heddle (1901) localities combined with abstracted data latent in an extensive personal library of topographical and geological maps in association with a nineteenth-century reprint collection. Further literature research draws upon personally owned antiquarian publications and modern mineralogical works. Local regional journals and library reference material are culled for historical information.

Subsidiary to mineralogical interests is agate collecting combined with cutting and polishing to enhance presentation. Classic Scottish agate material is well represented including one extraordinarily large agate (24 x 16cm cut surface) discovered at Dunure, Ayrshire, which formed subject matter for an article in the *Galloway Gazette* of 8 February 1993. David Anderson's mineral and agate collections are epitomised by a superb amygdale in basalt (17 x 12 x 8cm) found in Angus, which is lined with 2cm of blue-white banded agate and in-filled by flawless smoky quartz crystals. Selections from his rock, mineral and agate collections were exhibited at the McKecknie Institute, Girvan, during April 1993 and reported upon in the *Carrick Gazette* (2 April 1993) and *Ayrshire Post* (9 April 1993).

SIMON MARK INGRAM – born on 6 November 1959 in southeast London, a geologist and mineralogist by qualification and self-employed with business interests in gemstones, minerals, fossils and lapidary. Curiosity in minerals and fossils awakened during 1968 as a consequence of discovering echinoids and belemnites in the domestic driveway. Early interest was nurtured by his parents, though more intensively cultivated by Ernest Wilkinson (d.1983), a lapidary/mineral collector domiciled in Hull, Yorkshire. Encouragement plus inspirational guidance led to reading geology and mineralogy at Aberdeen University, from where he graduated in 1983.

Simon Ingram's collection, largely self-collected, expanded as a consequence of field excursions plus summer vacational work under the patronage of the British Geological Survey. Comprising in excess of 700 thumbnail to cabinet-size specimens, including over 500 from Scotland, the collection is partially accommodated in drawers, display cases and on shelving. Embodied in the collection are representations from the classic Scottish localities at Strontian (55 specimens) and Mull and Skye (25). Quarries at Loanhead, Beith, Ayrshire, and Ardownie, Dundee, respectively yielded aesthetic zeolite specimens (40) and agates plus deeply coloured geodic amethyst (150+). Non-Scottish suites embrace species from the northern Pennine orefield (100+) and Caldbeck Fells (35). An exchange contact, Jan Holt, residing at Gol, Buskerud County, Norway, generated 45 Norwegian specimens for the collection.

All specimens are labelled and fully catalogued using a chronological accession number system stored on a computer database. Comprehensive curatorial data includes specimen number, primary phase and chemical formula, secondary phase(s), date collected or acquired, description and location (Ordnance Survey sheet number plus grid reference).

In recent years numerous collecting trips, usually in the company of Gordon Todd and David Anderson, have been undertaken throughout Scotland, and solo during 1991 and 1992 to the northern Pennine orefield. Since 1991, Simon Ingram has subscribed to membership of the Edinburgh Geological Society. A diversification of interests into lapidary resulted in membership of the Scottish Mineral and Lapidary Club (Edinburgh) in 1993. Simon Ingram's published works include articles on pyrite at Goat Quarry, Fife (Ingram *et al.*, 1992); molybdenite at Glen Creran, Argyllshire (Ingram *et al.*, 1993) and agates from Ardownie Quarry, Monifieth, Tayside (Ingram, 1994). Simon Ingram also co-authored a paper detailing the Alva silver deposit (Moreton *et al.*, 1998).

CHRISTINE MATHIESON – born in Inverness, graduated in biology at Aberdeen University and undertook Postgraduate Librarianship Studies. Her interest in minerals developed some 20 years ago from fascination in preparing and photographing rock thin-sections. Highlights of her collection include a suite of phenakite specimens from Ben Macdhui, beryls from the Cairngorm mountains, mountain leather found at Cabrach, Aberdeenshire, and a 1.5cm free-standing apophyllite from Storr, Skye. All her collection is self-collected and details of each numbered specimen are entered into a card index. Species lists from each locality are additionally filed. The collection numbers some 850 specimens, mainly cabinet size, and 100 thumbnail and micromounts.

# THE MINERALS

## ANGLESITE

A lead sulphate of the barite group which characteristically forms in oxidation zones of lead deposits.

Under the heading 'Crystallized Sulphate of Lead', Sowerby (1806) records: 'I believe there is no place in the world, except Anglesey in North Wales, where this substance has been found; and all we have heard or seen of it came from the Parys mine.' However, Jameson (1805) reports an earlier analysis, by Klaproth, of lead vitriol from Wanlockhead, noting that it occurs in lead glance veins at Wanlockhead and at Parys Mount in Anglesey, and that Proust found the sulphate in Andalusian lead glance veins.

Allan (1834), under his description of 'Sulphate of Lead' (anglesite was named by Beudant in 1832), records:

The largest and most beautiful crystals of this mineral are found at the mines of Wanlockhead and Leadhills, often in transparent tabular-shaped individuals some inches in diameter .... Many of the ores of lead are unquestionably derived from the decomposition of galena, and none more distinctly so than the sulphate, which not only contains the same ingredients, but is frequently

Anglesite prisms, to 5 x 1.5 x 0.8cm, sheathed with cerussite. Specimen size 8 x 5 x 5cm, Susanna Mine, Leadhills, Lanarkshire. (Heddle Collⁿ 721.16.)

Striated anglesite prism 7cm high, with transecting crystals, Susanna Mine, Leadhills, Lanarkshire. (NMS G 1994.1.18, ex Carruth Collection.)

met with at Leadhills, either occupying the cavities of cubical crystals, or disposed on a surface of galena, which has all the appearance of having been acted upon by acids.

In the Leadhills-Wanlockhead orefield, anglesite adopts many forms (Lang, 1859) including occasionally stalactitic, although it usually occurs as long-bladed crystals. Temple (1954) recorded anglesite from many veins in the area, although the most productive have been the Susanna and Belton Grain veins. Commonly colourless or white to grey, anglesite may be yellowish, brown or black, the latter due to finely disseminated galena. Anglesite is frequently associated with cerussite, lanarkite or linarite. Galena may be pseudomorphed by anglesite, which in turn can be replaced by cerussite, caledonite, linarite and quartz (Wilson, 1921; Temple, 1956).

Noteworthy anglesite in the Scottish Mineral Collection are: a specimen, without matrix, of grey to greenish bipyramidal crystals, to 2cm, from Leadhills (Heddle Collⁿ 721.35; confirmed as anglesite by X-ray powder diffraction photographs 5435 and 5452); bladed fawn-coloured glassy crystals, 5 x 2.5cm, in cavernous quartz with corroded galena occurring on an Old Museum specimen, 21 x 18 x 10cm, which may originate from the Old College (Jameson's Museum) collection, no. 721.37; crystals, to 2.5cm, investing a specimen with galena in a cavity backed by barite, specimen size 20 x 14 x 8cm, Leadhills; and brownish cream-coloured crystals, to 2.5cm, in pulverulent and with massive galena, plus cerussite, Leadhills (Heddle Collⁿ 721.24 and 25).

## ARAGONITE

Calcium carbonate, a trimorph of calcite and vaterite which bestows its name to the aragonite group. Aragonite is less common, and more unstable, than calcite to which it frequently converts. It is found in near-surface, low-temperature rocks, and is commonly present in oxidised zones of ore deposits.

Excellent specimens have been recovered throughout the Leadhills-Wanlockhead orefield. Spicular radiating aragonite crystals (1cm) occur on scalenohedral calcite extracted from the Bay Vein (NMS G 1951.8.14 and 1955.16.1) and the Glencrieff Mine (NMS G 1926.2.6), Wanlockhead.

Complex growth of aragonite, with multiple terminations, 7cm long, on limonitic boxwork matrix, Leadhills, Lanarkshire. (NMS G 1881.38.11.1, presented by Dr J. Wilson.)

Radiating aragonite crystals up to 1.5cm richly encrusting a calcite galena sphalerite specimen, Bay Mine, Wanlockhead, Dumfriesshire. (NMS G 1951.8.13. Purchased from R. Brown.)

Milky coralloidal aragonite, with 4cm 'fingers', constitutes a 14cm Wanlockhead specimen (Heddle Coll[n] 277.42). Between the 80 and 120 fathom levels of the Glencrieff Mine, a cavity 6ft long, 6ft deep and 2ft wide [183 x 183 x 61cm] in the centre possessed aragonite- and cerussite-invested walls. 'All that the miners did was simply to pick up a few specimens that were lying loose, and then fire their remaining shots, thus destroying the whole lot' (Brown, 1919).

Aragonite is well represented from the Leadhills mines, where on one specimen (NMS G 1881.38.11) it invests a 7cm-sided partial galena cube. Two small pale-green radiating Leadhills aragonite specimens, with fans to 4cm (Heddle Coll[n] 277.40 and 41), had mistakenly been identified as strontianite (Walker qv, Torrie, qv). This colour variant (plumboaragonite) was reported (Heddle, 1882) to contain 1.73% $SrCO_3$. (The aragonite structure permits ready substitution by larger cations, eg strontium and lead). Blue-green radiating aragonite fans (2cm radius) may be located in dump material from Laverock Hall Vein, Leadhills (NMS G 1994.112.1).

Elsewhere found within the Unst is serpentinite, most notably at Hagdale chromite quarry as stellate and radiating groups (NMS G 1990.58.28). Also found at Swinna Ness and Nikka Vord. It is elsewhere found in volcanic rocks at Kincraig shore west of Elie, Fife, as pinkish bifurcating veins up to 3cm wide (NMS G 1878.49.529, Dudgeon Coll[n]), and in cavities within the old limestone quarry at Pertershill, Bathgate Hills, West Lothian.

## BARITE

Barium sulphate; barite gives its name to a group of orthorhombic sulphates. Barite occurs as a gangue mineral in low-temperature hydrothermal veins, and is also common in sedimentary rocks, especially limestones, and as cavity fillings in veins and lavas.

In Scotland barite has an extremely wide geographical distribution; accordingly only a few occurrences are briefly noted. Beveridge et al. (1991) present an overview of barite production throughout Scotland from the six principal vein deposits, and the stratabound barite sequence at Aberfeldy, Perthshire. MacGregor et al. (1944) describe the history and geology, and present

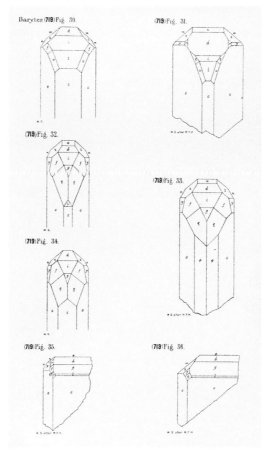

Drawings of barite crystals from the Great Vein, Glen Sannox, Arran. (From Heddle, 1901.)

Colourless barite prism, 9cm long, on a thin plate of calcite and barite with bitumen, no. 14 pit, Addiewell, West Calder, Midlothian. (NMS G 1880.23.)

production figures, for ten barite-producing districts in central Scotland. Only the wider, and more readily accessible, of several hundred barite veins recorded throughout Scotland have been exploited. Three of the six main vein deposits, at Strontian, Argyllshire; Myres Burn, Eaglesham, Renfrewshire; and at Cumberhead, Coalburn, Lanarkshire, were exploited in the 1980s, although to a lesser degree than the deposits worked prior to 1970 at Gasswater, Cumnock, Ayrshire; Muirshiels, Queenside Muir, Renfrewshire; and Glen Sannox, Arran. The latter three collectively produced 0.8-0.9 million tons of barite (Beveridge et al., 1991), with the Gasswater Mine alone yielding 0.5 million tons (Scott, 1967). At the Foss Mine, Aberfeldy, barite is won from a bedded massive deposit composed of anhedral to irregular barite grains 0.1-2mm across, which commonly exceed 90% by volume of the rock.

Barite is the only primary sulphate in the Leadhills-Wanlockhead district and occurs in most veins. Massive barite is associated with galena or sphalerite in the hanging and foot walls, or as vein infilling. Heddle (1901) listed many of the forms adopted by barite crystals in vugs. Yellow-tipped barite crystals from the Bay Vein, Wanlockhead, contain iron or cadmium derived from tiny sphalerite inclusions; elsewhere green and blue barite has been reported from the orefield (Temple, 1954).

95

Parallel growth of tabular barite crystals possessing yellow terminations, Wanlockhead, Dumfriesshire. Front of specimen measures 12 x 8cm. *(NMW 83.41 G.M8930; Bob King Collection (K3898). Courtesy Geology Department, National Museum of Wales.)*

Stepped brewsterite crystal to 4mm on a barite specimen 9.5 x 7 x 3cm from Whitesmith Mine, Strontian Main Vein, Strontian, Argyllshire. *(Heddle Colln 439.4.)*

Both Leadhills and Wanlockhead mines produced magnificent barite specimens. From the former a large example, 32 x 30cm and 14cm high, of lamellar crystals, creates a sphaeroidal aggregation (Heddle Colln 719.129); and from the Wembley Shaft a fine spray of crystals, to 1.5cm, invests galena (NMS G 1927.7.9). Cockscomb barite from Wanlockhead, on a 27 x 25cm base and 17cm high, is composed of tabular crystals to 12cm (NMS G 1952.5.310). Preserved within the Scottish Mineral Collection are bright yellow-tipped and yellow-edged milky-white to clear, glassy, tabular crystals up to 8cm across which form aggregates, from Glen Crievie, Wanlockhead (NMS G 719.69 and 79). Also accessioned are two rudimentary hemispheres, composed of lamellar plates, from Wanlockhead, which measure 19 x 17cm and 28 x 20cm across respectively (NMS G 1926.2.11 and 1952.5.308).

## BIRNESSITE

A sodium manganese oxide hydrate. Birnessite is found in exhalative sediments, thermal springs associated with volcanic activity, as a secondary mineral replacing primary manganese oxides, supergene enriched sediments, and as deep-sea nodules.

Originally recorded from Birness, Aberdeenshire (Jones and Milne, 1956), the species has since been reported from three additional localities in Scotland. At Birness a manganese pan in fluvioglacial gravel deposits contains grains ($<8 \times 4 \times 3$mm), predominantly birnessite and quartz. Elsewhere it is located in exhalative sediments near Noblehouse, Peeblesshire, and in manganese-cemented raised beach deposits at Luce Bay, Wigtownshire (Nicholson, 1983, 1988). It is also present as coatings, and as blue-black to black discrete grains in redbluish iron oxide deposits along joint planes and fractures within a trachyte sill in Craigmuschat Quarry, Gourock (Nicholson, 1988a).

## BISMUTH

A native element. Bismuth belongs to the arsenic group of minerals, occurring in hydrothermal veins associated with silver, cobalt, nickel, uranium and tin, and less commonly in pegmatites.

The presence of bismuth ($<2$mm) was noted by Miller and Taylor (1966) in small carbonate-quartz veins occupying north-west tear faults in the Criffel granodiorite aureole at Needle's Eye, Dalbeattie. The Scottish Mineral Collection is devoid of Heddle bismuth specimens, although Heddle (1901) reported bismuth with erythrite and native silver at Alva. Bismuth featured in the geochemistry of the latter deposit, for Parnell (1988) analysed a silver-bismuth-selenide (1µm) in pyrobitumen from the main spoil heap at Silver Glen, Alva. It also infills, with pyrrhotite and electrum, fractures in galena from the Corrie Buie veins, and in pyrite at Tomnadashan, Tayside (Pattrick, 1984). Additionally, Webb *et al.* (1992) reported bismuth in a W-Sn-Mo-Bi-Ag assemblage in a zinnwaldite-bearing granite at Glen Gairn, near Ballater, Aberdeenshire.

## BREWSTERITE

A strontium barium aluminium silicate hydrate belonging to the zeolite group, brewsterite forms in hydrothermal ore veins with calcite and barite.

The abundance of brewsterite at Strontian (the type, and only Scottish locality) strongly, though erroneously, suggests that the strontium zeolite is common. Realistically it is known from only Canada, France and Russia (Tschernich, 1992).

At Strontian, brewsterite forms glassy crystals up to 4-5mm (see Dyer and Wilson, 1988). In general it richly coats fault breccia, lines cavities, and was one of the last minerals to be deposited in the lead veins, even forming after mining operations. Associated minerals are harmotome, strontianite and, rarely, ancylite on calcite-barite and quartz brecciated gangue. Mining operations commenced in 1722 as shafts were dug, followed by intermittent operations until 1872 when mining ceased. A short burst of activity occurred between 1901 and 1904. North Sea oil exploration injected a new lease of life as the search for barite continued from the mid-1970s for a decade or so (Winrow, 1986). With so much sporadic mining activity, plus brewsterite availability, it is not surprising that the species 'appears to be common'.

A mineral from Vesuvius sent by Heuland to Brewster, upon examination by the latter, was named 'Comptonite', for Brewster (1821a) recognised it as

Brewsterite forms: *i* from Goodchild (1903); *ii* from Tschernich (1992); and *iii* from Akizuki *et al.* (1996).

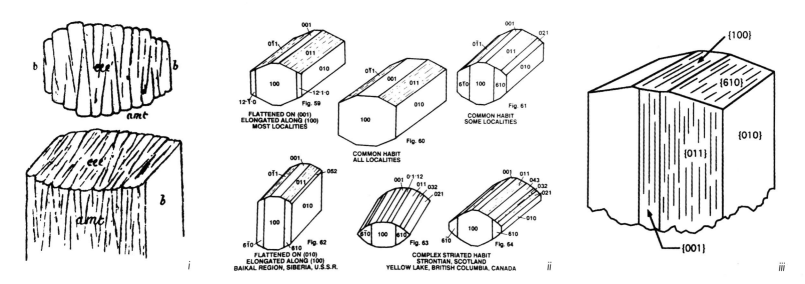

a new species (Heuland named it 'Apophyllite? New'):

> The crystals which I have examined, have the form of right prisms, nearly rectangular, with plane summits; or the same figure truncated on the lateral edges, so as to compose an eight-sided prism. This last form is the most common; but though some of the crystals are very perfect and beautiful, yet there is such an irregularity on the faces of the prism, that it is impossible to obtain very precise measurements of the angles. (Brewster, 1821a)

The above description is appropriate for brewsterite, although the mineral detailed was thomsonite and exemplifies early difficulties encountered in specific zeolite recognition.

One year later, Brooke (1822) states:

> A mineral from Strontian, which has been called in France *Primitive Stilbite*, and was at one time considered to be Apophyllite, is certainly a distinct substance.
>
> I have therefore given it the name of *Brewsterite*, on account of the many important discoveries connected with crystallography, which have resulted from the experimental researches of Dr Brewster. The primary form of the *Brewsterite* is a *right prism, .... whose bases are oblique-angled

*parallelograms*, M on T measuring 93°40', as deduced from the inclination of *a* on *a'*, *c* on *c'*, and *a* on *c*, .... I have not been able to cleave the crystals with certainty in any other direction than parallel to the plane P. Yet when an attempt is made to divide them perpendicularly to P, and parallel to T, the new surfaces exhibit traces of cleavage planes.

Goodchild (1903) notes:

> The figures of this species given in text-books certainly do not correctly represent the usual habit of Brewsterite. The specimens in the Scottish Mineral Collection show the distal face made up of a series of wedges with their thick and thin ends placed alternately, and with the exposed parts of each wedge rounded from side to side, smooth, and with a vitreous lustre.

Nawaz (1990) tabulated interfacial angles for seven separate published morphological studies of brewsterite and, in conjunction with his single crystal X-ray investigations, showed that the X-ray orientation is $c>a$ with *a* elongation. Some previous authors (Nawaz, 1990) demonstrated that the morphological orientation ($c>a$ with $c$ elongation) conflicted with the X-ray orientation ($c>a$, and *a*

elongation). Both vertical and horizontal striations on brewsterite led some workers to erroneously judge $c>a$ for $c$ elongation. Nawaz concluded that 'Brooke's (1822) data are remarkably accurate for their time although Brooke did not specify the orientation'.

Chemical analyses published by Connell (1830) Thomson (1836) and Mallet (1859) showed brewsterite to contain some 8–9% SiO, 6–7% BaO and approximately 1% CaO. Strunz and Tennyson (1956), utilising Mallet's 1859 average analysis figures, ascertained 16.09 Si + Al atoms and 2.03 Ba, Sr and Ca atoms per formula unit. This study showed that Mallet's analyses were high quality, for the empirical results are in excellent agreement with the accepted formula of $(Sr,Ba,Ca)Al_2Si_6O_{16}.5H_2O$.

Akizuki *et al.* (1996), using back-scattered electron imaging, showed that a brewsterite crystal from Strontian possessed compositional growth zoning with respect to the Ba:Sr ratio, suggesting that the composition changed during growth. They also showed that the chemical composition of the {011}, {610} and {010} growth sectors was different and that their atomic structures differed slightly.

Three generations of calcite crystals, prism and first and second generation of rhombs to 1.5cm across. Strontian, Argyllshire. Specimen size 12.5 x 6 x 4cm.
(Heddle Coll<sup>n</sup> 270.42.)

Colour-zoned calcite rhombohedron 3cm across, Middleshop Mine, Main Vein, Strontian, Argyllshire.
(NMS G 1994.50.3.)

Prismatic calcite overgrowths to 3cm on central rhombohedra with colour-zoned edges, Strontian, Argyllshire.
(Heddle Coll<sup>n</sup> 270.17.)

## CALCITE

Calcium carbonate. Calcite is trimorphous with aragonite and vaterite; it forms a series with rhodochrosite and gives its name to the calcite group. Calcite is widely distributed, not only as a sedimentary rock-forming mineral, but also in igneous, metamorphic and metasomatic rocks and hydrothermal veins.

Due to Scotland's complex geology, it is not surprising that calcite, one of the commonest minerals known, occurs in a great variety of geological environments. Heddle (1901) devotes approximately eight pages to Scottish calcite localities, the greatest coverage for any species in his two-volume work, and in which he presents 224 crystal drawings of various calcite forms.

Unlike other Carboniferous limestone tracts in the United Kingdom, considerable expanses

Complex doubly terminated scalenohedral crystal 17.5cm long on dolomite chalcopyrite sphalerite matrix, Glencrieff Mine, Wanlockhead, Dumfriesshire.
(NMS G 1927.6.1. Purchased from R. Brown.)

Scalenohedral group of calcite crystals to 5cm peppered with pyrite on sphalerite, Glencrieff Mine, Wanlockhead, Dumfriesshire.
(NMS G 1923.3.9.)

of limestone do not exist in Scotland. Muir *et al.* (1956), in their petrographic and chemical studies of Scottish limestones, reported that calcite occurred in limestones ranging from Lewisian to Cretaceous in age.

Exquisite, showcase calcite specimens procured during periods of economic exploitation stem from Leadhills-Wanlockhead and Strontian lead mines. Financial pressures in a difficult industry were not conducive to fine mineral specimen acquisition. Brown (1919), in describing specimens from the Glencrieff Mine, Wanlockhead, commented:

> Had care been taken, hundreds of valuable specimens might have been saved that were simply blown to pieces .... As for calcites, it has produced a fine lot – one specimen measured about 24 inches by 18 inches [61 x 45cm]. It had knife-like, edged crystals about 4 inches by 3 inches [10 x 8cm], with small crystals of galena and zinc blend between the crystals; others with dark centres showing the 'nail-head' enclosing the 'dog-tooth' varieties, and some of them capped and showed as if there was a space between the two forms of crystals. Upwards of 30 varieties of calcite have been found in this district.

Throughout the Leadhills-Wanlockhead ore-field calcite is commonly associated with ankerite, barite, chalcopyrite, galena and sphalerite. Temple (1956) concluded that calcite probably formed either by magmatic introduction of Ca and $CO_2$, or by reprecipitation of $CaCO_3$ derived from ankerite replacement by ore minerals, or by both processes. Many Wanlockhead calcites possess a pronounced pink blush due to manganese, for which the tenor of 0.76 wt% MnO was ascertained by Temple (1954) in a Glencrieff Mine sample. Plumbocalcite is present in several Wanlockhead veins, the variant containing up to 9.5 wt% $PbCO_3$ (Lacroix, 1885; Collie, 1889; Heddle, 1901), and with similar levels also in opaque to vitreous calcites from Leadhills. From an x-ray examination Temple (1954) obtained a slightly larger unit cell for plumbocalcite than for calcite, which is indicative of lead substitution for calcium in the atomic structure.

Many fine calcites originate from the Strontian mines, with some specimens revealing multi-stage development. Milky-white hexagonal prisms overgrew 'hexagonal prisms' from which the rhombohedral faces project through younger basal pinacoids. The cores possess three-rayed stars either centred on the trigonal axis, or off-centre. Dotted around the basal pinacoid periphery may be 'acute rhombohedra, or scalenohedra on their sides', thus producing a striking coronet effect. A possible explanation for a rhombohedron overgrown by its prismatic habit is that the rhombohedron was followed by either a rhombohedron or scalenohedron of the opposite sign. Deep-brown to pale-brown stepped rhombohedra with platy hexagonal calcites are also encountered.

Extension of manganoan calcite compositions up to $(Mn_{47.7}, Ca_{45.1}, Mg_{7.2})CO_3$, closely approaching rhodochrosite, was reported by Calvert and Price (1970) for calcite within manganese nodules from Loch Fyne, Argyllshire.

Apart from the pink and deep-brown calcites previously denoted, Scottish calcites may additionally be pale blue as in the coarse-grained (3-4mm) Glen Dessary marble, Inverness-shire (NMS G 1968.54.1), or salmon-coloured as in the decorative and ornamental Tiree marble, Argyllshire. From Bleaton Quarry, Kirkmichael, near Blairgowrie, Perthshire, a group of nail-head calcite crystals attain 6cm across (NMS G 1982.29.1, presented by G. Sutherland). A large specimen, 45cm long, 22cm wide and 16cm high, from Wanlockhead is heavily studded with 4cm nail-head calcite crystals on galena; the latter in one instance forms a 5cm cube (Heddle Coll[n] 270.814). Similarly a specimen 45 x 30 x 16cm from Wanlockhead presents pink calcite scalenohedra 6-8cm long on galena and sphalerite (Heddle Coll[n] 270.825). Comparable with the latter are two specimens from Leadhills (NMS G 1912.70.14 and 15, presented by Rev. W. Peyton, of Llandudno, North Wales). A flat, stalactitic calcite curtain 28cm long, and two stalactites 15cm long joined at the roots, from Little Rack Wick, South-west Hoy, Orkney (Heddle Coll[n], 270.745 and 733 respectively), represent an unusual calcite occurrence, as does similarly, from the same locality, a single stalactite 32cm long (Heddle Coll[n] 270.746).

## CALEDONITE

A lead copper carbonate sulphate hydroxide which forms in the oxidised zone of lead and copper-bearing ore deposits.

One of three new lead ore species from Leadhills noticed by Brooke (1820) was termed by the latter 'cupreous sulphato-carbonate of lead'. This species had been observed earlier from Wanlockhead by Sowerby (1809), who classed it as 'crystallized green carbonate of copper'. An analysis in Brooke (1820) reveals six parts of lead sulphate, four lead carbonate parts and three parts of copper carbonate, although the author qualified the results 'if the carbonate of copper be chemically combined, and not accidental'.

Doubly terminated caledonite crystal 4mm long, with leadhillite and barite, Leadhills, Lanarkshire.
*(NMS G 1991.31.61, ex Sutcliffe-Greenbank Coll[n].)*

*Etiquette manuscrite de de Bournon retrouvée sous un des types de lanarkite. (Arm. 16. Tir. 18. γ' 81 de la Collection de Louis XVIII).*

Brooke (1820) comments:

The crystals are generally very minute, and appear sometimes in small bunches, radiating from their common point of attachment to the matrix .... The fact that presents itself to our notice here, of so distinct a difference of crystalline form, (as opposed to leadhillite and lanarkite) produced by a change in the proportions only of the elements of the crystallised body, will tend to confirm the intimate relation that subsists between the chemical and crystallographical characters of minerals; and it appears to disprove M. Beudant's conjecture, that only the secondary forms of crystals are affected by a change in the proportions of their constituent chemical elements.

Although Brooke recognised the phase as a new species and commented upon Beudant's supposition, Beudant (1832) takes credit for naming caledonite.

Early analyses (Story-Maskelyne and Flight, 1874; Collie, 1889) revealed $CO_2$ to be either absent or present in very low amounts (approximately 1.5 wt%). In the 1874 paper its presence was attributed

Caledonite 3mm long associated with cerussite lanarkite and leadhillite, Leadhills, Lanarkshire. *(Heddle Coll$^n$ 739.10.)*

to cerussite impurity. Heddle's (1901) ideal formula for caledonite $(Pb_1Cu)_2OH_2SO_4)$ is likewise deficient in a $(CO_3)$ component. Berg (1901) showed $CO_2$ to be an essential constituent. (From the ideal formula $Pb_5Cu_2(CO_3)(SO_4)_3(OH)_6$, caledonite theoretically contains 2.73% $CO_2$.)

From its early discovery at Leadhills, caledonite was known only from that locality for many decades. Heddle (1901) noted caledonite occurred in the Susanna, and Leadhills Dod mines and at Wanlockhead in the Belton Grain and Bay veins. Temple (1954) recorded the species from, in addition to the above localities the Cove, Crawfurd's, East Stayvoyage, Laverock Hall, Marchbank, Old Glencrieff and Scar veins. Temple (1954, 1956) postulated that caledonite formed from solutions in which lead predominated over copper. In the sulphate sequence anglesite-caledonite-linarite-brochantite, formation of caledonite inhibited brochantite development, and vice versa.

Caledonite-bearing specimens in the Scottish Mineral Collection corroborate Temple's (1954) findings, for they are devoid of brochantite. Exceptionally fine deep blue-green crystals (to 7mm) associated with lanarkite come from the Susanna Mine (Heddle Coll$^n$ 739.4). Numerous blue-green hemispheres and rosettes (to 5mm) formed in a 5cm-wide cavity partially lined with barite and susannite, from the Susanna Mine (Heddle Coll[n] 290.2: previously identified by Heddle as auri-chalcite but confirmed by X-ray photography, nos. 2901 and 2904, as caledonite). As a multicrystal 'plate', to 1cm across, it traverses a small cavity in a cerussite crystal mass, Susanna Mine (ex J. Neeld Coll[n], R. Barstow, and L. and P. Greenbank Coll[n]: NMS G 1991.31.62), and also appears as a poly-crystalline caledonite crust, with linarite, on galena, Leadhills (NMS G 1908.13.6). Of the 30 caledonite specimens in the Scottish Mineral Collection only three stem from Wanlockhead.

## CELESTINE (Celestite)

Strontium sulphate, forms both a series with barite, and a member of the barite group. Celestine occurs predominantly in sedimentary rocks, especially evaporite deposits, as a primary mineral in hydro-thermal veins, and in amygdaloidal and veined basic volcanic rocks.

Pale-blue celestine veins in trap rock are exposed near North Berwick, East Lothian, where the mineral developed as diverging tabular masses up to 10cm (Heddle Coll[n] 720.1 and NMS G 1901.65.8). In an attempt to discover the cause of colour in three celestine samples, including two non-designated specimens from the Scottish Mineral Collection from North Berwick, Friend and Allchin (1940) detected gold, which showed a range 30-55ppm, and iron 5-11ppm. Greg and Lettsom (1858) re-ported celestine from a cutting in the foundations of George IV Bridge, and as small foliated masses in basalt at Calton Hill, Edinburgh. It appears at Clachnahary, near Inverness 'in veins traversing Old Red Sandstone conglomerate; blue, foliated, and, rarely, with crystalline form' (Heddle, 1901), and as blue fan-shaped radiations to 4cm in dolerite near St Andrews (Heddle Coll[n] 720.8), and close to St David's, Inverkeithing, Fife (NMS G 1957.7.16). Celestine was also found infilling a rectangular cavity 10 x 5cm in a small dyke near Lot's Wife, Auchenreoch Glen, Dumbarton (NMS G 1971.42.2). It is present as blue masses up to 10cm with white strontian-barite at the Muirshiels barite mine,

Renfrewshire (NMS G 1969.39), and was described from this locality by Todd and Laurence (1989).

Elsewhere, at Shierglas Quarry, Blair Atholl, Perthshire, it occurs as deep-blue crystals with white celestine on slicken-sided calc-silicate rock (NMS G 1995.74.1). Bright-blue bladed celestine, with crystals up to 4cm, has been reported by Wolfe (1996) from Trearne Quarry, near Beith. Within the Ratagain igneous complex six miles east of Glenelg, Inverness-shire, Nicholls (1950) discovered celestine, together with quartz, that formed thin veins which traversed nordmarkitic hybrid rocks. Celestine was also reported by Wilson (1921) in the veins at Strontian.

## CERUSSITE

Lead carbonate, a member of the aragonite group. Cerussite is typically found in the upper oxidised portion of lead deposits, where it occurs with angle-site and pyromorphite and is commonly associated with secondary lead and copper minerals.

Of widespread occurrence in Scottish lead deposits, albeit in minor quantities, cerussite together with pyromorphite, is the most common secondary mineral throughout the Leadhills-Wanlockhead area being more abundant in the former. In 1805 Jameson noted its presence within the orefield. Sowerby (1804) described lead carbonate as having

> often a great resemblance to carbonate and sulphate of barytes. It has, however, the advantage of weight, is generally more milky in its appearance, and is mostly shorter in the cross fracture; it is also softer.

The lead carbonate figures in Sowerby (1804, 1806, 1809) are largely based upon specimens from Leadhills and Wanlockhead. He received mineral specimens from Mr Laing of 'Wanlock Head mines, near Glasgow', and reported in 1804:

> Mr. Laing judiciously observed, that the sulphuret of lead, or galena, in most cases, where it is decom-posing to form carbonate of lead, has a blue tarnish. It sometimes also becomes dusty or crumbly.

Cerussite was named by Haidinger in 1845.

Jackstraw cerussite crystals to 1cm on limonitic matrix with pyromorphite, Belton Grain Vein, Wanlockhead, Dumfriesshire, specimen size 9.5 x 6 x 4cm

(NMS G 1951.8.35. Purchased from Robert Brown.)

Silky white cerussite crystals up to 2mm on minutely globular malachite, Wanlockhead, Dumfriesshire.

(Dudgeon Coll[n] 1878.49.535.)

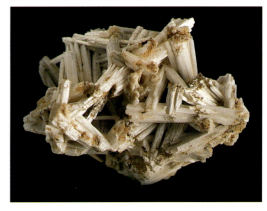

Mass of coarsely fibrous jackstraw cerussite crystals 6 x 4.5 x 1.5cm, Leadhills, Lanarkshire.

(NMS G 1991.31.53, ex Sutcliffe-Greenbank Coll[n].)

The occurrence of hydrocerussite with cerussite on the 120-fathom level of the New Glencrieff Vein, Wanlockhead, was described by Brown (1919). Cerussite appeared especially abundant between the 80 and 120 fathom levels: 'when a vug was opened up about six feet long, six feet deep, and two feet wide [183 x 183 x 61cm] in the centre the walls were simply covered with aragonite and cerussite.'

At the High Pirn Mine on the Belton Grain Vein, Temple (1954) records cerussite crystals to 3.8cm long formed in cavities within crystalline quartz. Cerussite may also be massive, occasionally stalactitic, and either creamy-white or black due to finely disseminated galena. It commonly coats or pseudomorphously replaces galena, the oxidation of which usually gave rise to anglesite followed by cerussite development. Occasionally cerussite replaces galena without the intermediate sulphate phase. Cerussite represents the starting point of numerous secondary mineral sequences present in the area and may be associated with caledonite, lanarkite, leadhillite, linarite, malachite and pyromorphite. Additionally cerussite may pseudomorphously replace anglesite, galena, lanarkite, leadhillite and pyromorphite (Wilson, 1921).

Cerussite was reported from the Struy lead mines, Strath Glass, Inverness-shire (Russell, 1946), where it developed as small patches throughout the galena-barite vein material at the Loch na Meine workings. It occurs here also as small colourless crystals, up to 5mm, of varied habit in pyromorphite-lined cavities. As a moderately common mineral varying from white to pale blue in colour, it exhibits a wide variety of forms in the slags at Meadowfoot smelter, Wanlockhead (Green, 1987). In the Cherty Rock at Stotfield, near Lossiemouth, Moray, cerussite encrusts both quartz and galena; it also forms crystals attaining 11mm in length with acicular-bladed crystals to 8mm. At this locality cerussite is associated with sphalerite, pyromorphite and phosgenite (Starkey, 1988).

The greater abundance of cerussite in Leadhills as opposed to Wanlockhead is reflected in the Scottish Mineral Collection holdings, for some 85% of cerussite specimens emanate from Leadhills. From here, noteworthy specimens are NMS G 1991.31.51, which contains 6mm glassy crystals surrounded by yellow-green pyromorphite on galena; straw-like creamy fasciculate groups to 2cm on ferruginous matrix (Heddle Coll$^n$ 281.65); and snowflake cerussite specimens (NMS G 1994.1.41 and 45) which attain 4.5 and 6cm across respectively. From the Susanna Mine is a mass of glassy idiomorphic crystals, to 5mm, with additional cream-coloured cerussite (NMS G 1994.1.37, *ex* J. Neeld Coll$^n$ no. 3218).

## CHENITE

A lead copper sulphate hydroxide secondary mineral developed in oxidised lead-copper ore.

Chenite was first discovered on a small Susanna Mine, Leadhills specimen (3 x 3 x 2.5cm), once in the world-famous Karabaczek Collection (Cassirer and Martin, 1979), and which possibly passed through the hands of the well-known mineral dealer J. Böhm of Vienna (Paar *et al.*, 1986). The chenite-bearing specimen is predominantly coarse-grained galena with included minor chalcopyrite. Chenite developed in a single cavity as twelve transparent to translucent, sky-blue prismatic crystals (<1 x 0.5 x 0.2mm). Secondary associates include caledonite, linarite, leadhillite, anglesite and mattheddleite. It is additionally present on a Susanna Mine specimen (purchased by W. H. Paar of Salzburg University, Austria) with susannite, leadhillite, caledonite, hydrocerussite, linarite, mattheddleite and lanarkite, and was subsequently located, with elyite and lanarkite, in slag material from an old dump at the site of the ruins of the

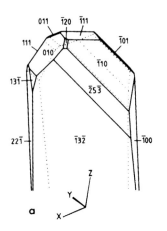
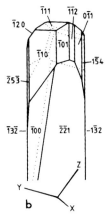
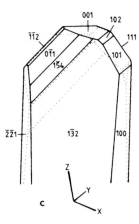
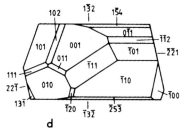

Front, side, rear and top views of a chenite crystal, drawings from Paar *et al.* (1986). *(Courtesy of the Mineralogical Society.)*

Glengaber smelting mill at Meadowfoot, Wanlockhead (Paar *et al.*, 1986). Chenite is also known from Horner's Vein dump material, Leadhills, where it forms light-greenish to blue bladed crystals to 0.5mm (Rust, 1994).

## COPPER

A native element most commonly associated with basic volcanic rocks, copper also occurs in secondary enrichment zones of copper sulphide/oxide deposits and is found within serpentinites and sandstones.

The most notable locality in Scotland is Boyleston Quarry near Barrhead, Renfrewshire, where native copper occurs as plates, nuggets and leaves. Moreover small crystals and grains may be disseminated within analcime, calcite and prehnite. Two additional modes of occurrence are stringers or, more rarely, as crystals in cavities. Heddle (1901) reported copper in delicate leaves and flat lumps up to 7oz (198gm). An irregular-shaped mass of native copper 10cm long, accessioned in the Scottish Mineral Collection (NMS G 1957.7.6), weighs 4.45oz (126gm), while a sheet (NMS G 1962.5) measures 16 x 12cm. Further descriptions of Boyleston and Loanhead quarries copper and that from Hartfield Moss, Renfrewshire, are furnished by Meikle (1989, 1989a and 1990).

From fluid inclusion studies of associated minerals, Hall *et al.* (1989) concluded that remobilised copper resulted from disseminated basaltic chalcopyrite in a hydrothermal system at Boyleston Quarry. Upon cooling and reduction, dissolved copper reprecipitated as the above forms in a sulphur-free environment.

Copper is known also as filaments within prehnite in Glenfarg lavas, Perthshire, and with prehnite, datolite and chalcocite from the same locality (NMS G 1902.70.4), also occurring as plates on basalt from the Gleniffer Hills near Paisley, Renfrewshire (NMS G 1862.27.1, 2 and 5). Plates up to 5cm occur on fracture surfaces and bedding planes of the New Red Sandstone at Ballochmyle, Ayrshire (Heddle Coll^n 15.29).

Seven additional native copper occurrences were noted by Heddle (1901), including the Unst serpentinite, Shetland Isles. In the serpentinite Prichard *et al.* (1987) described native copper associated with platinum-group minerals. Green and Todd (1996) describe chabazite, with bright metallic lath-like copper inclusions, from land-slipped basalt at Moonen Bay, Skye.

## CORUNDUM

Corundum, an aluminium oxide, belongs to the hematite group of minerals and occurs in pegmatites, contact zones, some silica-deficient igneous rocks and recrystallised aluminous xenoliths in igneous rocks.

Early records of Scottish corundum (Sowerby, 1804) are erroneous, for the mineral described was 'Red Schorle' from Achendoor, Aberdeenshire. No record of corundum from Scotland appears in Greg and Lettsom (1858), whereas Teall (1899) discovered corundum in contact rocks of the Ben Cruachan granite, Argyllshire. Teall described medium-grained, dark bluish-grey contact rocks composed of cordierite, andalusite, alkali-feldspar, oligoclase, biotite, pyrite and green spinel. Corundum was detected in the residue from a hydrofluoric acid digestion, and formed either irregular grains or crystals exhibiting combinations of the hexagonal prism, primitive rhombohedron and basal plane.

The first discovery of corundum (sapphire) on Mull occurred in 1912 during geological mapping by E. M. Anderson, in a rock from the bed of a tributary flowing into Abhuinn nan Torr (Thomas, 1922). Further discoveries around Carsaig, and to the north near Tiroran, eventually sourced the material to sapphire-bearing xenoliths in sills. Thomas (1922) classified the aluminous xenoliths into three distinct groups, with sapphire occurring in anorthite-corundum-spinel assemblages. Xenolithic corundum commonly occurs as deep-blue brilliant tabular crystals from 1mm to, rarely, 15mm across. Xenoliths in the gabbro of Haddo House, Aberdeenshire, contain corundum-spinel-

Corundum (sapphire) 4cm high and 2 x 2cm across in camptonite from Carishader, Loch Roag, Lewis, Outer Hebrides.
*(GLAHM M16684. Courtesy of the Hunterian Museum, University of Glasgow.)*

Sapphire, locality as above. The 'Saltire Sapphire', as this gem is now becoming known, has received official recognition from *The Guinness Book of Records* as the largest UK-cut sapphire. Cut from a 39-carat crystal, the stone weighs 9.7 carats. Private collection.

magnetite assemblages; in some instances corundum may form up to 36 wt% of the rock (Read, 1931).

Richey *et al.* (1930a) described sapphire-spinel xenoliths in hypersthene gabbro from the northern flanks of Glebe Hill, 1.5km north-west of Kilchoan, Ardnamurchan. Microscopically, the xenoliths contain green spinel, calcium-rich plagioclase and colourless to deep-blue corundum plates up to 2-3mm. The xenoliths are thought to have been derived by intense thermal metamorphism of bole by the hypersthene gabbro. (From this locality originate specimens NMS G 1984.38.1-5).

Large megacryst sapphires, attaining 46mm diameter, occur in a camptonite dyke traversing Lewisian rocks at Carishader Loch Roag, Isle of Lewis, Outer Hebrides (Jackson, 1984; Offer, 1995: the outcrop is a Site of Special Scientific Interest, thus closed to collecting). Corroded sapphire megacrysts average 2-3cm in length and form striated to distorted, truncated hexagonal pyramids and bipyramids. Within individual crystals colour variation is marked, although all are gem quality (NMS G 1982.24.4 and .7, 1984.9.2). From the same locality a deep-blue partially resorbed rudimentary crystal, 3cm across the prism faces, and likewise along the *c* axis, weighing 100gm, is on long-term loan to the National Museums of Scotland from a private collector. Corroded megacrysts of augite, apatite, sanidine and anorthoclase are present throughout the camptonite.

Idioblastic to ragged corundum is disseminated throughout silica-poor hornfelses of the Comrie area, Perthshire (Tilley, 1924) in andalusite-corundum-cordierite; corundum-cordierite; and corundum-cordierite-spinel assemblages. Similarly, in hornfelses developed around the Belhelvie basic and ultrabasic complex, Aberdeenshire (Stewart, 1946). A suite of emery-like rocks occur in a pronounced thermal aureole adjacent to a dolerite plug near Strachur, Argyllshire (Smith, 1965). Blue and green corundum-bearing rocks possess distinctive assemblages as the contact is approached. Minerals associated with the corundum include cordierite, sanidine, mullite, spinel, magnetite, pseudobrookite, ilmenite, hematite and rutile. Close to these emery rocks, although more distant from the contact, pelitic bands comprise cordierite, corundum, spinel, magnetite, pseudobrookite and sanidine assemblages (from this locality is specimen NMS G 1969.57.1).

An unusual pelitic hornfels at Coire nam Muc, Ardgour, Argyllshire, contains green spinel, hypersthene, phlogopite and blue corundum (Drever, 1940). Corundum is also found with hercynite, potassium feldspar and sillimanite in high-grade contact metamorphosed pelites of the Lochnagar complex, near Braemar, Aberdeenshire (Goodman and Lappin, 1996).

Apart from igneous, xenolithic and hornfelsic occurrences, Scottish corundum is found in metamorphic terrains also. Heddle (1892) featured a hexagonal sapphire crystal included in red andalusite from quartz veins cutting schists of Clashnaree Hill, Clova, Aberdeenshire. In the pelitic rocks 1.7km east-north-east of Clachtoll, Lochinver, Sutherland, a pod-form restite body bears an unusual mineral assemblage of mainly pink, but occasionally blue-tinged corundum (to 1.5cm) and staurolite (1cm) in a chromian muscovite matrix (NMS G 1983.1.1 to 7) (Cartwright *et al.*, 1985).

Of indeterminate origin, though possibly xenolithic, are sapphire plates to 4cm across in a turbid matrix from a boulder in drift, Glenbuchat, Aberdeenshire (NMS G 1994.29.1 and 2; collected and presented by G. Sutherland). Corundum was shown by Kneller (1985) to occur over a wide area in north-east Scotland as part of the regional low-pressure assemblage in silica-undersaturated pelites. The corundum formed in high potassium and aluminium pelites associated with calc-silicate rocks.

## COWLESITE

A calcium aluminium silicate hydrate of the zeolite group found in basic lavas.

Cowlesite is known from basalt at Kingsburgh House on the eastern side of Loch Snizort, Skye; chemical analysis reveals that Kingsburgh cowlesite differs little from the ideal composition (Gottardi and Galli, 1985). Cowlesite was found also in basalt at Ouisgill Bay, Duirinish, Skye, as white spheres associated with phillipsite in a specimen donated by N. Hubbard (NMS G 1994.27.2 – confirmed by X-ray diffraction, photograph no. 4701), and at Moonen Bay in Duirinish, where it forms 1mm intergrown spherules, and is the sole occupant of cavities in basalt (Green and Todd, 1996).

Edingtonite, crystal 2.5mm in cavity in basalt, Old Kilpatrick, Dunbartonshire.
*(Heddle Coll^n 452.8.)*

## EDINGTONITE

A barium aluminium silicate hydrate belonging to the zeolite group. Edingtonite occurs in volcanic rocks, hydrothermal ore veins and in late-stage hydrothermally altered syenites.

Haidinger (1825a) noted edingtonite on one of Thomas Edington's specimens collected in 1823 on Lord Blantyre's estate, one and a half miles (2.4km) from Old Kilpatrick, Dunbartonshire:

> It occurred most frequently at a quarry half a mile NE of Old Kilpatrick; with Prehnite at Bell's quarry near Bowling quarry; and also at a quarry five miles north of Old Kilpatrick, where two specimens in Prehnite were found. (Greg and Lettsom, 1858)

> The majority of the specimens in the Scottish Mineral Collection have the form of small phacoids, with two doubly-curved surfaces, which bend in directions at right angles to each other on opposite sides of the crystals, and which are bounded by four narrow plane surfaces, representing the unit prism *m*, (110); and by four other hemipyramidal planes or sphenoids, *p*, (111), in pairs, alternately above and below .... The habit is very remarkable, and quite unlike that of any other mineral. (Goodchild, 1903)

Dana (1892) reproduced Haidinger's (1825a) crystal drawing of edingtonite, whereas Greg and Lettsom (1858) presented a slightly different form, to which Heddle's (1901) crystal drawing shows a remarkable (although not identical) similarity. The curved morphology figured by Goodchild (1903) is dissimilar also.

Crystal drawings of edingtonite: *i* from Goodchild (1903); *ii* from Haidinger (1825a); and *iii* from Greg and Lettsom (1858).

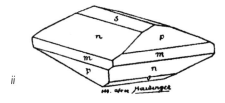

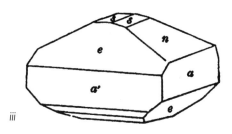

The first analysis of edingtonite by Edward Turner (reported by Haidinger, 1825a) revealed silica, alumina, lime and water, with an unsatisfactory deficiency of 11.22 wt%. From the association of edingtonite with harmotome in Old Kilpatrick material, Heddle (1855) conjectured that the mineral contained barium, which he confirmed by analysis. Hey (1934) demonstrated that Heddle's analysis was also not entirely satisfactory since the empirical formula $Ba_{0.83},Al_{2.14},Si_{2.97},O_{10.4}, H_2O_{3.34}$ suggests 10.4 oxygens and three water molecules instead of ten and four respectively (Fleischer and Mandarino, 1995).

After a lapse of over 169 years, Meikle (1994a) discovered edingtonite at Loanhead Quarry near Beith, Ayrshire. Meikle and Todd (1995) published a full description which revealed a low-barium, high-silica, optically positive, edingtonite associated with prehnite and rare diaspore found in a prehnite vein transected by a barite vein in Clyde Plateau lavas. Edingtonite-bearing specimens from Loanhead, apart from prehnite, contain analcime,

harmotome, thomsonite, natrolite, calcite and hematite. Haidinger (1825a) reported edingtonite associated with thomsonite, calcite and harmotome, whereas Heddle (1855) remarked that he never observed edingtonite with thomsonite.

## EPISTILBITE

A calcium aluminium silicate hydrate belonging to the zeolite group. Epistilbite typically forms in amygdaloidal basalts early in the zeolite crystallisation sequence when the silica content of the fluids, and pH, are high (Tschernich, 1992).

One of the rarer Scottish zeolite species, epistilbite was discovered during 1852 in two specimens at Talisker Bay, Skye, by Mr James Russell of Chapelhall, Airdrie. Greg acquired the two specimens for his collection and the Allan-Greg catalogue refers to the Russell specimens, of which one remains in the Natural History Museum collections (BM 95640). A report of a Field Meeting to Skye (September, 1880, *Mineralogical Magazine*, vol. 4, p. ix) led by the President (Professor Heddle) of the Mineralogical Society of Great Britain and Ireland mentions that:

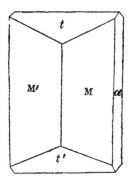

Epistilbite crystal drawing, drawn from a crystal in Greg's collection. Specimen discovered on Skye during 1852 by J. Russell. From Greg and Lettsom (1858).

the late Mr. James Russel here found (at Talisker) the only two specimens of *Epistilbite* which have yet been got in Britain; one of these is now in the British Museum, the other in Dr. Heddle's own collection.

Flesh-coloured, and milky, crystals 3-4mm long are the sole occupant of a cavity some 3 x 1.5cm in basalt from Talisker (Heddle Coll^n 440.1). Surprisingly, Goodchild (1903) cast doubt upon the authenticity of the Skye epistilbite. Epistilbite was also noted by Heddle (1901) on the shore below Beal, Skye, associated with analcime, and at Dearg Sgeir, Loch Scridain, Mull, with scolecite. King (1976) identified epistilbite with calcite and quartz in a quarry beside Loch Sligachan, Skye, and it was also confirmed by X-ray diffraction methods from specimens recently collected at Sgurr nam Boc, Sgurr nam Bairneach, and Talisker Bay (D. I. Green *pers. comm.*, 1995; Green and Wood, 1996).

## ERIONITE (*sl*)

A potassium calcium sodium magnesium aluminium silicate hydrate belonging to the zeolite group. Erionite is frequently found in volcanic and sedimentary rocks, especially where the latter resulted from volcanic ejectamenta, and is commonly intimately associated with offretite as epitaxial overgrowths on levyne in basic lavas.

Erionite forms creamy-coloured, silky fibrous aggregates associated with garronite in a cavity in vesicular basalt from Storr, Skye (NHM specimen BM 1968.61, identified in 1979). Adjacent cavities contain analcime, apophyllite, chabazite, heulandite, levyne, mesolite, stilbite and thomsonite.

It is found, in a rare case, as striated golden crystals, less than 1mm and possessing a distinct hexagonal outline, in vesicular basalt from a roadside cutting at Edinbane, Skye (Savage, 1995). Erionite-offretite outgrowths on levyne are noted under offretite.

Cubo-octahedral galena crystals 4 x 3.5cm (left) and 3.5 x 3cm (right) from Leadhills, Lanarkshire. *(Heddle Coll^n 45.66 and 45.37 respectively.)*

Cubo-octahedral galena crystal to 6cm, bearing Heddle's crystallographic notations, Leadhills, Lanarkshire. *(Heddle Coll^n 45.34.)*

## GALENA

Lead sulphide. One of the most extensively distributed sulphide minerals, and the principal ore of lead, galena occurs in numerous geological environments including in sedimentary rocks, especially limestones, veins, pegmatites, skarns, replacement bodies in metamorphic rocks, and in igneous rocks. Typically it is found in low-temperature veins, commonly associated with sphalerite, chalcopyrite, pyrite, silver minerals, tetrahedrite, calcite, quartz, barite and fluorite.

The economic and social history of Scottish lead mining is well documented (Cochran-Patrick, 1878; Wilson, 1921; and Smout, 1967), although extensive records remain to be researched. Within Scotland, galena has widescale geographic distribution, a measure of which may be appreciated from Landless (1985) – of the 216 mines throughout Scotland listed, 109 produced lead (galena). Principal lead-mining areas, in addition to the Leadhills-Wanlockhead orefield, include Tyndrum, Perthshire; Strontian, Argyllshire; Islay; and the Blackcraig Cairnsmore mines in Kirkcudbrightshire.

Many of the mines possess long exploitation histories, especially those of Leadhills, for monks of Newbattle Abbey mined lead at Glengonner

in 1239. During 1466 James Lord Hamilton was summoned by the Abbot of Newbattle for removing 1000 'stones' of lead ore from Friar's Moor at Leadhills (Cochran-Patrick, 1878). For 100 years, from 1660, ore exports from this area to Holland fluctuated widely. Commencing at 400 tons per annum, export peaked at 550 tons in 1690, thereafter plummeting to 100 tons in 1720. A dramatic rise occurred around 1748, to 500 tons per annum (Aldridge, 1988). Between 1700 and 1786 annual output from Leadhills mines varied from 750 to 1000 tons of ore. Similar tonnages were probably extracted from Wanlockhead mines (Wilson, 1921). Reasons for large fluctuations are not difficult to find: as wars raged in Europe, demand for lead became high. In total, approximately 500,000 tons of ore and lead bars were produced from the area. Although no silver minerals have been identified from the orefield, the argentiferous galena has yielded 25 tons of silver (Beveridge *et al.*, 1991).

All but a few of the 70 or so veins known from the Leadhills-Wanlockhead orefield produced ore. Details of individual veins are reported by Wilson (1921) and Temple (1954). The veins possessed a characteristic banded structure, or a breccia filling cemented by calcite and ankerite with interspersed

galena. In width the veins varied from mere stringers, to 5.48m. Ore stringers range from less than 2.5cm up to 1.22m wide with, exceptionally, 3.65-5.48m-wide pods of solid galena being encountered.

Temple (1956) recognised two periods of mineralisation: the first comprised quartz veins with traces of gold, pyrite, albite and muscovite, followed by a second lead-zinc mineralising epoch. Of the three major primary sulphides (galena, sphalerite

Crystal mass of galena, 7 x 7 x 5cm, Wanlockhead, Dumfriesshire. *(Dudgeon Coll^n 1890.114.104.)*

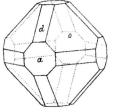
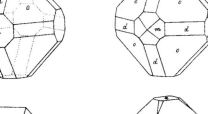

Galena crystal drawings, after Leadhills specimens, Lanarkshire, from Heddle (1901).

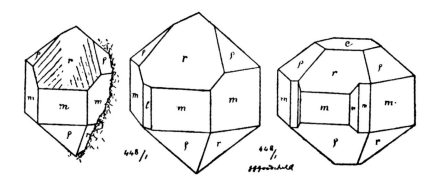

Gmelinite crystal drawings after Skye specimens showing complex rhombohedral and hexagonal habits, from Goodchild (1903).

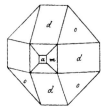
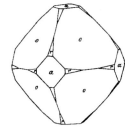

and chalcopyrite), galena developed in two stages – firstly, with sphalerite, although later than ankerite but earlier than calcite and barite; and secondly, following calcite and barite. Sphalerite occupies a similar position in the paragenetical sequence, whereas chalcopyrite may be earlier or later than galena and sphalerite formation, and a second generation of chalcopyrite is later than galena.

Emplacement of the deposit took place at temperatures over the range 281-143°C (Temple, 1956) and at depths of 609-1219m below the surface. Samson and Banks (1988), from fluid inclusion and stable isotope studies, obtained a temperature of approximately 150°C and concurred with Temple's view that the deposit formed from a brine modified by meteoric waters. (Over 60 different species have been identified from the deposit, many being derived from deep oxidation zones which bear no relationship to present-day water table levels.)

In the veins galena was commonly found in the foot or hanging walls, with the former predominating. Massive galena generally formed in veinlets and pods, although crystal faces developed in vugs. Heddle (1901) lists many of the magnificent forms found. Of Leadhills galena he writes:

The numerous low faces .... on Leadhills specimens occur on crystals of one or more pounds

weight, and so are beyond the sustaining power of an ordinary goniometer.

Cubo-octahedral forms, up to 5cm across, originate from both Leadhills and Wanlockhead mines. Scottish Mineral Collection specimens which demonstrate the above forms are represented by 'imperfect half crystals', of which the 'opposite side' to the crystal faces is a series of cleavage planes. Some Heddle specimens retain his notations, for example 'a d o' on a 4.5cm cubo-octahedron from Leadhills (Heddle Coll[n] 45.35), and 3cm and 5cm crystals (Heddle Coll[n] 45.39 and 45.32 respectively). From Wanlockhead one specimen displays dominant octahedral faces (4.5cm edge) truncated by a 2mm cube face (NMS G 1991.9.9) and a second with simple cubes to 11cm, plus calcite scalenohedra (3cm) and barite (Heddle Coll[n] 45.83).

## GARRONITE

A sodium calcium aluminium silicate hydrate belonging to the zeolite group. Garronite is found in amygdaloidal basalts.

Garronite-bearing lavas occur on the east side of Storr, Skye. Milky-white compact, radiating masses with concentric fractures normal to the

fibre length fill amygdales up to 0.8cm (NMS G 1977.31.1-4 and 1982.37.20). Also from the northern end of Moonen Bay as whitish-grey, fibrous compact masses infilling cavities, within basalt, to 1.5cm (NMS 1999.4.17, 18 and 21; confirmed by X-ray photographs, 5750 and 5751).

## GMELINITE (sl)

A sodium calcium aluminium silicate hydrate member of the zeolite group. Gmelinite is primarily found in basic volcanic rocks, and basalts.

Gmelinite is of extremely restricted occurrence in Scotland, the main locality being Talisker Bay, Skye, from where there are two representative specimens in the Heddle Collection (448.1 and 2). Specimen 448.1 exhibits milky to vitreous idiomorphic crystals to 5mm which line 24-25cm$^2$ of a cavity, whereas the second small specimen features a 1cm crystal. From the same locality, Mr B. Young collected vitreous flesh-red 1mm crystals in basalt (NMS G 1990.3.2). In all the above denoted specimens, gmelinite forms monomineralic linings to the cavities which are devoid of other species. It also occurs as imperfect glassy crystals, to 2mm, lining a small cavity in levyne-bearing amygdaloidal basalt from Echoing Craig, Quiraing, Skye (Heddle Coll[n] 448.3).

Although gmelinite might potentially be considerably more widespread throughout the Skye basalts, studies of Scottish zeolites distributions in other areas are few. Dyer *et al.* (1996) reported gmelinite associated with thomsonite from the southern extremity of Arran at Breadlebane.

## GOLD

A native element, forms a continuous series with silver -3C. Notable quantities of gold occur in hydrothermal veins, pegmatites, and contact metamorphic deposits, and it is also found in consolidated, and unconsolidated, placers.

The earliest records of Scottish gold extend back to 1125 and the reign of King David I of Scotland (1082-1153; ruled from 1124), who granted a title to the Holy Trinity Church of Dunfermline 'for a tenth of all the gold found in Fife and Fothrif' (Dudgeon, 1877). On 26 May 1424 an Act of King James I of Scotland (1394-1437; ruled 1424-37) decreed that of Scottish gold:

Gif ony myne of gold or silver be füdyn in ony Lordes landes of ye realme, and it be prwyt tht thre half-pⁿnys of siluer may be fynit owt of ye punde of leid, The Lordes of Parlimnt consentes tht sik myne be ye Kinges as is vsuale in uthir realmys. (Atkinson, 1619)

The Martin nugget, 41gm, found at Straitstep, Wanlockhead, Dumfriesshire.
(NMS G 1994.20.4.)

King James IV of Scotland (1473-1513; ruled 1488-1513) devoted considerable energies towards gold extraction, particularly in the Crawford Moor area. This area embraces the Leadhills-Wanlockhead orefield in the west and extends eastwards to Megget Water, Selkirkshire. In addition to Atkinson (1619), Scottish gold occurrences are extensively documented in Lauder Lindsay (1867), Porteous (1876), Cochran-Patrick (1878), Heddle (1901), Adamson (1988), Callender (1990) and Beveridge *et al.* (1991).

Apart from restricted 'vein' occurrences, gold in Scotland is widely disseminated throughout alluvial deposits as irregular grains, auriferous quartz fragments, and small nuggets. Alluvial gold was first discovered in Scotland by Gilbert de Moravia during 1245, at Durness (Richardson, 1974). Alluvial deposits at Leadhills-Wanlockhead, and Helmsdale, in Sutherland, are the most noteworthy. From the former area, gold extraction first made the district famous. King James IV raised finances to prospect the area and just as the workings became profitable he was killed, fighting on foot, at the Battle of Flodden (9 September 1513). At one period 300 men were employed for several summers. During the reign of King James V (1512-42; 'ruled' 1513-42), £100,000 sterling worth of gold was extracted by washing. Between 1538 and 1542 the district produced 1163gm of gold, to fashion a crown for King James V, and 992gm for his queen's crown. A considerable portion of the gold coinage of James V and Mary, Queen of Scots (1542-87) was minted from Leadhills-Wanlockhead gold, including the famous Bunnet pieces, so named after the 'capberet' favoured by the king, the first dated coin made from Scottish gold minted in 1539. Commercial gold working appears to have ceased post 1620.

In gold-rush history Scotland played a role. The discovery of gold during late 1868 at Kildonan, Sutherland, quickly led to exploitable quantities being discovered in the Suisgill and Kildonan burns. Inevitably the local papers, the *Northern Ensign* and the *John O'Groats Journal*, extensively reported the 1869 Kildonan gold rush. Over 600 hopeful prospectors descended upon the normally deserted glen (Callender, 1990). At the beginning

gold pegged at £4.50 per ounce, although during summer months prices fell to £3.40. The Duke of Sutherland refused to allocate additional prospecting ground, and with the herring season commencing, and weather deteriorating, prospector numbers plummeted to around 50. At midnight on 30 December 1869 Scotland's gold rush ended (Callender, 1990). A resurgence occurred during 1895, and 1911, although both ventures folded due to economic reasons. Two cross-shaped pendants, one owned by the Countess of Sutherland, are fashioned from Kildonan gold (Callender, 1990): the precious metal primarily went into jewellery, especially rings in the distinctive Sutherland design.

Even today the lure of gold continues, as the gold-producing rivers of history in the Leadhills-Wanlockhead area, Elvan Water, Short Cleuch, Long Cleuch, Glengonner, Windgate Burn and Mennock Water, and in the Suisgill and Kildonan burns at Helmsdale, still produce small quantities for the discerning recreational, enthusiastic panner.

Adamson (1988), in his book *At the End of the Rainbow*, endeavoured to include all validated gold-bearing sites throughout Scotland, and by panning at most listed localities in the northern, central and southern goldfields (tripartite demarcation by Lauder Lindsay, 1867) obtained gold. More recently Coats *et al.* (1991) demonstrated that alluvial gold is widespread over the central Ochil Hills, Stirlingshire and eastwards to the Firth of Tay. Electronprobe-microanalysis of 31 gold grains from Borland Glen, in the Ochil Hills, reveals a median composition of 92% Au and 6.3% Ag. In gold concentrates from the latter area, Davidson (1995a) identified rutheniridosmine. Studies on gold grains panned from Perthshire, Kildonan and Suisgill burns, and localities throughout the Southern Uplands, by Leake *et al.* (1993) revealed a distinctive inclusion assemblage signature which reflected the control of the source mineralisation. Apart from common species forming inclusions in the gold, the following are also reported as inclusions: berzelianite, clausthalite, cobaltite, galenobismutite, hessite, matildite, palladium, tetradymite and wittichenite (Leake *et al.*, 1993).

Scottish gold. Fig. 1: medal, 135gm, 10cm diameter, fashioned from Sutherland gold for Lauder Lindsay. Fig. 2: Martin nugget, Wanlockhead. Fig. 3: Gilchrist nugget, Kildonan, found during 1869. Figs. 4, 5, 6: from Leadhills-Wanlockhead. Fig. 7: from Kildonan (4-7 larger than actual size). Figs. 8 and 9: Kildonan nuggets (actual size) used as centrepieces of brooches and breast-pins. From Lauder Lindsay (1880).

NATIVE SCOTTISH GOLD

and is rich in gold with minor quartz (BM 87500). The Martin nugget (41gm), which was found at Straitsteps, Wanlockhead, and is entirely gold (NMS 1994.20.4), is named after Dr James Martin (1790-1875) from Leadhills, who joined the army medical core and served in Spain during the Peninsular War (1808-14). He returned to Britain during 1814 and graduated MD in 1826. Eventually he undertook medical practice in Leadhills. He died on 10 February 1875 in Edinburgh and is buried at Leadhills (Porteous, 1876). Mr R. Sutherland over many years made an unequalled contemporary collection of Scottish gold, including numerous nuggets in the 4-9.6gm range, plus an outstanding rich auriferous quartz nugget of 22gm. The latter probably ranks as the largest surviving nugget recovered since the discovery of the Martin and Kildonan nuggets.

Dudgeon (1877) relates that a specimen of auriferous quartz found at Wanlockhead 'was obtained by the late Professor Traill, about the beginning of the century and is now placed in the mineralogical collection in the Museum of Science and Art, Edinburgh'. There are auriferous quartz specimens from Leadhills in the Dudgeon collection, but unfortunately there is no register reference to a similar Wanlockhead specimen being presented by Professor Traill. Dudgeon's paper refers to a second auriferous quartz specimen from the area the Gemmell nugget, of somewhat dubious origin. Andrew Gemmell discovered the specimen 'the day after a barrow full of lumps of Australian gold-quartz had been wheeled over a rough track' (Heddle, 1901). The latter additionally reported 'its gold is of Australian, and not Scotch colour'. Two men from Wanlockhead worked in the Australian goldfields and returned to the village, although their opinions as to whether the Gemmell nugget originated locally were never sought (*North British Daily Mail*, 16 May 1878). The Gemmell nugget provenance problem could possibly be resolved with the application of gold fingerprinting methods which utilise laser ablation inductively coupled plasma mass spectrometry (Watling *et al.*, 1994) or electronprobe microanalysis methods. Gemmell unfortunately broke the specimen into several

Waterworn masses of placer gold, plus appreciable gold in quartz matrix, form nuggets of which, albeit small, there are numerous recordings from Scottish rivers. Lauder Lindsay (1880) presented an account of various Scottish gold nuggets, mainly auriferous quartz, including some considered erroneous (possibly from Australia, California, New Zealand and Nova Scotia) located in national and public museums throughout Scotland and England. A tabulation of Scottish nuggets was presented by Adamson (1988). The 'Turrick' nugget found around 1840 at Turreich near Amulree, Perthshire, and once owned by R. P. Greg, weighs approximately 65gm, measures some 3 x 2 x 1cm pieces, disseminating portions, by sale, to different parties. Dudgeon (1877) obtained all the major fragments and reassembled them into a mass measuring 10 x 7.5 x 6.3cm. The Gemmell nugget, exhibited at a Royal Society of Edinburgh meeting on 5 March 1877 (Dudgeon 1877a), is accessioned into the Scottish Mineral Collection (NMS G 1877.35).

Hutchinson (1951) reported a 'sensational find' on 6 June 1940 by the Leadhills prospector and retired lead-miner John Blackwood, who was working beside the Windgate Burn and found a nugget of 500 grains (32.35gm). The gold-bearing quartz measured 4 x 3 x 1.5cm. Russell (1944) furnished a detailed description, plus coloured photograph, in which he states 'thickly coated on one side and to a slight extent on the other with massive gold rising to form a solid tongue in one corner'. The nugget is housed in the Russell Collection (BM 1964R, 3549) in the Natural History Museum, London.

Scottish auriferous veins carry gold either as minute particles, or combined in a suite of minerals. Two areas of recent commercial interest are Cononish, near Tyndrum, and Calliachar Burn close to Aberfeldy, Perthshire (Pattrick *et al.*, 1988; Parker *et al.*, 1989). At Cononish economic quantities of gold with silver, are present in a pyrite-rich galena-bearing quartz vein. Gold is present either as 5-100µm grains located within fractures in the sulphides (at Tyndrum electrum, petzite and sylvanite are present). Considerable mining activity occurred although it has now ceased due to depressed gold prices on the world market. From the Calliachar Vein deposits, small quantities of gold have been smelted. The wheel turned full circle when, from Calliachar gold, replica James V ducat pieces mounted in paperweights were produced (*Sunday Picture Post*, 23 May 1993).

## GONNARDITE

A sodium calcium aluminium silicate hydrate member of the zeolite group. Gonnardite is found with thomsonite, natrolite and tetranatrolite.

Hemimorphic greenockite crystal, 3mm long on basalt, Bishopton Railway Tunnel, Renfrewshire. Specimen collected by S. J. Shand. (NMS G 2000.10.2.1.)

The first recording of Scottish gonnardite was by Mackenzie (1957), who discovered this species in vesicular lava at Allt Ribhein, Fiskavaig Bay, on the south side of Loch Bracadale, Skye. Here the vesicles may be filled with stilbite, heulandite and thomsonite, or with saponite. Some vesicles possessed a saponite core rimmed with zeolites: in two samples the zeolite was largely gonnardite. It was recently confirmed as small (1mm) off-white crystals from vugs in a dyke at Ard Fernal, Isle of Jura, with natrolite, mesolite and scolecite also (Dyer *et al.*, 1996), while at Milngavie it is found in amygdales with analcime, chabazite, natrolite, wairakite, thomsonite, laumontite and cristobalite (Evans, 1987).

Although gonnardite and tetranatrolite are difficult to recognise, gonnardite rarity may be misleading for it can easily be overlooked in thomsonite, natrolite and tetranatrolite assemblages (Tschernich, 1992).

## GREENOCKITE

Cadmium sulphide. Greenockite is found rarely as idiomorphic crystals in amygdaloidal basic lavas, and typically as earthy coatings on sphalerite.

Charles Murray Cathcart (1783-1859), the second Earl of Cathcart, whose earlier title was Lord Greenock, discovered the unusual mineral, which he gave to Jameson, in material excavated during construction of the Bishopton rail tunnel, Renfrewshire. Jameson (1840a) named it in his honour and placed greenockite in the 'Order Blende'. (Lord Greenock, who was interested in geology, had attended Jameson's natural history classes during 1830.) Similarly, Thomas Brown, of Lanfine (T. Brown, *qv*), who collected the mineral many years before, about 1816, arranged it among zinc blende in his extensive collection. Brown's catalogue, commenced 1832, contains an entry no. 749 stating 'Prehnite with Blende Kilpatrick'. An additional note opposite to the entry dated 1836

A rare thin-section, 7.62 x 2.54cm (3 x 1in), by M. F. Heddle, of greenockite in prehnite, Bishopton, Renfrewshire. Courtesy of Hunterian Museum, University of Glasgow.

again states Kilpatrick. When tunnel greenockite was actually discovered is unknown as records do not exist, although tunnelling work should not have commenced prior to March 1837 when the railway line Enabling Act became operative (Meikle, 1992a).

A pencilled note with Brown's no. 749 specimen (Hunterian Museum, register number M377) refers to it as the original find, the largest crystal measuring close to 1cm. Lord Greenock presented Professor Connell (1794-1863) of St Andrews University with a crystal sufficiently large (3.71 grains/0.24gm) to analyse. Connell's (1840) results published simultaneously with Jameson's (1840a) description showed 77.30% Cd and 22.56% S (total 99.86%). Thomson (1840) likewise analysed the greenockite and obtained 77.6% Cd and 22.4% S, whereas Forbes (1840) published the results of an optical study (over 160 years later, there are few greenockite analyses throughout the mineralogical literature). The Irish mineral dealer Patrick Doran (1781-1881) sold five greenockite specimens to Jameson in 1843 (Old College Register, p.278, item 8). Two specimens in the Scottish Mineral Collection, NMS G 68.1 and 68.11 and previously numbered 4 and 5 respectively, are possibly Jameson's purchased material.

In excavated material from the Bishopton tunnel that remained in a nearby field, Harwood (1951) discovered an incomplete greenockite crystal which measured 6 x 4mm. During 1974-75, when the Bishopton Bypass was under construction close to the railway tunnel, greenockite was again recovered.

Greenockite crystals are known from Boyleston Quarry near Barrhead (Heddle, 1887) and Loanhead Quarry, Beith and Erskine (Meikle, 1989a, 1992a), it is also present as citron-yellow earthy coatings on sphalerite from Wanlockhead.

## GYPSUM

Calcium sulphate hydrate. Gypsum is the commonest sulphate mineral and frequently forms large evaporite deposits. It occurs when sulphate-bearing solutions interact with limestone, as an efflorescence, encrustation, or concretions in clays and marls.

In the southern half of Scotland gypsum is widely reported (Heddle, 1901), albeit in relatively small quantities, for example: in caves, as stellate masses on shales and associated with alum shales, in fireclay, within quartz-lined druses in lavas, traversing sandstones, and in coal mines. A paucity of good-size crystals characterises Scottish gypsum. Eight kilometres east of Langholm, Dumfriesshire, a band 0.5cm thick was recorded in the Carboniferous Limestone Series (Lumsden et al., 1967). A bore at East Doura near Kilmarnock, Ayrshire, intersected gypsum comparable to that at Ballagan, Stirlingshire (Richey et al., 1930). In the Cementstone Group of the Carboniferous, around Hutton Castle 11.3km south from Eyemouth, Berwickshire, gypsum crystals are widely disseminated in shales, although principally concentrated as joint fillings, and veins. 'Ribs' up to 25mm thick are common in the topmost 50m of a bore, whereas at 106m a layer 75mm thick was logged. Gypsum concretions, and masses of flat pinkish gypsum, occur at intersections of many white veins (Greig, 1988).

An unusual anhydrite and selenite occurrence was reported by MacGregor (see Mitchell et al., 1960) in veins, vugs and druses of the Traprain Law phonolite, near Haddington, during the working life of the quarry. Brown (1925) includes gypsum in his minerals listing for the Leadhills-Wanlockhead area but the species is not described. Temple (1954) observed gypsum to form small, white elongated and striated crystals on ankerite from the Glencrieff Mine, Wanlockhead. In the latter area Green (1987) observed that gypsum formed stout prisms or long prismatic crystals on slag from the Meadowfoot smelter. Gypsum occurs as small granular crystals coating a sample of calcite and pyrite from the orefield (Williams, 1964).

Off-white botryoidal masses up to 1cm thick were found at the foot of Main Craigs, Ailsa Craig, Ayrshire (Harrison et al., 1987), and it also forms sugary white microcrystals in fossil cavities in Lower Carboniferous limestones of Trearne Quarry, Beith, Ayrshire (Todd, 1989). Of recent origin, as colourless prismatic crystals with cowlesite, it occurs in a sea-washed basalt cavity at Moonen Bay, Skye (Green and Todd, 1996). Williams (1965) reported small pockets of fibrous gypsum in calcite veins at Southerness and Arbigland, near Kirkbean, Kirkcudbrightshire.

David Corse Glen presented a fibrous gypsum specimen ('satin spar') 4cm across from Auchenreoch,

Dunbartonshire (NMS G 1879.37.1), and a specimen similar in size and appearance is derived from the Michael Colliery, Wemyss, Fife (NMS G 1994.15.25). From shales, a 1cm vein of 'satin spar' at Ballagan Glen in the Campsie Hills, Stirlingshire (Heddle Coll[n] 746.1). Salmon-pink gypsum forms thick bands, nodules, and beds of nodules with shaley divisions, in a specimen (35 x 20 x 9cm) from Edington Mill, Hutton, Berwickshire (NMS G 1971.26). Gypsum is present as glassy, colourless and iron-stained, poorly formed, small twisted crystals with goethite and jarosite on granite matrix from the underground workings of the Bellsgrove Mine on the Strontian Main Vein, Strontian, Argyllshire, and was restricted in occurrence to areas of decomposing pyrite (NMS G 1978.41). Gypsum forms a glassy outgrowth (2.5 x 1.5 x 1.5cm) from a brachiopod in limestone, Trearne Quarry, Beith, Ayrshire (NMS G 1989.40.9).

## HARMOTOME

A barium potassium aluminium silicate hydrate. Harmotome, a member of the zeolite group, possesses the same aluminium-silicon-oxygen framework, to which water molecules are attached, as phillipsite. Harmotome is defined as a member of the phillipsite group with greater than 50% barium (Tschernich, 1992). Overall, although considerably less common than phillipsite, harmotome occurs mainly in late-stage hydrothermal lead-silver veins, and in volcanic rocks where its derivation is due to hydrothermal alteration of barian feldspar.

Harmotome was defined by Haüy (1801) although the name was first used by Daubenton (1784) in his *Tableaux Méthodiques des Minéraux*. In the etymology of his 'Cross stone' species (after complex cruciform Marburg and Perier twins) Jameson (1804) refers to Strontian as 'the only other place where it has been observed in veins after Andreasburg and Oberstein'. Allan (1834) comments, 'The finest simple crystals of this mineral occur disposed on calcareous spar at Strontian in Argyllshire, where individuals of a pure white colour have been met with nearly an inch in

Euhedral harmotome 2cm, on barite calcite matrix, Strontian, Argyllshire. (NMS G 1998.44.5.)

diameter,' and also 'occasionally accompanying analcime in the amygdaloid of Dumbartonshire'. Haidinger (1825) reports 'simple crystals from Strontian in Scotland, are very generally known – It is very frequent in amygdaloid, as in various places in Scotland'. In addition to the localities previously mentioned, Greg and Lettsom (1858) noted harmotome from the Campsie Hills, and Corstorphine Hill (Edinburgh), and Heddle (1901) adds Bishopton Tunnel and Kilmacolm, Renfrewshire, and Bowling Quarry, Kilpatrick, Dunbartonshire.

Harmotome from Strontian ranks amongst the finest known (Tschernich, 1992). Specimens of exceptional quality stem from the lead mines (Corrantee, Whitesmith, Middleshop, Bellsgrove and Fee Donald mines), where harmotome developed upon calcite or barite, and rarely on brewsterite. All the harmotome crystals are twinned, with Morvenite, Marburg and Perier twins

being represented. The name 'morvenite' was first proposed by Thomson (1836) after the area. Goodchild (1903) reports:

Three varieties of Harmotome occur there [at Strontian] and are often to be found in the same specimens. One of these may be described as consisting of spindle-shaped crystals, as these are greatly elongated parallel to *b*, (010), and they may be said to taper off to each end from the middle. Crystals of this habit used to be regarded as simple, and were long known by the name of Morvenite, from the locality where they occur. They are usually of small size. A second habit is one in which the elongation is parallel to *a*, (100), and the crystals are rectangular in cross-section transversely to that, and are terminated at the longer ends by what look like low four-faced pyramids. This variety appears to be the commonest .... The third variety is that which is best known through the medium of the figures in text-books. It is cruciform in cross-section ....

Russell (1946) reported harmotome from the Struy lead mines on the west side of Strath Glass, Inverness-shire. Here, the species, though not abundant, forms crystals up to 5mm which exhibit various modifications of the characteristic complex cruciform twins. The harmotome is emplanted on sphalerite, barite, calcite and the feldspathic matrix. Cruciform harmotome coats fissures and cavities in a breccia zone enriched with hematite at a quarry 0.4km south-west of the southern margin of Loch an Arbhair, Ross-shire (Waterston, 1953; NMS G 442.36.43). At the Hilderston silver mine, near Bathgate, West Lothian, Marburg twinned harmotome (0.6 x 0.12mm) associated with witherite and barytocalcite, in barite vein matrix occurs on dump material (Meikle, 1994). Harmotome is found rarely associated with barite and prehnite in loose basalt blocks at Hartfield Moss, near Beith (Meikle, 1990). Marburg and Perier twinned harmotome (6 x 2mm) is associated with edingtonite formed in cavities within prehnite-thomsonite assemblages at Loanhead Quarry, Beith (Meikle, 1989a, and *pers. comm.*, 1996, and Meikle and Todd, 1995). It is known also from Touch Forest (chemically analysed crystals), Gargunnoch near Stirling, the Burn of Sorrow at Dollar Glen, Clackmannanshire, and Craignaught Quarry, Dunlop, Ayrshire (T. K. Meikle *pers. comm.*, 1996).

Scottish harmotomes from numerous localities require investigating chemically to ascertain compositional range.

Some 45 harmotome specimens in the Scottish Mineral Collection display excellent milky-white euhedral crystals. Many are from the Heddle Collection, although the total includes numerous high-quality examples collected over the past 15 years. A brown calcite specimen (29 x 22cm) emplanted with excellent crystals to 1.5cm is the largest harmotome-bearing specimen in the collection (NMS G 1888.50.7). This specimen was purchased, in a lot of ten fossil woods and minerals, from Joseph Bryson, an optician in Princes Street, Edinburgh (the lot cost £2). Noteworthy also are glassy harmotome crystals (1.2cm) lining a cavity in granite, rather than investing calcite or barite, from the Middleshop section of the Strontian Main Vein (NMS G 1984.47.9). Opaque pale-greenish 'simple' crystals (0.5cm) with curved prismatic faces, and elongated on *b*, differ markedly in appearance from normal harmotome at this locality (NMS G 1984.47, 12 and 13). On calcite, and after brewsterite, are crystals (1cm) from the Middleshop section of the Strontian Main Vein (NMS G 1994.50.1).

## IKAITE

A calcium carbonate hydrate originally discovered as columnar-shaped skerries just below sea level at low tide in the Ika Fjord, south-west Greenland (Pauly, 1963).

Numerous, odd-shaped crystals were dredged from the River Clyde muds near Cardross, Dunbartonshire, around 1888-89 during dredging trials, and formed the subject matter of early speculative articles, especially by Trechmann (1901), Gemmell (1903) and MacNair (1904). Elongated prismatic, and almost square to rhomb-shaped in cross-section, the leathery-looking crystals, with undulose faces, taper in a curved manner towards the coigns. A chemical study by MacNair (1904), who concluded that the crystals were pseudomorphs after gaylussite, revealed a calcium carbonate composition with minor magnesium carbonate and calcium phosphate. The crystals are calcite and Shearman and Smith (1985) demonstrated that ikaite ($CaCO_3.6H_2O$) was

parental to the pseudomorphs. Ikaite is stable at 3kbar at 2°C and 5.08kbar at 25°C (Marland, 1975) and precipitates as a metastable phase at low temperatures as the pressures do not prevail during deposition. The presence of these crystals (now calcite) in the stratigraphic record reflects past, very cold periglacial conditions.

On the Garvellach Islands in the Firth of Lorne, Argyllshire, Spencer (1971) observed quartz-dolomite segregations in the top of the exposed Dalradian limestone. The segregations pseudomorphed blade-shaped crystals as stellate groups several centimetres long. Spencer (1971) also demonstrated that sandstone-wedge features were ice wedge casts. Johnston (1995) found similar characteristics in a glaciomarine sequence in the Dalradian of Donegal, Ireland. The segregation structures are consistent with a widely accepted late Precambrian glaciation and thus probably represent ancient ikaite.

# JULGOLDITE

A calcium iron aluminium silicate hydroxide hydrate. This chemically complex mineral belongs to the pumpellyite group and forms two series, with iron pumpellyite and magnesium pumpellyite.

Heddle (1901) regarded small, black crystals occurring in pectolite from a Ratho Quarry, near Edinburgh, as aenigmatite, whereas X-ray diffraction examination of the original Heddle specimen (NMS G 343.2) revealed the mineral to be julgoldite. A second Heddle pectolite specimen (NMS G 435.27) from Auchinstarry, Kilsyth, contained lustrous, black plumose clusters (<2mm) of julgoldite (Livingstone, 1976). At both localities the julgoldite-bearing pectolite formed rare pockets in quartz-dolerite. The 1976 paper represented the second recording of julgoldite.

Radiating spray of thin blades, 3.5mm long, of julgoldite on pectolite, Auchinstarry Quarry, Kilsyth, Stirlingshire.
*(Heddle Coll^n 435.27.)*

# LANARKITE

Lead oxysulphate. A rare mineral generated during galena oxidation.

Brooke (1820) designated a new species from Leadhills as 'sulphato-carbonate of lead' and described the mineral as:

> Colour whitish, bluish, and greenish-grey, sometimes approaching to apple-green. The crystals I have seen are seldom distinct, always minute, and aggregated together lengthwise, presenting a character approaching to fibrous.

The lanarkite primary form was perceived by Brooke (1820) to be a right prism whose base is an oblique-angled parallelogram of 59° 15' and 120° 45'. Schrauf (1877) published a complete description of lanarkite morphology based upon crystals from the Susanna Mine, Leadhills. Richmond and Wolfe (1938) revised Schrauf's morphological data on crystals from the same locality held within the Harvard University mineral collection and published new cell parameters.

On the basis of specimens in Count de Bournon's collection Haüy (1813) chemically categorised the species as lead carbonate. Greg and Lettsom (1858) reported carbonate-bearing analyses also and commented that it was found

> … in diverging aggregations, from 1 to 2 inches in length, at Leadhills, in Lanarkshire, associated with Susannite and Caledonite. It is the rarest of the lead-salts met with at Leadhills, which is its only locality in the United Kingdom.

Pisani (1873), Story-Maskelyne and Flight (1874), and Collie (1889) ascertained that earlier analyses were on impure material. Although Beudant (1832) named lanarkite, Allan, in his work *Manual of Mineralogy*, which appeared two years later, still referred to the mineral as sulphato-carbonate of lead. He did comment, however, that 'it dissolves in nitric acid without perceptible effervescence'.

Heddle (1901) recorded that lanarkite occurred associated with susannite and caledonite, and rarely with cerussite, at the Susanna Mine. Brown (1919) discovered lanarkite with leadhillite, on the old heaps from Margaret's Vein, at the old dressing floors on the Crawfurd's Vein, and as very fine crystals from the New Glencrieff Vein, Wanlockhead, and it has also been found in the Cove and Straitstep veins, Wanlockhead, and Hopeful Vein, situated equidistant between the Wanlockhead and Leadhills villages (Temple, 1954). In the latter work the author noted that on one specimen from Whyte's Cleuch, Wanlockhead, lanarkite developed directly from galena. Green (1987), in his study of slags from the Meadowfoot smelter, Wanlockhead, discovered that lanarkite formed radiating sprays, transparent prismatic crystals, and white to pale-blue botryoidal aggregates.

The Scottish Mineral Collection houses 45 specimens curated under the name lanarkite, for the species is adjudged to be the most important phase on the specimens. Lanarkite is also present, as an associate, on specimens of caledonite, leadhillite, susannite, macphersonite and mattheddleite. All 45 lanarkites are from Leadhills, with one exception, a small specimen from Pibble Mine, Creetown, Kirkcudbrightshire (NMS G 1974.24.2). This latter specimen (collected and presented by J. Williams, and confirmed by X-ray diffraction – SMC X-ray photograph library) reveals 2mm lanarkite crystals in a cavity associated with anglesite in corroded galena (Williams, 1970).

Leadhills lanarkite crystals are either glassy, or dull in appearance due to a surface coating. Two diverging prismatic 'multi-crystal' groups, to 5cm, possess glassy interiors with creamy-grey coated surfaces (Heddle Coll^n 737.3; the surface coating is leadhillite as confirmed by X-ray diffraction). Pale-green bladed crystals, to 2cm, are partially encrusted with susannite (NMS G 1991.31.27), and also in association with galena plus yellow pyromorphite from the Susanna Mine (NMS G 1991.31.29). Lanarkite also appears on cerussite with yellow pyromorphite, 'fibrous' crystals to 2.5cm with leadhillite (NMS G 1991.31.35) and similarly as glassy pale-green interleaving 2.5cm prisms with caledonite (NMS G 1995.88.1, *ex* 1994). (The 1991 registration number refers to specimens from the Greenbank-Sutcliffe collection.)

Abdul-Samad *et al.* (1982) redetermined the free energy of formation at 298.2°K for lanarkite

Divergent spray of 'fibrous' lanarkite crystals 5 x 3.5cm, Susanna Mine, Leadhills, Lanarkshire.
*(NMS G 1991.31.32, ex Barstow, Sutcliffe-Greenbank Coll<sup>n</sup>s.)*

Divergent lanarkite prisms 2 x 0.3cm, specimen size 3.5 x 3 x 2.5cm, Leadhills, Lanarkshire.
*(NMS G 1991.31.38, ex Sutcliffe-Greenbank Coll<sup>n</sup>s.)*

Divergent lanarkite prisms 2.5 x 0.5cm, coated with leadhillite and cerussite, specimen size 7 x 4.5 x 3.5cm, Leadhills, Lanarkshire.
*(NMS G 1991.31.35, ex Sutcliffe-Greenbank Coll<sup>n</sup>s.)*

Lanarkite crystal drawings from Heddle (1901).

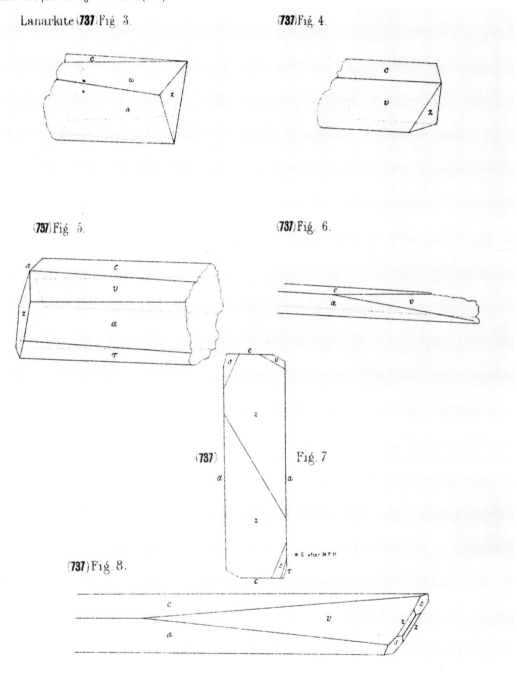

Lanarkite (737) Fig. 3.

(737) Fig. 4.

(737) Fig. 5.

(737) Fig. 6.

(737) Fig. 7.

(737) Fig. 8.

W. G. after M. F. H.

from the Susanna Mine, Leadhills, and incorporated the new value into thermodynamic calculations in order to elucidate equilibrium assemblages involving anglesite, hydrocerussite, leadhillite and lanarkite. They found that lanarkite is thermodynamically stable only at low activities of the bicarbonate ion in aqueous solutions. Lanarkite rarity was therefore attributed to low activity levels rarely being achieved in nature. (In carbonate-bearing vein systems bicarbonate ion activity was commonly high. The activity of a compound is a measure of its *effective* concentration, or concentration as modified by the effects of the solvent and other dissolved substances.)

## LEADHILLITE

A lead sulphate carbonate hydroxide and a trimorph of susannite and macphersonite. Leadhillite is a rare secondary mineral found in the oxidised zone of lead deposits.

Count de Bournon appears to be the first author who erected, chemically, this mineral as a separate phase, which he designated *plomb carbonaté rhomboidal* (Haüy, 1813). Haüy enunciated:

*Elle vient de la mine de plomb de Leadhill, en Ecosse, où elle est accompagnée de fort beaux cristaux de plomb carbonaté de l'espèce précédente, et très fréquemment de plomb phosphaté d'un beau jaune orangé: je ne l'ai jamais observé venant d'aucun autre endroit; elle est fort rare à Leadhill même.*

*La couleur de cette substance, dans les cristaux petits et très-transparents, qui est l'état le plus habituel sous lequel elle se présente, est un brun jaunâtre tirant un peu sur le vert; mais dans les cristaux un peu plus grands, cette couleur est plus pâle, et devient d'un gris sale, un peu verdâtre. Ces cristaux ont un éclat très-brillant.*

[It came from the lead mine at Leadhills, in Scotland, where it was found together with very beautiful crystals of lead carbonate of the preceding species, and very often with lead phosphate of a beautiful yellow-orange: I never observed it in any other place; it is very rare at Leadhills itself.

The colour of this substance, in the smaller and very transparent crystals, which is the commonest of the forms in which it presents itself, is a slightly greenish yellowy brown, but in the slightly larger crystals, this colour is paler and becomes a dirty grey and more greenish. These crystals have great brilliance.]

A few years after Haüy, Brooke (1820) chemically designated the mineral as sulphato-tri-carbonate of lead. Mohs (Haidinger, 1825) termed the mineral axotomous-lead-baryte and included a crystal drawing – one of twelve published a year later by Haidinger (1826a), who had previously studied specimens from Thomas Allan's and Brewster's collections in Edinburgh.

Count de Bournon, who owned a fine, extensive collection (22,000 specimens of which 10,000 were isolated crystals), left France in 1792 a ruined man, due to confiscation of his property, and took refuge in London. As a royalist faithful to the Bourbons he did not return to France until the restoration of Louis XVIII in 1815. By that time he became a member of the Royal Society of London and, as a co-founder, had been instrumental in establishing the first Geological Society in the world, the Geological Society of London, in 1807. De Bournon tutored Louis XVIII in mineralogy and sold his collection to the king, later becoming its Director, with Beudant (1787-1850) as Assistant Director. During his sojourn in London, de Bournon, throughout the Napoleonic era, continued to exchange specimens with Laumont and Haüy (Schubnel, 1987). In Paris Beudant (1832) erected leadhillite as a species, simultaneously with lanarkite and caledonite.

Leadhillite is a rare mineral, though widely dispersed throughout the Leadhills-Wanlockhead orefield. Heddle (1901) writes:

Leadhills; has been found in all of the veins, but in largest quantity in the Susanna Mine (Wilson). It sometimes occurs alone; more frequently with Cerussite, or with this and Lanarkite. Very rarely Aurichalcite is its associate. In the veins of the Leadhills Dod it accompanies Caledonite, or Caledonite with Linarite, Chrysocolla, Malachite and Cerussite ....

Wanlockhead; has been found in all the veins, except the Belton Grain Vein.

Bryce Wright (c.1880) illustrated a translucent, honey-yellow coloured crystal, procured by himself during 1873, and noted that it is now in the celebrated collection of W. S. Vaux, Esq., of Philadelphia. Correspondence from the Academy of Natural Sciences, Philadelphia (Cooper, *pers. comm.*, 1995) reveals that the now dulled crystal measures 4 x 3 x 1.5cm. Additional information could not be solicited for the collection is currently regarded as essentially closed (Fischer, *pers. comm.*, 1995). (Bryce-Wright's sales catalogue, c.1880, priced leadhillite at 10/- to £10.)

Brown (1925) records that many fine specimens of leadhillite were found during the previous years at the Susanna and Hopeful veins. In his examination of the orefield, Temple (1954) discovered leadhillite to be present in 13 out of 57 veins, including the Belton Grain Vein. His work revealed that leadhillite formed prior to lanarkite as the solutions responsible for secondary mineral development changed from carbonic to sulphuric, although the general trend was in the opposite direction.

Ninety-four leadhillite specimens are housed in the NMS Scottish Mineral Collection. Generally, leadhillite-bearing secondary lead-mineral assemblages embrace lanarkite, caledonite, cerussite and pyromorphite. Rarely, macphersonite, mattheddleite and susannite are associates. Leadhillite frequently co-exists with cerussite, or lanarkite only. Leadhillite crystals (0.4mm) developed directly on unaltered galena in quartz, this specimen being devoid of other secondary phases (from Brown's Vein, Glengonner, Leadhills; Heddle Coll[n] 734.15), and richly invests an 8 x 4cm cavity in barite with cerussite (1.5cm) from the Susanna Mine, Leadhills (Heddle Coll[n] 734.31). Cream-coloured leadhillite (1cm), intimately associated with cerussite on galena from Leadhills (NMS G 1994.1.78; Carruth Coll[n]), mineralogically reflects changing sulphuric to carbonic conditions of formation. As 'truncated pyramidal' crystals (to 4mm), leadhillite occurs with lanarkite (2cm) and pyromorphite on quartz from the Susanna Mine, Leadhills (NMS G 1991.31.9). This specimen comes from the collection of Joseph Neeld (1789-1856)

Tabular pseudohexagonal leadhillite crystal, 6.5mm across, Brown's Vein, Glengonner, Leadhills, Lanarkshire. *(Heddle Coll[n] 734.27.)*

Nest-shaped mass of large, subparallel leadhillite crystals to 2cm across with caledonite and barite, specimen size 9 x 7.5 x 4cm, Leadhills, Lanarkshire. Sold by Bryce-Wright to the British Museum (BM 41118) in 1867.
*(NMS G 1991.31.1, ex Sutcliffe-Greenbank Coll[n]s.)*

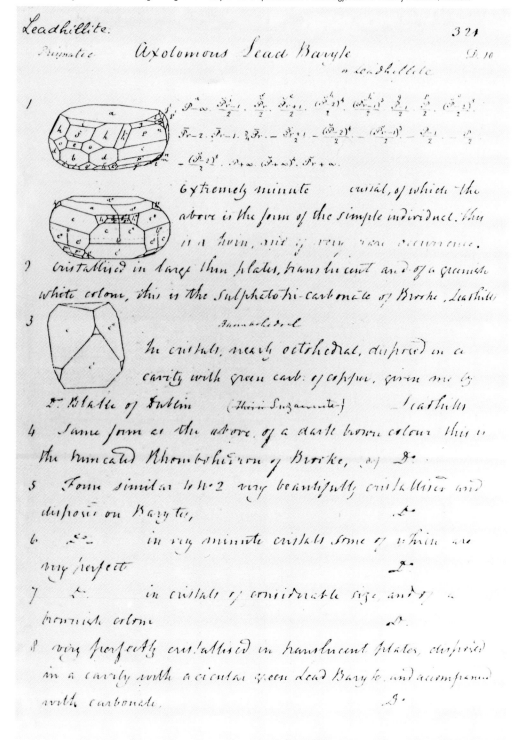

and retains the original Neeld number, 3616. Neeld's mineral collection, accumulated mainly between 1825 and 1830, was purchased in 1974 by B. Lloyd, who retains the original Neeld catalogue which is watermarked 1830. Unusually, leadhillite appears as pseudo-hexagonal prisms (5mm) in ferruginous quartz gangue from dump material on the western side of Wanlock Dod, Whytes Cleuch, Wanlockhead (NMS G 1981.47.1–3). From an old dump of the Hopeful Vein, Temple (1954) also recorded pseudo-hexagonal leadhillite.

Leadhillite chemical analyses reveal minimal variation, although Collie (1889) and Russell *et al.* (1984) demonstrated slight changes in sulphate, carbonate and hydroxyl contents; the latter authors ascribed the differences to mutual substitution within the leadhillite structure. However, Temple (1954, 1956) reported a leadhillite from the Hopeful Vein to contain 0.5% chromium.

## LEVYNE (sl)

A calcium sodium potassium aluminium silicate hydrate belonging to the zeolite group. Levyne occupies vapour cavities in basaltic rocks.

Greg and Lettsom (1858) reported opaque flesh-red levyne at Hartfield Moss, near Beith, Renfrewshire, although from their published figure (p.177) conceivably the mineral may be gmelinite. Levyne is found in Tertiary basalts at Storr, Isle of Skye, where it forms milky-white twinned, 1–2mm crystals (Greg and Lettsom, 1858; Tschernich, 1992). It is frequently the sole zeolite in cavities, while those nearby contain a rich variety. Levyne is also present as a boxwork of crystals with offretite outgrowths, in vesicles up to 1.5cm diameter, at Quiraing, Skye (Livingstone and Macpherson, 1983; NMS G 1982.37.12 and 13).

Green and Todd (1996) reported levyne as an uncommon occupant of small cavities in sea-worn boulders at Moonen Bay, Duirinish, Skye. The species is present in two distinct forms: thin opaque hexagonal plates with epitaxial outgrowths of erionite-offretite; and thick, transparent, hexagonal crystals possessing large pinacoid faces

Linarite crystals to 0.8cm with cerussite, Susanna Mine, Leadhills, Lanarkshire. *(NMS G 741.8.)*

bounded by smaller pyramids. Crystals of the first habit reach 1cm exceptionally, although commonly less than 2mm. Again monomineralic levyne-lined cavities are characteristic, or rarely with associated analcime. Levyne is also found in basalts at Ouisgill Bay (NMS G 1990.3.1) to the north of Moonen Bay. Of over 400 zeolite localities on Skye, King (1976) discovered that levyne occurred at 4% of them.

## LINARITE

A lead copper sulphate hydroxide developed in the oxidised zone of lead-copper ore bodies.

Excellent linarite crystals are known from Lady Anne Vein, Brown's Vein, and the Susanna Mine at Leadhills. At Brown's Vein where linarite developed within cavitied quartz-rich specimens, few secondary minerals are present. Heddle (1901) contrastingly reports that numerous secondary lead minerals of the orefield are associated with linarite at the Susanna Mine (except susannite), in the mines of Leadhills Dod, and at the High Pirn and Belton Grain mines, Wanlockhead. Temple (1956) deduced that linarite formed in response to changing lead – copper ratios in solutions which deposited sulphates in the sequence anglesite – caledonite – linarite – brochantite.

Specimen NMS G 1963.16.20, on which an old label states 'Blue Carbonate of Copper Wanlockhead', is similar to one figured by Sowerby (1817), who reports that 'the present rare specimen is preserved in the cabinet of G. Laing, Esq., and comes from Wanlockhead mine'. (Linarite was named by Glocker, 1839.) Sowerby believed that the Wanlockhead linarite was the earliest recording of the then chemically defined mineral in Great Britain.

Linarite is additionally known from Pibble Mine, near Creetown, where associates are secondary lead minerals, galena, chalcopyrite and malachite.

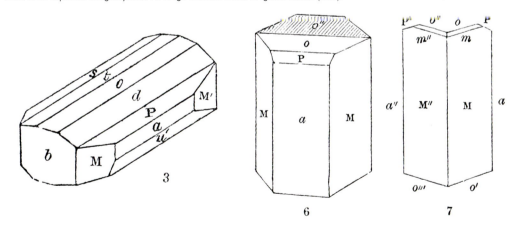

Linarite crystal drawings based on specimens from Leadhills, Lanarkshire. No. 3 'represents a crystal in M. Brooke's collection,' and 6 and 7 'represent a large crystal in Mr Greg's collection'. From Greg and Lettsom (1858).

## LUDWIGITE

A magnesium iron borate, after which the six-membered ludwigite group is named. Ludwigite developed in high-temperature, complex magnetite – diopside – forsterite skarn deposits.

First reported in Scotland by Tilley (1948) from the Loch Kilchrist, Broadford area, Skye, ludwigite is present in complex-zoned boron and fluorine skarns generated at granite-dolomite limestone contacts. The principal locality at Kilbride is a ludwigite-chondrodite-magnetite skarn which carries accessory fluoborite and szaibelyite (Tilley, 1951). It is also found in marbles at nearby Camas Malag.

## MACPHERSONITE

Lead sulphate carbonate hydroxide, a trimorph of leadhillite and susannite. Macphersonite is an exceedingly rare secondary mineral developed during galena oxidation.

Macphersonite crystal group 2 x 0.7cm, Leadhills, Lanarkshire
(NMS G 1991.31.21, ex Sutcliffe-Greenbank Collⁿs.)

Type macphersonite was simultaneously described from Leadhills Dod, Leadhills, Lanarkshire, Scotland and the Argentolle Mine, near Saint-Prix, Saône-et-Loire, France (Livingstone and Sarp, 1984). In Leadhills material the leadhillite-susannite trimorph is pale amber to tan coloured *en masse* and colourless in cleavage fragments. Leadhills macphersonite projects into a mattheddleite-lined cavity in quartz (Heddle Collⁿ 721.34) as a composite, dumb-bell shaped crystal associated with leadhillite, susannite, cerussite, caledonite and pyromorphite.

The co-existence of leadhillite, susannite and macphersonite on the same small specimen poses complex genetic problems which are not yet understood. A mineral is only stable under certain restricted physical and chemical conditions which can be represented graphically by a stability field. Should conditions change instability emerges, and the mineral transforms into a different phase which becomes stable within an adjacent field. However, a mineral from one stability field can develop *very close to and on the unexpected side* of a common boundary. This phase is said to be existing *metastably*. Possibly one or two of the trimorphs may be co-existing metastably, or each is a stable phase existing under conditions where all three fields meet. Leadhillite, susannite and macphersonite all form at an intermediate stage when cerussite (alkaline) develops from anglesite (acid). As the trimorphs contain both carbonate and sulphate ions critical physical and chemical conditions must prevail for their genesis. That these conditions are rarely achieved is reflected by the scarcity of the trimorphs when one considers the vast number of altered lead deposits throughout the world containing cerussite and anglesite (Livingstone, 1993). The recently determined atomic structures of macphersonite (Steele *et al.*, 1998) and susannite (Steele *et al.*, 1999), when compared with that of leadhillite (Giuseppetti *et al.*, 1990), reveals many similarities. Within the layered structures of the three polymorphs principal differences lie in the sulphate layers which possess different sequences of up- or down-pointing sulphate tetrahedra.

## MATTHEDDLEITE

Lead silicate sulphate chloride. Mattheddleite is a silicate isostructural with members of the apatite group, and late-stage secondary mineral developed in oxidised zones of lead ore bodies.

Originally described from a single Leadhills specimen (Livingstone *et al.*, 1987), mattheddleite has subsequently been identified on numerous specimens from the orefield. It forms colourless to white encrusting masses of glassy hexagonal crystals (<0.3mm) capped by hexagonal pyramids. On the type specimen (Heddle Collⁿ 721.34) mattheddleite is associated with macphersonite, leadhillite, susannite, caledonite, cerussite and pyromorphite. A mattheddleite-bearing specimen in the Catty Collection, Maidstone Museum, Kent, and erroneously labelled 'Caldbeck Fells', contains lanarkite, cerussite, hydrocerussite, leadhillite and caledonite. Additionally found at Horner's Vein, Leadhills, with queitite, cerussite, susannite and pyromorphite (Jackson, 1990) and with chenite (Paar *et al.*, 1986). Mattheddleite, which is distinctive morphologically, occurs on dump material at Cove Vein, and Wilson's Vein, Wanlockhead (Meikle, *pers. comm.*).

The crystal structure of mattheddleite, which has very recently been determined (Steele *et al.*, 2000), reveals an apatite structure in which Pb replaces Ca and equal amounts of Si and S replace P. The X anion is OH + Cl and for the crystal used in the structural determination (from NMS G 1878.49.488.2) strong OH-Cl variation was noted with an OH-rich core and Cl-rich exterior.

## MORDENITE

A calcium sodium potassium aluminium silicate hydrate belonging to the zeolite group. Commonly found in geothermal wells, also in basalt, tuffaceous breccia, dolerite, and rhyolite in geothermal areas.

One of the less common Scottish zeolites, mordenite appears to be first recorded from Scotland as brick-red, needle-shaped, radiating crystals (= ptilolite) in quartz-filled cavities of

Scanning electron microscope photograph of mattheddleite, from Leadhills, Lanarkshire. Photograph of Dr R. S. W. Braithwaite's specimen no. RSWB84-125, which is part of one belonging to Dr J. Jung. Courtesy of Dr R. S. W. Braithwaite.

0.1mm

Mordenite tufts to 2cm associated with stilbite, Sgurr nam Boc, Skye. (NMS G 1994.11.54.)

## MULLITE

Aluminium silicate. Mullite, a high-temperature refractory mineral, occurs in metamorphic rocks of the sanidinite facies, especially in fused pelitic xenoliths (buchites) or in high-grade thermal metamorphism of aluminous material.

Bowen *et al.* (1924), in their investigations of equilibrium relations of alumina and silica, showed that there is only one compound of those oxides stable in contact with liquid in the binary system. The compound showed a remarkable similarity with natural sillimanite and the authors reported that the compound was not only stable at very high temperatures where liquid occurs in the binary system, but also at temperatures as low as $1000°C$ or lower. They remark:

> In view of this fact it seemed to us that the compound could scarcely fail of occurrence in natural rocks though here again probably determined as sillimanite. The most promising material for the occurrence of the 3:2 compound $[3Al_2O_3.2SiO_2]$ appeared to be the so-called sillimanite buchites or fused argillaceous sediments occurring rather frequently as inclusions in Tertiary intrusives of the Western Isles of Scotland.

volcanic rock at Dumbarton, Dunbartonshire (Bøggild, 1922; the specimen is housed in the Copenhagen Museum). Utilising single crystal X-ray techniques Waymouth *et al.* (1938) identified fibrous crystals from Aros, Mull, Argyllshire as mordenite (BM 47614). Mordenite also developed as an alteration product of a glassy groundmass in a quartz-fayalite pitchstone dyke in the shore at Tomore, Arran (Harris and Brindley, 1954).

Associated with stellerite, or singularly, mordenite occurs in Old Red Sandstone lavas at Todhead Point, 9km south of Stonehaven (Dyer *et al.*, 1993). Also as pink, fibrous cavity infillings, occasionally with heulandite, and tufts and veins in highly altered amygdaloidal volcanic rock from a gas pipeline excavation at Tinkletop, Ballindean,

Inchture, Perthshire (NMS G 1977.37.1-14, and 1977.38.1-3, latter from the same locality collected and presented by the late Dr R. R. Taylor via the Scottish Mineral and Lapidary Club, Edinburgh). Additionally, pink ramifying veins of mordenite to 1cm are associated with quartz in altered volcanic lava from the slopes of Black Craig, Gargunnock, near Stirling (NMS G 1978.40.1 and 2). It is found as silky-white, highly fibrous interwoven tufts to 2cm projecting into cavities associated with stilbite, or quartz, from Sgurr nam Boc, Skye (NMS G 1994.11.48-56). (The NMS specimens have been identified by X-ray powder diffraction and/or infrared techniques.) Green and Wood (1996) reported mordenite from Sgurr nam Boc as an associate of epistilbite.

The work of Thomas (1922) paved the way for Bowen *et al.*'s (1924) mullite discovery in fused phyllites at Seabank Villa and Nun's Pass, near Carsaig, south-western Mull. Cameron (1976) studied the sillimanite-mullite relationships in buchites of the Rudh' a' Chromain sill, Loch Scridain, Mull. He found that sillimanite, and 'relatively large' mullite which possessed a faint lilac pleochroism due to 0.7% $TiO_2$ in the structure, were associated with cordierite-rich areas. Cameron (1976) concluded that the xenolithic bulk compositions probably represent liquid products of partial melting of underlying Moine rocks.

Adjacent to a dolerite plug at Sithean Sluaigh, Strachur, Argyllshire a series of emery-like rocks contain mullite-corundum-magnetite-hercynite-pseudobrookite-sanidine, and mullite-cordierite-hercynite-magnetite-pseudobrookite assemblages (Smith, 1965). The assemblages are considered to represent aluminous residues from pelitic/psammitic phyllites after a granitic fraction had been selectively refused and removed. Smith and McConnell (1966), using electron-diffraction methods, studied mullite and sillimanite which formed needles 0.3 x 0.003mm that had developed from muscovite breakdown during thermal metamorphism at Strachur. They reported that mullite, which co-existed with sillimanite, developed a stabilised structure probably due to introduction of small amounts of ferric iron from hematite inclusions in muscovite grains. Additional results on the Strachur mullite-sillimanite association were published by Smith (1969), who showed that sillimanite, and a 'mullite-type phase', resulted from the decomposition of muscovite in the outer part of the aureole.

## OFFRETITE

A calcium potassium magnesium aluminium silicate hydrate belonging to the zeolite group. Offretite has a chemical composition similar to erionite.

Of a somewhat more restricted occurrence than erionite, offretite frequently intergrows with the former in vesicular basalts. It forms fibrous outgrowths, less than 0.1mm thick, on glassy levyne

(NMS G 1982.37.14-19), occupying cavities up to 1.5cm in basalt at Quiraing, Skye (identified by X-ray diffraction, optics and electronprobe-microanalysis, Livingstone and Macpherson, 1983).

Offretite has been tentatively identified, from optical evidence, as minute crystals occurring either parallel, or normal, to titanite crystal faces, or as tufts, in lavas at Hartfield Moss, Renfrewshire (Meikle, 1990), and is present on micromount specimens from Loanhead Quarry, Beith (BMS no. 1036 [Meikle] and no. 1358 [Braithwaite], *British Micromount Society*, catalogue, 1994).

Green and Todd (1996) reported erionite-offretite epitaxial outgrowths a few tenths of a millimetre wide parallel to the *c*-axis of levyne in basalt at Moonen Bay, Skye.

## PHENAKITE

Beryllium silicate. Phenakite occurs in granite pegmatites, greisen, and hydrothermal veins.

Colourless glassy imperfect crystals (1.5cm) of phenakite occur in quartz-amazonite-pegmatite veins in syenite near Sgor Chaonasaid, north of Ben Loyal, Sutherland (Heddle Coll[n] 397.1 and 397.12). It also appears in a stream sediment concentrate on the south-western flank of Cam Eas in the Cairngorms (Bowie, 1962), and is present as ivory-coloured polycrystalline outgrowths on granite from the north-east slopes of Ben Macdhui, Cairngorms.

## PHILLIPSITE (*sl*)

A potassium sodium calcium aluminium silicate hydrate with or without barium. In phillipsite, a member of the zeolite group, potassium, calcium, sodium and barium atoms are linked to an aluminium-silicon-oxygen framework to which water molecules are attached. Less than 50% of the exchangeable cations are barium. Phillipsite and harmotome possess the same framework structure and form a continuous chemical series in exchangeable cations from calcium-rich to barium-rich.

Rhombohedral phenakite crystals forming outgrowths, to 1.8cm on orthoclase. Ben Macdhui, Cairngorms, Aberdeenshire. (NMS G 1982.70.1.)

Although phillipsite is a common zeolite in volcanic rocks, altered tuff beds and deep sea sediments, it is rare throughout the Scottish Tertiary basaltic province. King (1976) discovered that phillipsite occurred at only four locations out of over 400 examined on Skye, one locality being the northern slopes of Healabhal Mhor (McLeod's Table North) in Duirinish. Further Skye occurrences have been reported by Green and Todd (1996) at the northern end of Moonen Bay where it is usually the sole occupant of cavities up to 5cm. Here the phillipsite, which contains appreciable barium, forms drusy white to orange-brown crystalline crusts. Also to the north of the latter locality at Ouisgill Bay, where vitreous milky phillipsite crystals attain 2 x 3mm, lining a 2.3cm-long cavity (NMS G 1994.27.1, presented by N. Hubbard).

Outwith Skye, Dyer *et al.* (1996) reported phillipsite from a large dyke at Bennan Head, Arran. From a dyke on the beach of Glas Eilean island, Jura, the same authors discovered wellsite (an intermediate composition for which the name has now been discredited, Coombs *et. al.*, 1998)

Phosgenite crystals to 1.5mm on ferruginous flinty siliceous matrix, Lossiemouth, Moray.
*(D. I. Green Coll^n.)*

Spiky encrustation of tapering pyromorphite crystals to 1mm forming spheroidal groups, Leadhills, Lanarkshire.
*(NMS G 1994.1.30, ex Carruth Coll^n, NC74.)*

Pyromorphite prisms to 2mm diameter, Leadhills, Lanarkshire.
*(Heddle Coll^n 550.20.)*

with harmotome. Phillipsite was found as an inclusion in a grey apatite crystal, and forming brownish cream crusts on a sapphire-bearing camptonite dyke from Carrishader near Uig, Loch Roag, Isle of Lewis (Ross and Cromarty). Electron-probe-microanalysis on three separate grains yielded $K_2O$ 5.3-8.2%, $Na_2O$ 0.9-2.3%, CaO 6.2-7.7%, BaO 0.3-0.4%, $Al_2O_3$ 23.7-24.3% and $SiO_2$ 44.8-47.2%. Further phillipsite-bearing specimens from the camptonite dyke reveal that sub-idiomorphic crystals, up to 0.5mm, line elongate cavities (NMS G 1988.11.1-5 and 1994.28.3).

## PHOSGENITE

Lead carbonate chloride. Phosgenite forms a secondary mineral in the oxidation zone of lead deposits and, by alteration of galena plus other lead minerals, under surface conditions.

Greg and Lettsom (1858) first reported the occurrence of phosgenite in Scotland: 'Mr. Wright has quite recently found this mineral, in minute crystals, with galena and quartz, at Lossiemouth lead-mine in Elgin, Scotland.' After a period of 130 years Starkey (1988) confirmed the occurrence at Lossiemouth, detailing phosgenite crystals (<5 x 4mm) associated with pyrite, cerussite, pyromorphite,

sphalerite and galena in quartz-lined cavities of the Cherty Rock formation. Starkey (1988) ascribed phosgenite origin to atmospheric exposure and galena-seawater interaction in the inter-tidal environment.

## PYROMORPHITE

A lead phosphate chloride secondary mineral of the apatite group. Pyromorphite forms primarily after galena, and mainly in the oxidised zone of lead deposits.

Although pyromorphite is widely distributed, albeit in small quantities, as a secondary phase in many Scottish lead deposits, the main concentration predictably occurs in the Leadhills-Wanlockhead orefield. In the latter particularly good localities are Belton Grain Vein, at the High Pirn Mine, Wanlockhead and the Sarrowcole Vein, Leadhills. Massive pyromorphite, often pseudomorphously replacing galena, is common and ranges in colour from black, through green and yellow to white, while crystals display green, yellow and red to orange colours. Sowerby (1809) notes:

The yellow phosphates of lead of Wanloch-head mines, Scotland, are found coating Galena in the Bellan-grain vein, from 20 to 30 fathoms below

the surface, but gradually disappear at greater depths. From this mine our specimens came by favour of G. Laing, Esq.

Allan (1834) remarks:

The varieties from Leadhills are more remarkable for the richness of their colours than the beauty of their crystalline forms, being generally aggregated, grouped in rosettes, forming superficial coatings, and otherwise indistinctly defined. The Reverend W. Vernon has ascertained orange phosphate of lead from Leadhills to contain about one per cent. of the chromate, to which admixture he attributes the splendid tinges of that variety.

Surprisingly Heddle (1901) does not present a crystal drawing of pyromorphite in spite of Greg and Lettsom's (1858) statement that 'a specimen in Dr Heddle's collection, however, which in colour surpasses the chromate, is in large crystals'.

Temple's (1954) researches showed that pyromorphite from the orefield fell into two distinct structural groups. X-ray powder diffraction patterns revealed that crystals differ structurally from the massive pyromorphite, although some crystals associated with the latter fall into the second group. The first group is devoid of calcium whereas the second characteristically contains between 7 and 9.5% CaO with a corresponding

Part of a pyromorphite area 4 x 2.5cm, on a specimen bearing caledonite and linarite, from Leadhills, Lanarkshire. The specimen was formerly in the collection of C. Bement in the American Museum of Natural History, New York, Bement having purchased it from Bryce-Wright during 1871. *(NMS G 1991.31.66, ex Sutcliffe-Greenbank Coll⁰s.)*

reduction in lead. Numerous pyromorphite X-ray diffraction patterns from Scottish Mineral Collection material confirm the distinction detected by Temple (1954), although calcian pyromorphite possesses a line at 8.45Å (not recorded by Temple) which is absent in normal pyromorphite patterns. However, a line at approximately 8.5Å is present in X-ray powder patterns of approximately half of the apatite group members. Electronprobe-micro-analysis results concur with those reported by Temple (1954), for a narrow range of 8 to 10% CaO was detected. Collie (1889) analysed a black, botryoidal mineral from Leadhills that contained $15.8\% Ca_3(PO_4)_2$ (or 8.6% CaO) and named it calcium vanado-pyromorphite. Heddle (1901) concluded that the material analysed by Collie was a pyromorphite in which calcium replaced lead, and vanadium replaced phosphorus. Examination of 'colleite' (black hemispheres from the Belton Grain Vein, Wanlockhead) in the Scottish Mineral Collection (Heddle Coll⁰ 552.10) reveals it to be mottramite.

Pyromorphite may pseudomorphously replace galena, leadhillite, calcite or barite (Wilson, 1921).

*Left:* Pyromorphite from Leadhills, Lanarkshire; mineral paintings by Richard Cust prepared between 1830 and 1836. 'The phosphate is crystallised but the crystals are so small they cannot be expressed in a drawing.' The paintings are arranged in families according to the system of Robert Jameson; these from volume 4, one of 5 volumes. Edinburgh University, Special Collections MS 2739-43.

*Below:* 'Cairn Gorum crystals' and three line drawings, plus sketch of a fluid inclusion, are from crystals 'in the collection of G. Laing, Esq. of Edinburgh … The others are in my own cabinet'. From Sowerby (1806).

## QUARTZ *var.* Cairngorm

Silicon dioxide. Cairngorm, a yellowish-brown to black variety of quartz, occurs in veins and geodes, especially within granites, and basic to intermediate lavas. It is occasionally found infilling agates. There are millions of silicon atoms in a quartz crystal but every one million has a few thousand replaced by aluminium atoms. Natural radiation acting upon aluminium-bearing quartz transforms it into the cairngorm variety (Heaney *et al.*, 1994). Dana (1837, p.341) is credited with erecting the name 'cairngorum' for the smoky-brown quartz of the type found in the Cairngorm mountains, Scotland.

John Stoddart (1801), when in the company of Mr and Mrs Farquharson of Invercauld, Aberdeenshire, related:

Our leisure was partly occupied with the examination of a well-selected cabinet of natural curiosities, the most remarkable of which was a large

124

'Cairngorm' smoky quartz, slice revealing colour zoning, 4cm across, Aberdeenshire.
*(Heddle Coll^n 210.77.)*

'Cairngorm' smoky quartz crystal 10.5 x 5cm, Ben Avon, Cairngorms, Aberdeenshire.
*(Dudgeon Coll^n 1878.49.101.)*

crystal of the Cairn Gorum kind, nearly two feet long, and found within Mr. Farquharson's domain on Ben-y-Bourd. These beautiful minerals are produced, in some abundance, on all the surrounding mountains . . . .

The above-mentioned crystal remains in the possession of the Farquharson family and is currently on display in Braemar Castle. It measures 53.3cm in length, is 22.8cm high, and possesses a girth of 66cm, weighing 23.6kg (J. S. Blackett, *pers. comm.*, 1995).

Cairn Gorum crystals have been known for some years, and are said to have first caused the lapidaries to settle in Aberdeen, where they have been constantly employed in cutting them for seals, ring-stones, &c. (Sowerby, 1806).

On 6 September 1850, Queen Victoria (1819–1901) ascended Ben a' Bhourd, and close to the summit noted: 'We came upon a number of "cairn-gorms", which we all began picking up, and found some very pretty ones' (Duff, 1980). A number of cairngorms collected by Victoria and Albert are in the collections of Osborne House, Isle of Wight. (Osborne House was Queen Victoria's home at the time of her death.) Within the *Osborne Catalogue* (1876), item no. 104 the 'Highland Inkstand', now at Balmoral, is described as 'formed of specimens of granite, cairngorms, quartz, deers' teeth … &c'. Cairngorm featured in a sale by Bryce Wright (c.1880, p.12) and details appeared in a *Catalogue of Minerals, Geological, Conchological, & Archaeological Specimens, &c.*:

CAIRNGORMS – A large series of these beautiful Scotch Stones, including an enormous specimen, undoubtedly the largest cut stone known, weighing 52oz., or 7878 carats.

From the Cairngorm mountains Heddle (1901) reports 'crystals sometimes attain to a weight of 40lbs or more'.

Notable localities, all in Aberdeenshire, are Murdoch Head (Heddle Coll[n] 210.798, large crystals to 20cm); Stirling Hill, Peterhead (Heddle Coll[n] 210.808; NMS G 1896.470, a donation by Miss R. Orrock of a 15 x 9cm crystal); Culblean and the Loch Avon area, the latter also being a Heddle locality (Heddle Coll[n] 210.23) in the Cairngorms. It occurs additionally in geodes with fluorite and calcite at Kempock Point, Gourock, Renfrewshire (Dudgeon Coll[n] 210.142) and in granite on the east side of Goat Fell, Arran (NMS G 1960.19.1-8).

## SCAWTITE

A calcium silicate carbonate hydrate generated by thermal metamorphism of limestone by basalt and gabbro.

Scawtite was first reported from Scotland by Agrell (1965), as occurring in Jurassic limestone forming a screen between two ring dykes at Kilchoan, Ardnamurchan. Two complex metamorphic events took place, with the first creating minerals characteristic of the decarbonation series described by Bowen (1940). With the introduction of water and fluorine during the second phase, earlier-formed minerals, including spurrite, became unstable. Scawtite developed from spurrite as scaly aggregates, especially associated with ettringite. It is also found with tacharanite and xonotlite in narrow fracture-filled veins in gabbro at Binhill Quarry, Cairnie, Huntly, Aberdeenshire (Livingstone, 1974).

Sheaves of scotlandite crystals to 1mm, Wanlockhead, Dumfriesshire. (*J. R. Knight Coll[n] JRK 0979.*)

## SCOTLANDITE

Lead sulphite: the first naturally occurring sulphite.

The scotlandite-bearing barite sample (5 x 5 x 3cm) found at the Susanna Vein, Leadhills, probably originated from an old collection. More recently it successively passed via the late Richard Barstow (1942-82), Howard Belsky of New York, and Dr D. H. Garske, one-time resident in Illinois, to W. H. Paar of Salzburg University, Austria. A large cavity within the sample encrusted by secondary lead minerals contains anglesite, lanarkite, lead-hillite, susannite and scotlandite. The latter is found in vuggy anglesite and forms yellowish single spear-shaped crystals (<1mm) which are sometimes arranged in fan-shaped aggregates. Minute pyromorphite inclusions impart the yellowish colour, whereas a second colour variant, devoid of pyromorphite, is whitish to water clear (Paar *et al.*, 1984).

Grey-coloured scotlandite clusters (<0.5mm) with susannite and mattheddleite are present on a galena-bearing lanarkite-rich specimen (10 x 5 x 3cm) from the Susanna Mine within the Scottish Mineral Collection (NMS G 1994.1.59). This specimen, originally in the A. G. Moss collection, no. M809, came to the National Museums of Scotland via Simon Harrison Minerals, N. Carruth and L. and P. Greenbank collections.

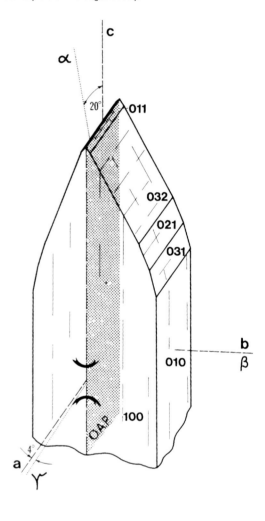

Idealised drawing of scotlandite crystal. From Paar *et al.* (1984). Courtesy of the Mineralogical Society.

## SILVER

A native element which forms a series with gold, occurs both as a primary constituent of some hydrothermal veins, and in ore deposit oxidation zones.

Silver has been reported from numerous localities throughout Scotland. The most noteworthy silver deposits are located at Alva in the Ochil Hills, Clackmannanshire, and Hilderston in the Bathgate Hills, West Lothian. Elsewhere silver, as an associate mineral in tiny quantities, is present in a wide variety of mineral assemblages and geological environments.

Native silver, dendrite 5mm long, Alva silver mine, Silver Glen, Clackmannanshire. *(S. Moreton Coll[n] SM2.)*

Approximately 3km north-east of Bathgate is the Hilderston silver mine. Meagre surface expressions betray little of past mining activities. Meikle (1994) expounded the mine's long history from 1606, when the silver vein was first discovered, until site investigations during 1993. The most comprehensive account of the Hilderston silver mine is given by Cadell (1925), who quotes extensively from previous accounts and compiles information on the original seventeenth-century workings and final nineteenth-century workings. The owner of the ground, Sir Thomas Hamilton of Binnie and Monkland, took out a mineral lease from the Crown in January 1607. Shortly afterwards the mine recorded a profit of £500 sterling per month based upon a silver price of 4s-6d per ounce. When this was reported to the Privy Council, King James VI immediately nationalised the mine which was taken over by the state in 1608 (Stephenson, 1983). Atkinson (1619) described the first ore, discovered by a coal miner prospecting on his own for a new coal seam, as:

redd-mettle; [nickeline?] no man thereabout ever saw the like. It was raced with many small stringes, like unto haiers or threadds.

A second similar piece was obtained from the vein:

And the vaine thereof, out of which I had it, was once two inches thicke .... the mettle thereof was both malliable and toughe. It was course silver, worth 4s 6d the ounce weight .... Untill the same redd-mettle came unto 12 faddomes deepe, it remained still good; from thence unto 30 fathome deepe it proved nought.

Ore from Hilderston was processed in mills erected upon the Water of Leith (the surrounding area becoming an Edinburgh district named Silvermills).

From old mining records Stephenson (1983) deduced the nature of the vein mineralisation which revealed two metallogenic episodes. In the barite vein the first episode of chalcopyrite, galena, pyrite and sphalerite was followed by silver, nickel and cobalt mineralisation. Examination of recently collected dump material (Meikle, 1994) revealed silver in barite as isolated irregular grains up to 0.25 x 0.16mm. Other minerals detected include annabergite, erythrite, acanthite, amalgam (Ag,Hg) barytocalcite, cinnabar, harmotome, maucherite, mercury, witherite and unidentified nickel-cobalt sulpharsenides.

An epigenetic hydrothermal silver deposit near Alva, Clackmannanshire, was deposited in the Devonian Ochil Volcanic Formation during Late Carboniferous times and sited predominantly on a fault breccia.

The silver-bearing vein was first discovered in December 1714 on Sir John Erskine's estate in what is currently known as Silver Glen. In the Erskine–Murray papers (NLS MS 5099), folio 170 elaborates upon the mine discovery and extent of the workings:

The vein was discovered from the appearance [at the surface] of a small pile of spar [*sic*] .... When the miners followed the spar a small string of ore 1½ inch wide with which the miners went down almost perpendicularly for 4½ feet all in day on the face of the mountain.

Small-scale working continued until the summer of 1715 when the vein suddenly produced a bonanza, some 40 tons of ore being extracted over a period of approximately three months. In February 1716 mining activity ceased. Sir John Erskine left the area to join the Jacobite rebellion and proceeds from the mine helped fund the insurrection.

A Royal Commission was granted to Dr Justus Brandshagen from the Harz silver mines to inspect the Alva Mine. Instructions at Hampton Court are dated 3 September 1716 and countersigned by The Right Honourable The Lords Commissioners of His Majesty's Treasury, and direct Brandshagen to report on the mine and his activities. (Brandshagen, 1717; Treasury Record T64/235, report dated Edinburgh, February 5, 1716/7):

That after I had bestowed considerable Time Pains and Expenses in cleaning this Mine and in building the Furnaces for assaying and melting and that everything was got in readiness for the Tryals, I wrote to Edinburgh and advertised The Earl of Lauderdale Mr Haldane and Mr Drummond of the same, desiring they might return to Alva to Witness the breaking off of the Ore, and making the Assayes, as Directed by His Royal Highness's Instructions, and they accordingly arrived at Alva upon the 2nd of January in the Morning.

That upon the said 2nd of January The above Commissioners went with myself, Thomas and James Hamiltons to the Mine and thereafter viewing the Mine and the Vein of Ore therein, it was agreed to break off Ore in Six Several Places of the Vein (and which Places are marked in Draught or the Description of the Vein with the Numbers 1, 2, 3, 4, 5, 6) and also it was agreed that we should lay aside part of Each of these Six Several Parcels of Ore for the Tryals, and part thereof also to send to London to Your Lordships, so accordingly at the Commissioners Sight, The Hamiltons broke off the Ore at the Six places of the Vein following Tryal at No 1 … where the vein is five inches broad.

(The Earl of Lauderdale held the position of General of the Edinburgh Mint, and Mr Drummond, a goldsmith, the Warden. Mr Haldane was the brother-in-law of Sir John Erskine.)

Selection of the six assay samples occurred over approximately 12ft (3.6m) of the vein, which ranged from 2ins (5cm) wide to one whereby the main body of the vein attained 15ins (38cm). According to Brandshagen the widest position of the vein became exposed in January 1717 measuring 2ft 8ins (81cm) which included 15ins (38cm) of ore (the black vein). The main ore body measured some 24ft (7.3m) vertically, 11ft (3.4m) horizontally with a vein up to 3ft (0.9m) wide (anon. miner, 1753; Moreton, *pers. comm.*). The vein commenced just below the surface and continued to a depth of 29ft (8.8m).

After Sir John's pardon for his part in the rebellion, mining recommenced in 1719 until ore became quickly exhausted. At the height of activity the weekly output of silver ore averaged £400 in value with a gross of some £40,000-£50,000. Silver was fix priced at 5s-7d per troy ounce (Newton, 1716; in

Hall and Tilling, 1976) hence a sale of about 5 to 6 tonnes of silver metal took place (Moreton, 1996).

In St Serf's Parish Church, Alva, are two plain communion cups dated 1767, which bear an Edinburgh hall mark with the initials KD of the silversmiths Ker and Dempster. The Latin inscription reads:

Sacris

in Ecclesia S. Servani

apud Alveth

A.D. MDCCLXVII

Ex Argento indigena

D.D. Cg.

Jacobus Erskine

[Sacred in the church of St Serf by Alva AD MDCCLXVII. Made from indigenous silver, James Erskine]

An unmarked silver font plate is believed locally to be fashioned from Alva silver (R. C. Snaddon, *pers. comm.*, 1994).

During the spring of 1758 mining resumed (Williamson, 1758) and in that year a mass of cobalt ore was discovered lower down the glen. Sir John Erskine died in 1739 and his son sold the estate to Charles Erskine (or Areskine) – an uncle who later became Lord Justice Clerk. It was he who re-opened the mine in 1758. On 17 January 1759 the leading Scottish chemist Joseph Black of Glasgow wrote about the cobalt ore (anon. miner, NLS MS 5098, f.49):

of that kind which affords the saffer or blue for porcelaine …. I shall procure some trials to be made by painting some delftware and comparing the colour with saffer [sic] used here.

The Alva cobalt ore produced a beautiful deep blue colour and was extensively used in porcelain manufacture at Prestonpans, East Lothian. History of the Alva cobalt ore in relationship to porcelain was elaborated by Turnbull (1997).

Edward Wright MD, in a letter dated 9 August 1760 to Emanuel Mendes da Costa (NLS MS 5098, f.89), wrote a revealing description of the Alva Mine:

About a month ago I had the pleasure of going through all the Mines on that Estate, in company with James Erskine Esqr. one of the Barons of Exchequer [sic] I just mentioned. There I particularly observed the old works out of which the rich silver ore I formerly wrote you about was got, which brought the proprietor many thousands of pounds. It was not a continued vein but a large belly of ore, in the form of nearly a double cone, thick in the middle and gradually dying away at both ends; though it is not improbable but the narrow filon? or thread in which it terminated may lead to another belly of the same kind, which the Baron, who is very curious in mines, designs to try.

Secreted at the rear of the Erskine-Murray papers are three smelt products. Two small silvery-bronze coloured ingots measure 2 x 0.75cm and 1.5 x 0.75cm; the third, in five fragments, is bluish-black and probably derived from copper ore.

Sowerby (1811) illustrated Alva silver:

This specimen of Crystallized Native Silver, accompanied by Flowers of Cobalt, was sent me from Alva mine in Stirlingshire by my kind friend formerly mentioned in this work by the name of G Laing, Esq. It is a very useful specimen, as exhibiting a very elegant arrangement of the primitive *nuclei*, if I may so call these minute octaëdrons …

Francis *et al.* (1970) report that the main workings are sited on a mineralised fault breccia containing calcite, pyrite, chalcopyrite, arsenopyrite, argentite, galena, erythrite and hydrocarbon botryoids. Both Heddle (1901) and Wilson (1921) report argentite, although recent work by Moreton *pers. comm.* and Moreton *et al.* (1998) and in the National Museums of Scotland casts considerable doubt on argentite presence in the ores. Dump excavation yielded appreciable quantities of primary ore in which argentite has not been discovered. Native silver dendrites in dolomite attain 3cm but more commonly range to 1cm in recently collected specimens. Safflorite is commonly associated with native silver and is the dominant cobalt ore mineral. The Alva silver mine produced the richest bonanza of native silver ever recorded in Britain (Moreton, 1996). The old copper mines in

Mill Glen, just north of Tillicoultry, Clackmannanshire, worked around 1750, also contained silver and cobalt (Wilson, 1921). Silver ore is said to have been found just to the west and north of Castle Campbell, approximately 1.6km north of Dollar, Clackmannanshire, 'but that it did not answer the expence of working it' (Watson, 1795).

Rare silver in prehnite is reported from Boyleston Quarry near Barrhead, and Loanhead Quarry, Beith (Meikle, 1989, 1989a) and at Hartfield Moss, Renfrewshire (Meikle, 1990). At the first locality silver exhibits simple and twinned cubic crystals up to 1 x 0.4mm, elongated bipyramidal crystals and filliform growths. Loanhead silver is present as well developed crystals to 1.8 mm, aggregates and filliform masses to 0.3mm with associated native copper and covellite. At Hartfield Moss silver crystallised as modified cubes to 0.2mm and filliforms to 1.4mm with square cross-section to 0.4mm.

Todd and McMullen (1991) described silver from calcite-barite vein material occurring in the Mannoch Hill and Nutberry Hill areas near Lesmahagow, Lanarkshire. Here it is present as cleavage face coatings in sphalerite, as plates, as small irregular aggregates of cubic crystals and as well-developed cubic crystals (0.5 - 2.7 mm). Silver is also reported associated with electrum, hessite, petzite and sylvanite from veins at Tyndrum, Perthshire (Pattrick *et al.*, 1988), and is found with altaite, coloradoite and hessite in the quartz, chalcopyrite, galena and pyrite Urlar Burn veins near Aberfeldy, Perthshire (Pattrick *pers. comm.*, 1990). Native silver is locally present between laths of molybdenite and within masses of molybdenum ochre at Glen Gairn, Aberdeenshire (Webb *et al.*, 1992).

## SPHALERITE

Zinc iron sulphide. Sphalerite belongs to the sphalerite group and is trimorphous with matraite and wurtzite. It is the most common zinc mineral and is intimately associated with galena in lead-zinc deposits, occurring in contact metamorphic deposits, low-temperature veins, pegmatites and sedimentary rocks.

Throughout Scotland galena-bearing veins carrying lesser quantities of zinc ore are widespread (Wilson, 1921) but most abundant in the western regions of the Southern Uplands and in the Dalradian. As sphalerite is widely distributed mention need be made of only several localities.

Before 1880 sphalerite was not gathered at Wanlockhead; however, in that year 39 tons were produced and output increased to 1000 tons per annum. No zinc ore was produced at Leadhills (Wilson, 1921). Sphalerite in the Leadhills-Wanlockhead orefield occurs either as aggregates of massive ore devoid of galena and chalcopyrite or as dark-brown to almost black crystals. Where the latter are large, and in association with calcite scalenohedra, specimens adopt a striking appearance. Wanlockhead mines produced by far the highest quality sphalerite-calcite specimens in Scotland.

Elsewhere sphalerite was common in the mines at Strontian Argyllshire, Tyndrum Perthshire; Pibble, Blackcraig and Woodhead mines in Kirkcudbrightshire; on the island of Islay; and in the complex ore at the Louisa antimony mine, Glendinning, Eskdale, Dumfriesshire, where considerable portions of the vein material contained 'crystals

Twinned sphalerite crystal 3.2 x 2.5cm with quartz and calcite scalenohedra, from between the 160 and 200 fathom levels, Glencrieff Mine, Wanlockhead, Dumfriesshire. Presented by F. N. Ashcroft to the British Museum (Natural History) in 1920. BM 1920.483. Natural History Museum, London.

and crystalline sphalerite – mainly "Black Jack" but a little "Honey Blende" also' (Williams, 1965). A more recent occurrence of sphalerite was reported by Coats *et al.* (1980, 1984) from the large stratabound barite and sulphide deposits near Aberfeldy, Perthshire. Here the sphalerite (including a manganoan variety) is associated with galena, pyrite and pyrrhotite in a barium-rich mineralised zone in Dalradian metasedimentary rocks. A possible origin for the deposit envisages the introduction of a metal-rich hydrothermal brine onto the sea floor under anaerobic sulphate-reducing conditions.

## STELLERITE

A calcium aluminium silicate hydrate belonging to the zeolite group. Scottish stellerite is found only in basalts although the species is known from a wide variety of rock types and geological occurrences (Gottardi and Galli, 1985).

Stellerite is difficult to distinguish from stilbite to which it is closely related structurally and chemically. Stilbite (monoclinic) and stellerite (orthorhombic) both possess similar optical properties, thus making stellerite validation difficult. Gottardi and Galli (1985) demonstrated slight differences in their X-ray powder diffraction data. Diagnostic characteristics lie in their alkali content, for stilbite contains sodium as an *essential* cation, whereas only traces are present in stellerite.

Dyer *et al.* (1993), utilising X-ray powder diffraction, thermal analysis, scanning and analytical electron microscopy, recorded stellerite for the first time from Scotland, at Todhead Point, 9km south of Stonehaven (Grampian Region). At this locality stellerite is present with mordenite in olivine basalt. Heddle (1901) published two analyses of calcium stilbite from Dunbartonshire, with 11.5 wt% and 9.8 wt% CaO, although sodium was not reported. Stellerite occurring as 3cm wheat-sheaf crystals associated with laumontite and calcite has been confirmed from Moonen Bay, Duirinish, Skye (Green and Todd, 1996). At another Skye locality, Edinbane, at the head of Loch

Greshornish, Savage (1995) described stellerite from vesicles in basalt; neighbouring vesicles contained erionite, stilbite, heulandite and calcite.

## STIBNITE

Antimony sulphide. Generally formed at low temperatures in hydrothermal veins, replacement deposits and hot springs.

Excellent stibnite specimens originate from the Fountainhead Mine, Hare Hill (The Knipe) located on the Marquess of Bute's estate approximately 5km south-south-east of New Cumnock, Ayrshire. At 518m above sea level the mine occurs in a small 'granite' mass where a vertical vein, 30-45cm wide, was worked from a level driven some 9-15m into the hillside (Flett, in Dewey, 1920). The stibnite workings are located within a late-Caledonian granodiorite mass, 1.6km in diameter, intruded into an Ordovician accreted turbidite terrain (Boast *et al.*, 1990). There is no mention of the mine in the New Cumnock Parish section of the *Statistical Account of Ayrshire* (anon., 1842) although several trials for lead had been made in the 'Nippes', whereas Simpson and MacGregor (1932) relate that it operated prior to 1860. The mine was probably worked by the mid-1840s, as an entry in the Old College Museum register (NMS Geology and Zoology Department archives) for 1845-46 states: 'Large specimen of Grey Antimony Ore, from Hare Hills, near New Cumnock, Ayrshire.

Plumose aggregates of stibnite prisms to 7cm in quartz with creamy stibiconite, specimen size 19 x 14 x 7cm, Hare Hill, New Cumnock, Ayrshire.
*(Dudgeon Coll^n 1890.114.72..)*

Presented by the Most Noble the Marquis of Bute to College Museum.' Ore transportation occurred over rough terrain by horse-drawn sledge to Kirkconnel, whereafter it was shipped out via rail (Brownlee *pers. comm.*, 1993). Minerals detected by Boast *et al.* (1990) in the zoned As-Sb-Cu-Pb-Zn assemblage, which contained significant gold grades, include arsenopyrite, galena, argentian-tetrahedrite, pyrite, chalcopyrite, stibnite, gold, bournonite and pyrrhotite. Secondary antimony minerals in the ore include rare senarmontite as octahedra (<0.5mm), valentinite and stibiconite (Macpherson and Livingstone, 1982). In June 1989 the area around the mine, including the ore dump, became a Site of Special Scientific Interest.

Wilson (in Dewey, 1920) records stibnite from the Louisa Mine, Glendinning, Dumfriesshire, where the vein was discovered in 1760 although not worked regularly till 1793. Jameson (1805) described the vein as 'quartz, and calc-spar; the ores grey antimony, brown blende, fine-grained lead glance, and iron pyrites .... The ore of antimony is the radiated grey antimony'. A detailed description of the history and mining activity at the Louisa Mine was given by McCracken (1965). During 1918 Arthur Russell collected dump material at the Louisa Mine, from which Smith (1919), utilising goniometric and chemical techniques, validated semseyite. Cunningham (1843) described the presence of stibnite:

> In the limestone quarry of Maisley, near Keith, a vein of prismatoidal antimony, or antimony-glance, has long been known to exist .... The antimony is accompanied with masses of green and purple fluor-spar ....

## STRONTIANITE

Strontium carbonate. Strontianite, a member of the aragonite group, occurs in low-temperature hydrothermal veins, and is found associated with barite, celestine and calcite. It also occurs in geodes or concretionary masses in limestone, and less frequently as a gangue mineral in sulphide veins.

Strontianite occupies a unique position in the annals of Scottish mineralogy. It became the first new species from Scotland, and from it the element strontium was derived. Sulzer (1791) named the species, although 17 years elapsed before Davy (1808) extracted the silvery-white metal.

Brewster and Jameson (1822) published a 'Notice of Mineralogical Journeys, and of a Mineralogical System; by the late Rev. Dr John Walker', in which they mention:

> Between the years 1761 and 1764, I [Walker] found in those mines (at Leadhills-Wanlockhead) the Strontianite; the Ore, and the Ochre of Nickel; the Plumbum pellucidum of Linnaeus …. All these were here, for the first time, discovered in Britain ….
>
> After examining all the coasts from the Shore of Assynt, to the Isle of Sky, I there parted with the cutter. I then traversed the countries of Glenelg, Kintail, Glenshiel, and several districts of Lochaber; examined Morven, and the mines of Strontian. There I found several rare minerals, and particularly that singular substance, since called the Strontianite, in great plenty; though I had observed it but very sparingly, three years before, in the Mines at Leadhills ….

In a footnote to the statement, Brewster and Jameson (1822) comment:

> It is not generally known, that at one period, small quantities of strontites were found at Lead Hills; and the fact in the text proves, that to Dr Walker the merit is due of having determined mineralogically that Strontites was a new mineral species. Dr Hope afterwards, by the *discovery* of the strontitic earth, added to the interest of the determination of Dr Walker, and proved that strontites was also a new chemical species.

However, Heddle (1901) remarks:

> Strontianite was discovered by Walker in 1764; but he had, in 1761, noticed it at Leadhills, on specimens unquestionably brought by miners from Strontian [but see J. Torrie].

Prior to 1791, strontianite had attracted mineralogist's attention around 1787:

Strontianite fan, 11cm radius, Strontian, Argyllshire.
(NMS G 1989.4.1.)

The mineral is found in the lead mine of Strontian in Argylshire. It was brought to Edinburgh about six years ago in considerable quantity. It was generally received as the aerated barytes [Abstract published in 1794 (Hope, 1794) of a paper read by T. C. Hope on 4 November 1793 – the full paper was published by Hope (1798); both were published in the *Transactions of the Royal Society of Edinburgh*].

In the 1798 paper Hope states:

> The mineral, of which I have the honour to lay an account before the Society, was brought to Edinburgh in considerable quantity about six years ago by a dealer in fossils, though indeed it had found its way, long before this period, into one or two collections.

Strontianite spheroids to 7mm on barite; specimen size 8 x 7 x 5cm from Muirshiels barite mine, Lochwinnoch, Renfrewshire.
(J. G. Todd Coll[n] no. BG-6.)

Jameson (1804), in his *System of Mineralogy*, remarks:

Dr Hope, in his masterly memoir published in the Edinburgh Philosophical Transactions for 1790, first made us acquainted with the peculiar earth which this genus [strontians] contains. It afterwards engaged the attention of other able chemists, particularly Kirwan, Klaproth, Pelletier, and Vauquelin.

In the second edition Jameson (1816) issued a similar statement:

The peculiar earth which characterises this mineral was discovered by Dr Hope, and its various properties were made known to the public in his excellent memoir on Strontites, inserted in the Transactions of the Royal Society of Edinburgh for the year 1790.

Although Hope had admirably demonstrated that barium salts differed markedly from those of the new peculiar earth, Jameson incorrectly attributed the discovery to Hope. It was Dr Adair Crawford, and his assistant, Mr Cruikshank, who first demonstrated that a difference existed:

In the year 1784 I made several experiments and observations on the medicinal properties of the Muriated Barytes, from which I concluded, that it might probably possess considerable powers as a deobstruent ....

The muriated barytes exhibited [administered] in St. Thomas's Hospital since the month of May 1789, was obtained by the decomposition of the heavy spar. Having procured some specimens of a mineral which is sold at Strontean, in Scotland, under the denomination of aerated barytes, I was in hopes that the salt might be formed with less difficulty by immediately dissolving that substance in the muriatic acid.

It appears however from the following facts, which have been verified by the experiments of my assistant, Mr. Cruikshank, as well as by my own, that this mineral really possesses different properties from the terra ponderosa [barium earth] of Scheele and Bergman. (Crawford, 1790)

Recently Crawford (1748-95) was immortalised mineralogically by Khomjakov *et al.* (1994), who erected the new species crawfordite, $Na_2Sr(PO_4)(CO_3)$, in honour of his discovery.

The finest Scottish strontianite specimens emanate from the lead mines at Strontian from where clove-brown and apple-green, fan-like masses have been extracted. Whitesmith Mine produced an exceptional apple-green specimen with 12cm-long radiating glassy prisms (NMS G 1878.49.532.3; Dudgeon Coll[n]), whereas a brown specimen 8cm radius (NMS G 1878.49.532.2; Dudgeon Coll[n]) bears the Fee Donald Mine location. An adhering label on a 8cm-radius brown fan (NMS G 1878.49.532.1 Dudgeon Coll[n]) states 'Carbonate of Strontites, Strontian', suggesting possible derivation from a very old collection. Heddle (1901) reported that brewsterite is a rare strontianite associate although harmotome frequently co-exists. Again, brewsterite is associated with apple-green strontianite, galena and calcite (NMS G 1994.135.106).

Todd (1989) reported strontianite from Trearne Quarry, near Beith, as small white spherical aggregates located in Visean limestone brachiopod fossils. Less commonly, strontianite forms subparallel bundles of elongated crystals forming tufts to 0.5mm. Especially in the brachiopods, associates include calcite, dolomite, quartz and colourless fluorite cubes. From the same quarry Wolfe (1996) reported free-standing strontianite crystal groups to 3cm in a celestine-bearing limestone block. Small limestone solution cavities at Allt nan Uamh, Inchnadamph, Sutherland, contain cream to pinkish, 1-2mm, strontianite crystals forming coarse efflorescent-like growths (NMS G 1995.121.1-3; presented by J. G. Todd).

Todd and Laurence (1989) figured strontianite as globules with cockscombe barite and quartz in cavities formed within massive barite at the Muirshiels barite mine, Renfrewshire. Most of the strontianite specimens are generally a calcian variety, the most calcium-rich developing a highly elongated pyramidal habit. Additionally found in a similar situation (NMS G 1968.56.1-14) where one specimen (no. 10) presents coalescing globules (7mm) covering approximately 15cm². In the majority of these specimens the globules are composed of tabular strontianite crystals which exhibit a pseudo-hexagonal aspect due to cyclic twinning. The twinned crystals are stacked upon each other producing a columnar effect.

Heddle (1883) listed 14 minerals, including strontianite, from a syenite boulder described as 'a perfect mineral casket', on the slopes of Beinn Bhreac, Tongue, Sutherland:

In diverging crystals, which form spherical masses in a cavity in the granitic vein, disposed upon Rock Crystal and Amazonstone .... Also in the mass of the syenite near the granitic vein, in small cavities, which were lined with crystals of hornblende and felspar. (Heddle, 1901)

From this locality, a 6cm-diameter cupola-shaped mass of 1cm radiating crystals was extracted (Heddle Coll[n] 280.14).

Strontianite, with calcite, dolomite and galena, fills veins and interstices in breccias and small faults near the Point of Ness, Stromness, Orkney (Mykura, 1976). Here the strontianite intimately associated with barite forms earthy brown to white masses (Heddle Coll[n] 280.3, 4, 10 and 15). (This material is the stromnite of Traill, 1806, 1823; Heddle, 1901.) Strontianite has a localised occurrence as an accessory mineral in the Glen Dessary syenite (Fowler, 1992).

## SUSANNITE

A lead sulphate carbonate hydroxide trimorphous with leadhillite and macphersonite. Susannite is found rarely in oxidised lead deposits.

Susannite, of considerably rarer occurrence than leadhillite, and from which earlier distinction proved difficult, was only known to occur at the Susanna Mine, Leadhills (Greg and Lettsom, 1858), until Heddle (1901) added a second locality, a vein in Leadhills Dod. Brown (1925) comments:

And I am pleased to say that this is one of the many rare minerals that have been re-discovered, after a lapse of 50 years, by Mr John Weir and Mr John Blackwood, of Leadhills, and myself at Susanna and Hopeful veins, Leadhills.

Susannite crystals, to 2mm, with cerussite and leadhillite, from the Susanna Mine, Leadhills, Lanarkshire. The specimen, BM 1964 R 12139, was purchased by Arthur Russell in 1934 from Gregory Bottley and Co. Courtesy of the Department of Mineralogy, Natural History Museum, London.

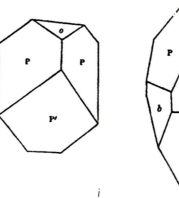
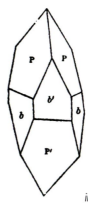
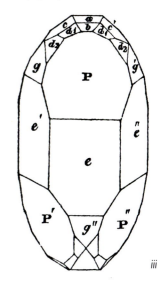

Susannite crystal drawings: *i* and *ii* from Greg and Lettsom (1858); and *iii* from Phillips (1823).

*i*

*ii*

*iii*

Brooke (1820) chemically designated the then-unnamed susannite and leadhillite as sulphato-tri-carbonate of lead, and distinguished crystals of the former as they possessed acute rhombohedral forms with interfacial angles of 72°30' and 107°30'. Both Brewster (1820) and Haidinger (1826a) undertook morphological studies of the two phases and Haidinger erected susannite as a distinct species in 1845 (Haidinger, 1845).

Kenngott (1868) regarded susannite as pseudo-rhombohedral leadhillite but concluded that it may represent a primary occurrence of the uni-axial polymorph of leadhillite obtained by heating that species. Palache *et al.* (1951) commented that conclusive evidence of this was lacking. Mrose and Christian (1969) demonstrated that leadhillite, when heated to 200°C, produced an identical single crystal precession X-ray photograph to that of rhombohedral susannite from Leadhills, and remarked that their X-ray powder patterns are indistinguishable. Utilising the latter, Livingstone and Russell (1985) demonstrated that the two species could be distinguished.

Susannite collected at the Susanna Vein, Lead-hills, is associated with leadhillite, lanarkite, pyromorphite, cerussite, caledonite, and rarely plattnerite. (Heddle Coll$^n$ 734a, 4, 5, and 11, NMS G 1991.31.77 and 1994.128.1 – the latter from the

Bertrand collection, Paris, 1878). [Émile Bertrand (1844-1909), a French mineralogist, designed the 'Bertrand lens' to observe interference figures in convergent polarised light.] Equant greenish-brownish crystals to 2mm invest a Leadhills specimen (NMS G 1994.1.71), originally in an old French collection which bears a label dated 1818. Susannite is associated with macphersonite and leadhillite, together with cerussite, caledonite and pyromorphite, on a small Heddle specimen (Heddle Coll$^n$ 721.34) occurring at Leadhills Dod (Livingstone and Sarp, 1984). Recently the structure of susannite has been determined by Steele *et al.* (1999) utilising a crystal from a specimen in the Scottish Mineral Collection (NMS G 1991.31.77).

## TACHARANITE

Calcium aluminium silicate hydrate. Tacharanite is restricted to amygdales in basaltic rocks.

Fine-grained amygdales in melanocratic olivine dolerite near Portree, Skye, contain varying mix-tures of thomsonite, mesolite, xonotlite, gyrolite and calcite. Tacharanite with tobermorite, sur-rounded by gyrolite and mesolite, forms cores to some amygdales with cryptocrystalline tacharanite containing gyrolite rosettes (Sweet, 1961). Fibrous

tacharanite associated with scawtite and xonotlite found in narrow veins cutting gabbro at Binhill Quarry, Cairnie, Huntly (Livingstone, 1974) was analysed, utilising electron microscopy and diffrac-tion, by Cliff *et al.* (1975). Seven separate fibrous tacharanite bundles yielded an average composition which revealed an appreciable aluminium content of 5.5% $Al_2O_3$ (corresponding to 2.4 Al atoms) whereas calcium gave 12.1 atoms (see *tacharanite*, ideal formula, Appendix 1). Tacharanite was also recorded by Walker (1971) from Mull basalts.

## TITANITE (Sphene)

Calcium titanium silicate. Titanite is a widespread accessory mineral of igneous rocks, especially inter-mediate and acid plutonic rocks, and nepheline syenites. In metamorphic rocks titanite occurs in schists and gneisses rich in ferromagnesian minerals, and numerous other metamorphic assem-blages. (In keeping with current terminology, except where a direct quotation appears, 'titanite' is substituted below for 'sphene'.)

Recordings of Scottish titanite are located in numerous publications relating to igneous and metamorphic petrological studies. Deer (1935) reported titanite as a common accessory, with

Titanite (sphene) crystal drawings based on crystals extracted from marble at Shinness, Sutherland. From Heddle (1882).

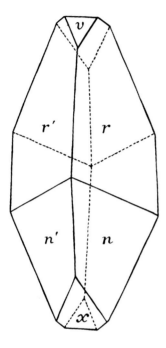

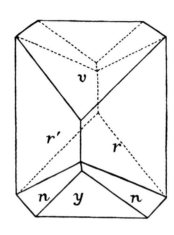

enclosed in the Ardgour feldspathic granulite and more rarely in the pelitic gneiss. Titanite is a very common accessory mineral in the aegirine granulites of Glen Lui, Aberdeenshire (McLachlan, 1951) where it varies in appearance from small, colourless blebs to characteristic greyish lozenge-shaped crystals. In the associated quartz-albite pegmatitic veinlets titanite frequently displays reddish patches around an early core of idiomorphic titanite crystals.

Throughout the Lewisian ortho and para gneisses of the Outer Hebrides, titanite is a common accessory mineral, sometimes attaining up to 6 vol.%, for example in the meta quartz micro diorite dyke at Garry a-Siar, Benbecula (Fettes *et al.*, 1992).

Yellow euhedral titanite, up to 6mm, is present both as an accessory mineral in the granite, and skarn, at Corrie Buidhe, Glen Creran, Argyllshire (Ingram *et al.*, 1993).

Clastic grains of titanite, together with mainly magnetite and zircon, form thick streaks in the Torridonian rocks of Skye and north-west Scotland (Peach *et al.*, 1907). In spite of rutile and anatase being present in heavy mineral concentrates generated from Scottish Carboniferous rocks, titanite was not reported (Bosworth, 1912) although Harding *et al.* (1984) noted its presence in locally derived stream sediments on St Kilda.

Noteworthy titanite specimens in the Scottish Mineral Collection are deep brown, vitreous 4-7mm crystals in granite/granodiorite from the Criffel-New Abbey area, Kirkcudbrightshire (Heddle Coll[n] 510.64; Dudgeon Coll[n] 1878.49.369). And also from near Tongue, Sutherland, a brown crystal, 1.5 x 0.5cm, in feldspar (Heddle Coll[n] 510.1). Clove-brown, resinous to vitreous idioblastic titanite crystals, 3 x 2cm, are present in a meta acid pegmatitic matrix from a roadside quarry at North Ford, North Uist, Outer Hebrides (NMS G 1976.11.1 and 2, collected and presented by T. K. Meikle). Also as a 4 x 3cm highly fractured imperfect crystal in quartz with epidote, garnet and molybdenite from Corrie Buidhe, Glen Creran, Argyllshire (NMS G 1995.136.7); and as colourless to creamy-brown crystals, to 7 x 5mm, with biotite books, standing proud of a flat rock surface in a

apatite, in the minor intrusions of the Cairnsmore of Carsphairn igneous complex, whereas Nicholls (1950) found titanite, mainly associated with biotite, in the diorites of the Glenelg-Ratagain igneous complex. On St Kilda in the Glen Bay and Na h-Eagan granites, titanite adopts an unusual sheaf-like habit of radiating fibres (Harding *et al.*, 1984). Basalts at Hartfield Moss, Loanhead and Boyleston quarries, Renfrewshire, contain colourless to pale greyish-green titanite crystal aggregates to 1.2mm across in vesicles (Meikle, 1990). Titanite is a common amygdale and matrix mineral in the Clyde Plateau lavas, and becomes more abundant as the burial metamorphism increases (Evans, 1987).

Heddle (1901) cites numerous locations for titanite in various rock types including granite, granite veins, syenitic granite, syenite, gneiss, diorite and primitive limestones. Heddle (1882) writes:

The *Sphenes* of Shinness are certainly the finest in Britain, at least as regards their form. One specimen also was the largest crystal which I have seen from Scotland. This was a crystal of simple

form, which we found exposed on the roadside. As this crystal seemed to be somewhat loose in its matrix of sahlite, it was measured, and found to be two inches in length, by one in width; it was of a pale-yellow colour. In 'dressing' the specimen the crystals started in fragments from the stone .... In general however, the sphenes are not half the size of peas. With the exception of the before-mentioned crystal, they are all of a dark hair brown colour.

Butler (1962) described schists composed essentially of biotite and titanite at many localities in the psammitic groups of the Moine Series in Ardnamurchan. Modally titanite may constitute up to 40 vol.% of the rocks although euhedral crystals are rare; similarly in the Lower Psammitic Group of Morar and South Knoydart. A titanite-rich band associated with an amphibolitic sill occurs at Fionn Allt, 6.4km west-north-west of the head of Loch Shin, Sutherland (Read, 1931a). Heavy mineral bands up to 1.27cm thick are fairly common in the Moine granulites (Phemister, 1936, 1948) and in Ardgour, Argyllshire, Drever (1940) reported titanite-magnetite assemblages in fragments specimen from Ternemny Quarry, Knock, Banff-shire (NMS G 1995.143.1). Additionally a clove-brown, lozenge-shaped crystal, 2.7 x 1cm, in a biotite/quartz garnet matrix from close to the mica prospect, Strathpeffer, Ross and Cromarty (NMS G 1995.84.1).

## TOBERMORITE

A calcium silicate hydroxide hydrate of which Scottish occurrences are confined to amygdaloidal basalts.

Heddle's (1880) original tobermorite description revealed meagre information regarding mode of occurrence 'totally filling small druses in the cliffs of the shore immediately to the north of the pier of Tobermory', and 'in the cliffs near the lighthouse, north of Tobermory, and along the shore towards Bloody Bay', Mull. At the first locality Heddle described tobermorite as:

> very pale pinkish white; it is translucent; it is frequently surrounded by a thin zone of pale blue massive mesolite. I have never found it in any druse which was not filled to its centre with it, therefore I have seen no trace of crystalline form.

Heddle (1901) refers to the host as 'basic eruptive rocks'. He also collected tobermorite from Gribun, Loch na Keal, Mull (Heddle Coll[n] 435.b.2 and 3) and Walker (1971) recorded additional numerous tobermorite localities on the island.

Sweet (1961) discovered tobermorite formed inner cores to numerous amygdales, some tacha-ranite-bearing, in olivine dolerite near Portree, Skye; other Skye localities include 'in a quarry out of which the pier at Dunvegan had been built' (Heddle, 1880) and from Sgurr nam Boc, Loch Eynort (Heddle, 1893), material which:

> Has a large plumose structure: the tufts are quite independent of each other in their offsets, but are thereafter interwoven with each other. Very arborescent and branching.

Heddle's material from this locality was examined by Claringbull and Hey (1952; BM 1937.

1490), McConnell (1954; BM 1952. 35) and Gard and Taylor (1958). Utilising electron microscopy, electron diffraction, X-ray and dehydration techniques, the latter two authors discovered that the 11Å basal spacing did not shrink upon dehydra-tion to the 9.3Å variant. Other Skye specimens include material from Dunan Earr An Sguirr, Loch Brittle (Heddle Coll[n] 433.1); Geoda Tuill, Loch Eynort (Heddle Coll[n] 435a.1); and near the Allt Coir a' Ghobainn River close to Drynoch (Livingstone, 1988) where the tobermorite is associated with reyerite, in basalt.

Currie (1907) reported reddish-white massive tobermorite with mesolite from Ardtornish Bay, Morven, Argyllshire.

## TOPAZ

An aluminium silicate fluoride hydroxide. Topaz is commonly found as an accessory component, and formed by pneumatolytic, hydrothermal and metasomatic processes in granite and granite pegmatites. It also occurs in gravel and alluvium deposits derived from granite massifs.

Jameson (1809a) detailed topaz from various Scottish localities although he demonstrated that some reports were erroneous. Of one greenish-white topaz crystal from Aberdeenshire Jameson (1809a) records:

> The form of the largest crystal is an oblique eight-sided prism, deeply bevelled at the extremity; the bevelling planes set on the acute lateral edges; the proper edge of the bevelment truncated, and the angle formed by the meeting of the bevelling plane and the lateral planes bevelled. The surface of the crystal is rough, owing to attrition.

The above crystal exceeded 7oz (198gm) where-as a second, from the same tract, was in excess of 1lb 3oz (539gm). Mr Farquharson of Invercauld possessed fragments of the latter crystal and Stoddard (1801) notes that topaz appeared along-side cairngorm quartz in the 'well-selected cabinet of natural curiosities' belonging to Mr Farquharson. (Currently only a large cairngorm crystal survives, Blackett *pers. comm.*, 1995). Jameson (1809a) based

his physical and crystallographic description of Scottish topaz on specimens from mineral cabinets, collectors or lapidaries. In the same paper he predicted that the 'Aberdeenshire topaz will be found in drusy cavities or veins in granite, and in considerable quantity'.

Sowerby (1811) figured a Cairngorm topaz from the collection of Thomas Allan:

> It is the more curious, as part of it is of a fine light blue colour, and part cinnamon-coloured, with a beautiful soft glowing warmth; and the disposition of the two colours seems almost to explain the nature of the crystallization.

Greg and Lettsom (1858) remarked that 'magnificent topazes have occurred in the Cairn-gorm district, occasionally in transparent rolled crystals, and masses of a pale blue colour 3 or 4 inches in diameter'. They also comment that the University of Edinburgh possessed some of the finest Scottish topazes. In contrast to rolled topazes Heddle (1901) featured the complex morphology exhibited by crystals from Beinn a Bhuird, Aberdeenshire. Topaz has also been reported at Machrihanish Bay, Kintyre, Argyllshire, in loose granite blocks transported by ice from Arran (Heddle, 1901; and McCallien, 1937); it is also recorded by the latter authors from the island, and by Macgregor *et al.* (1972) at Glen Shant and Brodick Bay as equant crystals up to 2mm lining drusy cavities in microgranite (Heddle Coll[n] 397.2, 4 and 5). Tindle and Webb (1989) detailed an intensely silicified quartz-topaz-muscovite matrix hosting niobian wolframite in a vein cutting zinnwaldite-bearing granite at Glen Gairn, near Ballater, Aberdeenshire; from the same locality came opaque 'feldspar-like' topaz crystals 1.5 x 0.6cm(NMS G 1987.15.26).

## VANADINITE

A lead vanadate chloride belonging to the apatite group occurring in oxidised zones of lead-bearing mineral deposits.

Vanadinite was named by Kobell (1838) after its chemical composition for vanadium had just been discovered in 1830 by Sefström (Mellor, 1929):

> In the course of last winter [ie 1830] my attention was directed to a mineral from Wanlockhead, which I obtained from Mr Rose as an arseniate of lead, to some varieties of which mineral it has a strong resemblance. On analysis, however, it proved different; and though I failed in obtaining *pure* oxide of chromium, yet I suspected at first that it contained a considerable quantity of chromic acid. Repeated attempts, however, to obtain a pure oxide having failed, and many properties manifesting themselves which chromium does not exhibit, I came at length to the conclusion that it contained a new metallic substance. This substance I was engaged in examining when the letter from Berzelius in the *Annales de Chimie*, came to hand, and showed me that my new metal was the Vanadium of Sefström. My stock of the mineral was very small, and it was from a portion of it, weighing only seven grains [1 gramme = 15.43 grains], that I prepared the compounds of Vanadium which I soon after exhibited to the Royal Society of Edinburgh ....
>
> The Vanadiate of lead, or minerals containing the metal in combination with lead, I have met with in two forms differing very much from each other ....
>
> This mineral has hitherto been found only in one mine at Wanlockhead, and only in one spot about six fathoms in length, where the vein had been subjected to a violent disruption. The mine has been unwrought for the last five or six years, and the only specimens to be met with are among the rubbish of the old workings. A supply of these specimens may be obtained from Mr Rose, mineral-dealer, South Bridge. (Johnston, 1831)

Vanadinite prisms: the largest measures 5.5mm in length, richly covering one surface of a specimen 8.5 x 5 x 5cm, Wanlockhead, Dumfriesshire.
(NMS G 1994.1.53, ex Carruth Colln NC21.)

Complex prismatic vanadinite crystals to 8mm on ferruginous matrix, Belton Grain Vein, Wanlockhead, Dumfriesshire.
(Heddle Colln 552.8.)

Vanadinite globules to 4-5mm on chalky plumboan apatite and pyromorphite, specimen size 9 x 4.5 x 3cm, Wanlockhead, Dumfriesshire.
(NMS G 1991.31.83, ex Sutcliffe-Greenbank Colln's.)

Allan (1834) records under VANADIATE OF LEAD:

> It is there found [at Wanlockhead] in small globular masses sprinkled over calamine, or in thin coatings on the surface of that mineral. Isolated and perfect crystals are rare, but occasionally the larger globules exhibit traces of six-sided prisms.

Thomas Brown (qv), in his mineral catalogue, records some 25 or so specimens of 'vanadiate of lead' with calamine, green phosphate, or manganese, from Wanlockhead. An appended note states:

> M Descloizea of Paris looked over these specimens 2nd Oct. 1845. He said that only the *reddish* coloured crystals were vanadiate of lead and that the shining *yellow* or opaque yellow were arseniate or arsenic phosphate. It is quite certain that Descloizea is mistaken in forming this distinction and that the shining yellow crystals are vanadiates as well as the brown ones.

Exquisite vanadinite specimens originate from the High Pirn Mine on the Belton Grain Vein at Wanlockhead although a misleading account by Greg and Lettsom (1858) states:

> Has been found both formerly and again lately, among the old heaps at the Hegh-pirn of the Susannah Mine at Wanlock Head, in Dumfriesshire [The Susanna Mine is at Leadhills, Lanarkshire] on common and cupreous calamine; very

rarely distinctly crystallised, usually occurring in small globular aggregations. The largest crystals not more than a quarter of an inch across. Very perfect ones are in Mr. Brown of Lanfine's, and in Dr. Heddle's collections.

Vanadinite forms cream, yellowish-brown, orange to dark-brown resinous globules (to 7mm), is rarely idiomorphic, and frequently invests pyromorphite. Waxy orange-brown globules (to 3mm) on apple-green pyromorphite coalesce to cover an area some 30cm² (NMS G 1994.1.52; ex Sutcliffe-Greenbank Coll$^n$). Two specimens (NMS G 1923.3.21, and 1926.2.16) originate from the New Vein, Glencrieff Mine, Wanlockhead. Elsewhere throughout the Leadhills-Wanlockhead orefield, Temple (1954) records vanadinite from the Brow, Brown's, Dobbies, Hopeful and New veins.

Considerable attention, chemically, has been devoted to Wanlockhead vanadinite; R. D. Thomson (quoted in Greg and Lettsom, 1858), Frenzel (1881) and Collie (1889) published analyses. Frenzel (1881) demonstrated that the vanadinite contained both calcium and phosphorus at about the 3% level with calcium considered to be replacing lead. An electronprobe-microanalysis study (Livingstone, 1994) revealed Wanlockhead vanadinite possessed up to 4.8% $P_2O_5$ and 1.4% CaO in the structure, low vanadium levels correlating with high phosphorus contents, and with Ca replacing Pb it suggests a coupled Ca-P substitution. (The Ca and P levels are not in the same ratio as those of the plumboan apatite on which the vanadinite developed.)

Extremely rare vanadinite has been discovered associated with mottramite on prehnite and calcite, in reddened tuffs, of the Clyde Plateau lavas at Loanhead Quarry, Beith (Meikle, 1989a).


## VESZELYITE

A copper zinc phosphate hydroxide hydrate, usually associated with secondary copper minerals in supergene deposits.

The first British occurrence of veszelyite was reported by Green (1990) from dump material on

Veszelyite showing two generations of growth. Lustrous veszelyite on the coign of a slightly altered 2.5mm earlier crystal. Painting (acrylics) by the author based on a photograph by M. P. Cooper of a N. Hubbard specimen from Straitstep, Wanlockhead, Dumfriesshire.

Straitstep Vein, Wanlockhead. Veszelyite forms rare lustrous, transparent to translucent dark-blue octahedral crystals up to 5mm. Sharp, pale-blue chrysocolla pseudomorphs, from 2–9mm, are more abundant than the veszelyite that in some cases forms a core. Associated secondary minerals include hemimorphite, either as euhedral crystals to 0.5mm, or more commonly as botryoidal encrustations; malachite, brochantite and aurichalcite. Also similar to above, NMS G 1991.5. 1 and 3, 6–13, and 15–21, collected from the same dump which surrounds possibly the Old Waygate Shaft.

In the 1740s the lease at the southern end of Straitstep Vein was held by Ralph Wightman, an Edinburgh merchant, and Dean of Guild. In 1744 Wightman opened up the old workings of Straitstep Vein and discovered good ore. About a year later Wightman died; although the Old Waygate

Shaft was sunk during his time, it was not completed until 1769 (G. Downes-Rose *pers. comm.*, 1989).


## VIVIANITE

A ferrous iron phosphate hydrate after which the vivianite group is named. Vivianite is frequently found in gossans of metallic ore deposits and widespread in clays, especially when associated with bones, decayed wood or other organic matter.

Cryptocrystalline vivianite is disseminated through carbonaceous clays from an old lake bed at the head of Haugh Burn, Cauldhame, Linlithgow, West Lothian, and partially invests red deer bones from the same locality (NMS G 1967.63.16 and 17 respectively). Similarly, it is found in clay and bones from the Royal Park, Edinburgh (NMS G 597.8) and as crystals to 5mm in bones from Kilmaurs, Ayrshire (NMS G 1969.7). It forms a blue coating on fossil fish and disaggregated scales from the Banniskirk Quarry, Caithness (NMS G 1911.18.2). Deep-blue, fasciculate, tabular vitreous crystals to 1.5cm were found in bones from a crannog at Loch Lee, Ayrshire (Heddle Coll$^n$ 597.1).

Vivianite developed as a secondary mineral on a suite of transition metal phosphates (including johnsomervilleite, graftonite, jahnsite, phosphosiderite, rockbridgeite and mitridatite) in two small metamorphic segregation pods in kyanite-sillimanite Moine gneiss at Loch Quoich, Invernessshire (Livingstone, 1980).


## WAIRAKITE

A calcium sodium aluminium silicate hydrate of the zeolite group. Wairakite is found in currently active geothermal fields and basaltic rocks which may, or may not, have suffered hydrothermal or burial metamorphism.

Wairakite is an uncommon zeolite and only two occurrences are noted in Scotland (Evans, 1987), both within the Carboniferous Clyde Plateau lavas of the Midland Valley (at Loanhead Quarry, Beith, and Milngavie). Electronprobe-microanalysis reveals a chemistry close to the ideal composition with minor sodium ($Na_2O$ 0.21-0.34 wt%). At Loanhead Quarry, wairakite mantled by analcime crystallised in a hydrothermal aureole adjacent to a Tertiary dyke cutting lavas which underwent zeolite facies burial metamorphism in the analcime-natrolite zone. A contact zone of carbonate metasomatism contains rankinite. With increasing distance from the dyke zones of epidote, prehnite-thomsonite and thomsonite are recognised, with wairakite forming in the latter (Evans, 1987). Metamorphism by a palaeo-geothermal plume occurred at a fluid pressure of 2kbar at 350-420°C (Evans, 1987), figures which concur with those suggested by Livingstone (1989) for hydrothermal garnet (grossular-andradite) formation after thomsonite at the same locality.

A porcellaneous calcian analcime-bytownite intergrowth within a reyerite-tobermorite (11Å) amygdale assemblage in Tertiary olivine basalt was discovered near Drynoch, Skye (Livingstone, 1988, 1989a). Within the calcian analcime considerable Na-Ca variation is evident, with a composition (Na 60%, Ca 40%) representing the highest calcium site occupancy found at this locality. This intermediate composition supports the case for a continuous analcime-wairakite solid solution series. Wairakite in the intergrowth may possibly have been depleted in calcium as a sodium-rich analcime pervades the intergrowth.

## WITHERITE

Barium carbonate, a member of the aragonite group of minerals. Witherite occurs in low temperature hydrothermal veins associated with barite, calcite and galena.

Not of frequent occurrence, the first Scottish witherite recording was by Brown (1919) who wrote:

*Witherite* was found in the early months of 1918 in the west branch of the new Glencrieff mine, along with quartz, calcite, barytes, and galena, about 200 fathoms from the adit level, in a vug or cavity four fathoms deep, 15 feet [4.5m] long, and about 3 feet [0.9m] wide in the centre. Some of the quartz specimens were about four feet [1.2m] in diameter, and had slipped from the hanger wall. They were jammed with large coarse calcites and barytes, which apparently had slipped from the foot wall. Some of the smaller quartz specimens were of a spongy nature. As the most of the *witherite* specimens were found near the bottom of the vug, and barytes near the top, it appeared as if the change from the sulphate to the carbonate of barium was not so distinctly evident near the top of the vug as it was near the bottom. Most of the specimens found were of a pseudomorphous nature; some of them showing the barytes going away, and witherite forming up. Everything went to show that the flow of the thermal waters had been downwards. The two globular specimens shown were found about three feet [0.9m] apart on matrix of quartz, on the ledges or foot wall near the bottom of the vug. The larger specimen was damaged in the taking off. It measured about 7 inches [17.8cm] in diameter; the smaller one about 5 inches [12.7cm]. Both have a coating of minute crystals.

To see them hanging in the cavity gave you the impression that you were looking at two large fungi. Nearly all the specimens found were white or grey in colour, except one large one, which was broken to pieces in the taking out. It was red, and resembled a Tam o' Shanter bonnet, both in size and shape. Another, about 4 inches [10.2cm] in diameter, had four galena crystals embedded halfway into the witherite, having apparently dropped on to the witherite when it was in a plastic state. Another specimen I have shows both the sulphate white and the carbonate grey in the process of change. It is of a spongy nature, and if you strike the crystals with a pencil they sound like a musical instrument. Another was mixed up with calcite, quartz and galena. It looked as if the witherite had run down amongst them and cemented them altogether. Without doubt a great many specimens were destroyed by the first blast. The miners said that when they were removing the *debris* after the blast it sounded as if they were shovelling china or bell metal. As witherite, with this one exception, had never been found in Scotland, I thought at first sight that the two globular specimens were *plumbo aragonite*, until I tried hydrochloric acid on them.

The ground has been worked up for about six fathoms from the top of the chamber, but no further discovery of witherite has been made, and since the galena is exhausted in that part of the mine, the working has now been abandoned.

Brown (1925) reported '*Witherite* has for the second time been found in the 80 fathoms level, west branch of Glencrieff Mine, at about 80 fathoms, and also at 140 fathoms north of the winding shaft'. From 80 to 40 fathoms level abundant barite pseudomorphed witherite crystals whereas globular witherite, about 1in (2.5cm) diameter, may cover barite, calcite and pyrite. At the 80 fathoms level witherite formed a 9in (23cm) diameter mass, which simulated a globe, with a barite cap ¼in (0.6cm) thick. A second reporting of witherite was additionally made by Brown (1925) from the Wembley Shaft, Leadhills. Temple (1954, 1956) reports that witherite also occurred in the Wembley Shaft, Leadhills, where it formed massive crystalline aggregates. A qualitative spectrochemical trace element analysis of the latter revealed the presence of Ag, Mg, Ca, Sr, Zn, Al, Si, Pb and V.

Witherite has been discovered in three samples from the Aberfeldy exhalative barite deposit in Perthshire. Here the witherite occurs as small inclusions in pyrite and contains up to 10 mol.% $SrCO_3$ (Moles, *pers. comm.*, 1993). Also as platy, elongated terminated crystals (<4 x 1.4mm) and pseudohexagonal tabular crystals with barytocalcite in dump material at the Hilderston silver mine (Meikle, 1994).

There are 17 witherite specimens in the Scottish Mineral Collection, all derived from Leadhills-Wanlockhead, eight of which were obtained via donation, and purchase, from Robert Brown of Wanlockhead. Of the remainder, five are accessioned, including three from the Wembley Shaft, Leadhills. One specimen, *ex* R. Brown (NMS G 1926.2.21), reveals embedded galena crystals (up to 4 x 2 x 2cm) in a partial witherite hemisphere, with barite, and accords with Brown's description of galena 'dropped on to the witherite when it was in a plastic state'. Six accessioned specimens display hemispheres of 5cm or greater.

Wulfenite bipyramids to 0.15mm, encrusting granitic matrix, Gairnshiel, Aberdeenshire. *(British Micromount Society Coll[n] BMS 0860.)*

A 1cm-wide vein of xonotlite in gabbro, from the Bin of Huntly Quarry near Huntly, Aberdeenshire. *(NMS G 1975.29.2.)*

## WULFENITE

A lead molybdate. Wulfenite is typically found in oxidised zones of lead ore deposits.

Throughout Scotland wulfenite is a rare mineral, although Green (1986) intimated that its readily recognisable characteristics under the stereomicroscope contributed to an increasing number of localities being discovered. Heddle (1901) reported wulfenite from only one locality: Lauchentyre Mine, Kirkcudbrightshire. For 46 years Lauchentyre Mine remained the only published Scottish locality until Russell (1946) added the second: Struy Mines, Inverness-shire. Green (1986) also noted wulfenite occurring on dumps at Broad Law, Leadhills, Lanarkshire, and Pibble Mine, Kirkcudbrightshire. Rust (1983) reported wulfenite from Susanna Mine dumps, Leadhills. Not all Scottish wulfenite is derived from lead ore deposits, as granitic environments at Pass of Ballater and Abergairn yield this species. In addition to the Lauchentyre and Pibble Mine localities above, Rothwell and Mason (1992) added Caulkerbush in the Clatteringshaws Loch area, and Drumruck Mine, where wulfenite can be found in Kirkcudbrightshire.

A chemical study of wulfenite from numerous Scottish localities (Livingstone, 1992) revealed little substitutional variation. Granitic environment wulfenite possessed higher tungsten levels than that from lead deposits. Compared to the above, two tabular crystals from Wanlockhead exhibited wide variation in molybdenum–tungsten substitution which ranged from molybdenian stolzite (White's Cleuch) to tungstenian wulfenite (Margaret's Vein).

## XONOTLITE

Calcium silicate hydroxide. Xonotlite is found chiefly as veinlets in serpentinite, and also in contact zones. It occurs as a phase in the $CaO$-$SiO_2$-$H_2O$ system along with tobermorite, gyrolite and okenite.

Heddle (1882) reported 'xonaltite' as being:

… first found by Mr. Rose and myself, along with *gyrolite*, near Kilfinnichan, Loch Screden (Scridain) Mull. More lately it was found by myself at Gribon, opposite Oronsay; and also on the north shore of Loch na Keal, in the same island.

The above represents the first British recording for xonotlite. Within Binhill Quarry, Cairnie, near Huntly, Aberdeenshire snow-white, highly fibrous veins to 1.5cm wide cut gabbro (NMS G 1970.58.4), and at the same locality it also occurs in narrow prehnite veins associated with scawtite and tacharanite (Livingstone, 1974). Heddle (1901) analysed 'wollastonite' (theoretical composition 48.27% CaO and 51.73% $SiO_2$) found as hard, compact bluish-white veins (to 2cm) possessing a fibrous structure from Allt Gartally, Milton, Glen Urquhart. X-ray powder diffraction identification (photograph no. 791) of Heddle's analysed specimen (329.1) confirms this material as xonotlite (theoretical composition 47.07% CaO, 50.42% $SiO_2$ and 2.51% $H_2O$).

Xonotlite is also associated with gyrolite at Killiemore, Loch Scridain (Heddle Coll[n] 435b.1, X-ray photograph no. 840) and found in olivine dolerite near Portree, Skye (Sweet, 1961) where it is associated with tacharanite, gyrolite, tobermorite and zeolites.

Walker (1971) ascribed xonotlite formation towards the base of a thick lava pile in Mull to a region of relatively high temperature. Associated with Mull xonotlite are okenite, pectolite, tobermorite, reyerite and tacharanite.

# THE MINERAL ALBUM
COMMON AND UNUSUAL SCOTTISH SPECIES

## ALMANDINE
Single crystal, 6cm diameter, Colvister, North Yell, Shetland.
*(Heddle Coll[n] 370.19.)*

## ANALCIME
Partial crystal 6cm diameter on matrix, Kilpatrick, Dunbartonshire.
*(Heddle Coll[n] 450.97.)*

## ANALCIME
Trapezohedral crystals to 3.5cm on matrix 13 x 11 x 7.5cm, Talisker, Skye, Inverness-shire.
*(Dudgeon Coll[n] 1878.49.391.)*

## ANDALUSITE
Prism, 7 x 1.5cm, in muscovite biotite plagioclase gneiss, Glen Clova, Aberdeenshire.
*(Heddle Coll[n] 398.17.)*

APOPHYLLITE
Crystal 1.8 x 1cm,
on phonolite,
Traprain Law,
East Lothian.
(NMS G 1991.31.69.
Presented by Rear
Admiral B. Brooke.)

APOPHYLLITE
Crystals to 2cm in basalt, Moonen Bay, Skye, Inverness-shire.
(M. McMullen Collⁿ.)

BARYTOCALCITE
Specimen from the Hilderston silver mine, Bathgate, West Lothian. The field of view is 10 x 8mm.
(S. Moreton Collⁿ.)

BERYL
Broken prism 4.5 x 2 x 2cm,
on quartz with muscovite,
Struy Bridge, Inverness-shire.
(Heddle Collⁿ 344.18.)

**BOTALLACKITE**
The crystal, on quartz, measures 2.5mm, Castleton Mine, Lochgilphead, Argyllshire.
*(D. I. Green Coll^n.)*

**CHABAZITE**
Crystals to 1.5 x 1.5cm, with stilbite and calcite, Kilmacolm, Renfrewshire. Specimen size 17 x 13 x 6cm.
*(Heddle Coll^n 447.80.)*

**CHALCOPYRITE**
Interpenetrant and single tetrahedra to 1.2cm on dolomite, Wanlockhead, Dumfriesshire.
*(NMS G 1991.31.69 ex Sutcliffe-Greenbank Coll^n.)*

**CLINOATACAMITE**
Crystals to 0.25mm, Rhuba a'Mhill, Islay.
*(D. I. Green Coll^n.)*

**CLINOZOISITE**
Prisms to 4 x 0.5cm, in quartz, Laggan, Dulnain Bridge, Moray-Inverness-shire border.
*(Heddle Coll^n 406.16.)*

**DOLOMITE**
Rhombs to 2cm, Black Craig Mine, Newton Stewart, Kirkcudbrightshire.
*(Heddle Coll^n 271.18.)*

**DOLOMITE**
Large cleavage fragment 7 x 5 x 4.5cm, with talc, North Cross Geo, Unst, Shetland.
*(Heddle Coll^n 271.9. Analysed specimen.)*

**ELBAITE**
A 1cm crystal in lepidolite matrix, Glenbuchat, Aberdeenshire.
*(D. I. Green Coll^n.)*

145

## EPIDOTE
Plumose aggregates to 5cm with minor actinolite, pyrite and calcite, Pitscurry Quarry, Pitcaple, Inverurie, Aberdeenshire.

*(NMS G 1980.32.2.)*

## EPIDOTE
Manganese-bearing epidote ('withamite') in 1cm cavity within andesite, Glen Coe, Argyllshire.

*(Dudgeon Coll[n] 1878.49.253.)*

## FLUORITE
Cubes, 0.7cm edge, on dolomite which lines cavities in lava, Gourock, Renfrewshire.

*(NMS G 1963.19.2.)*

## GOETHITE
Compact fibrous mass, fibres to 1.5cm, with yellow quartz, specimen size 11 x 8 x 6cm, Garleton Hills, East Lothian.

*(NMS G 257.51. Geological Survey Coll[n].)*

**HEMIMORPHITE**
Glassy, globular encrustation (field of view 12 x 10mm) from the Bay Vein, Wanlockhead, Dumfriesshire.
*(Dudgeon Coll[n] 1890.114.1404.)*

**HEMIMORPHITE**
Reniform encrustations, left 7.5 x 6 x 4.5cm and right 8 x 6 x 4cm, Leadhills, Lanarkshire.
*(Heddle Coll[n] 423.21 and 423.14 respectively.)*

**HEMIMORPHITE**
Prismatic crystals, to 1mm, forming encrustation, Glencrieff Mine, Wanlockhead, Dumfriesshire.
*(NMS G 1923.8.3. Purchased from R. Brown.)*

**HEULANDITE**
Crystals to 1cm, with chlorite, on basalt, Kilmacolm, Renfrewshire.
*(Heddle Coll[n] 438.23.)*

147

ILMENITE
Plates 7 x 0.5cm, in quartz, Loch na Lairige, Killin, Perthshire.
(NMS G 1981.10.1. Presented by R.T. Gardner.)

KYANITE
Blades 6 x 0.6cm, in quartz, from Botriphnie, Banffshire. (Heddle Coll[n] 400.18.)

KYANITE
Mass of kyanite 20 x 16 x 9cm with blades to 13cm, 1.6 km south-west of Scalla Field Summit, Shetland.
(NMS G 400.63. Presented by H. M. Geological Survey.)

MAGNETITE
Octahedra, 1.5cm centre edge, in chlorite schist, Pundy Geo, Point of Fethaland, Shetland.
(Heddle Coll[n] 237.54.)

148

**MANGANITE**
Cavity lining, 1cm crystals, specimen size 12 x 7 x 4cm. Granham Towie, Aberdeenshire. *(Heddle Coll^n 258.3.)*

**MESOLITE**
Spheres to 5mm diameter on thomsonite prisms to 5cm, with minor stilbite, Gryfe Tunnel, near Greenock, Renfrewshire. *(Heddle Coll^n 455.18.)*

**MICROCLINE**
Crystal of amazonite, with Heddle's notations, 5.5cm across, 5cm high and 2-3cm front to rear, from Beinn Bhreac, Tongue, Sutherland. *(Heddle Coll^n 315.18.)*

**NATROLITE**
Fans 6.5cm, in a largely natrolite specimen measuring 19 x 14 x 7cm, Boyleston Quarry, Barrhead, Renfrewshire. *(NMS G 1957.8.4.)*

**PECTOLITE**

Pseudomorphous masses, up to 3cm radius, after analcime, specimen size 10 x 10 x 5cm, Ratho, Midlothian.

*(Heddle Coll$^n$ 330.25.)*

**PREHNITE**

Botryoids to 1.5cm, specimen size 7 x 6.5 x 5cm, Bishopton Railway Tunnel, Renfrewshire.

*(Heddle Coll$^n$ 411.64.)*

**PREHNITE**

Botryoids to 1.5cm, with calcite and analcime, in a cavity within dolerite, Salisbury Crags, Edinburgh, Midlothian.

*(Heddle Coll$^n$ 411.100.)*

**PYRITE**

Cube, 2cm, in slate, Birnam Quarry, Dunkeld, Perthshire.

*(D. G. Anderson Coll$^n$.)*

## ROMANECHITE
On fragments of Old Red Sandstone, from near St John's Head, Lead Geo, Hoy, Orkney.
*(Heddle Coll<sup>n</sup> 269.45.)*

## RUTILE
Prisms and plates in quartz with chlorite, specimen size 8.5 x 5.5 x 3.5cm, Loch na Lairige, Killin, Perthshire.
*(Hunterian Museum, GLAHM-M16621.)*

## SCOLECITE
Fan, 10cm, specimen size 17 x 14 x 11cm, Maol nan Damh, Ben More, Mull, Argyllshire.
*(NMS G 1920.12.1. Collected by E. M. Anderson of H. M. Geological Survey.)*

## STILBITE
Sgurr nam Boc, Skye, Inverness-shire. The bow-ties attain 6.3cm and the fan 8.8 x 6.3cm; complete specimen measures 45 x 15cm.
*(M. McMullen Coll<sup>n</sup>.)*

## TALC
Books, 6cm, on serpentinite, Queyhouse Quarry, Quoys, Unst, Shetland.

*(NMS G 1990.58.2.)*

## THOMSONITE
Radial crystals 6.5cm, on analcime with calcite, Loanhead Quarry, Beith, Ayrshire.

*(J. G. Todd Collⁿ.)*

## THOMSONITE
Terminated 1cm crystals, in a thomsonite-lined 6cm cavity with grossular-andradite and analcime, Loanhead Quarry, Beith, Ayrshire.

*(NMS G 1981.18.2. Presented by T. J. K. Meikle.)*

## TOURMALINE
Prisms to 8 x 3 x 3cm in quartz-orthoclase matrix, Rubislaw, Aberdeenshire.

*(Heddle Collⁿ 426.48.)*

152

## TREMOLITE

The amphibole covers some 28 x 26cm, with prisms up to 1cm wide, from Shinness, Sutherland. This specimen was described by Heddle (1882) as 'one of the finest minerals ever found in Scotland. Nearly a foot in size in all directions, the mass is covered with parallel bundles of glancing fibres of the most delicate tenuity; these flash with a brilliant but chastened light, of more than silvery whiteness. No natural object devoid of colour could possibly be more beautiful.'

*(Dudgeon Coll$^n$ 1878.49.213.)*

## VESUVIANITE

Striated partial crystal 3.5 x 2 x 1.5cm, Dalnabo, Glen Gairn, Aberdeenshire.

*(Heddle Coll$^n$ 393.19.)*

## ZEUNERITE

Associated with bismuthinite in vein at Needle's Eye, Dalbeattie, Kirkcudbrightshire. Field of view 12 x 8mm.

*(T. Neall Coll$^n$.)*

## ZOISITE

Specimen size 12 x 11 x 5cm, Easter Laggan, Dulnain Bridge, Moray-Inverness-shire border.

*(D. G. Anderson Coll$^n$.)*

# APPENDIX I

## GLOSSARY OF SCOTTISH MINERAL SPECIES

In 1901 M. F. Heddle published a systematic and an alphabetical list of Scottish mineral species (Heddle, 1901), although the former was revised slightly and published in the same year (Goodchild, 1901).

Throughout the 100 or so years that have elapsed, the number of known valid Scottish mineral species has increased rapidly, in a period during which there have been great changes in mineralogical identifications. Consequently some names in Heddle's 1901 *Mineralogy of Scotland* are not included in the *Glossary of Scottish Mineral Species 1981* (Macpherson and Livingstone, 1982) as they no longer represent valid species. Further, in the context of modern mineralogical nomenclature, there seems little point in including varietal and synonymic lists. During the compilation it became apparent that not only is there a widespread lack of knowledge as to which mineral species occur in Scotland, but that there is also a need for greater precision in mineralogical nomenclature and in accuracy of identification.

Entries are mainly compiled from Macpherson and Livingstone (1982), Livingstone and Macpherson (1983), and Livingstone (1993a), with additional data from Scottish Mineral Collection records. Bold type is used to highlight valid Scottish species, whereas group names, and minerals of lower status, are represented in italics. For species acceptance the works of Fleischer and Mandarino (1991, 1995) and Clark (1993) were followed.

For the compilation of this Appendix, changes in pyroxene nomenclature (Morimoto, 1988) have been accommodated resulting in some 'species' in previous compilations being down-graded. As a result the total number of Scottish species in this Appendix now stands at 552.

### TABLE 2

Number of Scottish minerals discovered by decade
1901-90
(species known at 1901 equals 162)

| | | | |
|---|---|---|---|
| 1901-10 | 8 | 1951-60 | 22 |
| 1911-20 | 5 | 1961-70 | 35 |
| 1921-30 | 9 | 1971-80 | 100 |
| 1931-40 | 18 | 1981-90 | _132_ |
| 1941-50 | 28 | total | 357 |

In this compilation amphibole nomenclature follows that of Leake (1978) rather than the simpler, slightly less rigorous classificatory schemes and nomenclature of Leake *et al.* (1997). Under the new scheme analyses recalculations may result in name changes; the entry for crossite would disappear as this name is now recommended for extinction. Zeolite nomenclature closely follows the recommendations of Coombs *et al.* (1998), thus herschelite, previously listed (Livingstone, 1993a), becomes chabazite-ĸ.

### ACANTHITE

$Ag_2S$                    Monoclinic

As tiny platelets on native silver within, and on, prehnite from Boyleston Quarry, Barrhead and Hartfield Moss, Strathclyde. Also within quartz-hübnerite vein wall rock at Gairnshiel Bridge, Glen Gairn, Aberdeenshire.

### Acmite = AEGIRINE

### ACTINOLITE

$Ca_2(Mg,Fe^{+2})_5Si_8O_{22}(OH)_2$         Monoclinic

In actinolite-chlorite-epidote-amphibolite, 0.6km south-east of Loch Gair Hotel, Loch Gair, Argyllshire, and elsewhere.

### AEGIRINE

$NaFe^{+3}Si_2O_6$                Monoclinic

Found in an aegirine-nepheline-syenite pegmatite dyke cutting the ultrabasic rock of the Conc-na-Sròine mass, 640 metres south of Ledmore, Sutherland.

### AEGIRINE-AUGITE

$(Na,Ca)(Fe^{+3},Fe^{+2},Mg,Al)Si_2O_6$     Monoclinic

Within calcareous rocks which contain diopside and aegirine-augite, in the Ledbeg River, 360-460m north-east of Ledbeg, Sutherland.

### AENIGMATITE

$Na_2Fe_5^{+2}TiSi_6O_{20}$              Triclinic

A local accessory mineral in the epigranites of the Marsco area, Skye, Inverness-shire.

### AIKINITE

$PbCuBiS_3$                 Orthorhombic

Small inclusions ≈ 10μm in pyrite, with chalcopyrite, tetrahedrite-tennantite, calcite and quartz at the Tomnadashan copper deposit, Perthshire.

### *Akaganeite*

$\beta\text{-}Fe^{+3}O(OH,Cl)$              Tetragonal

Possibly present in a thin iron-pan, 1-3mm thick, formed from basalt at Killiechronan just south-west of Salen, Mull, Argyllshire.

## AKERMANITE

$Ca_2MgSi_2O_7$     Tetragonal

Colourless or faintly brown crystals in thermally metamorphosed Jurassic limestone at Kilchoan, Ardnamurchan, Argyllshire.

## ALBITE

$NaAlSi_3O_8$ (An 0-10)     Triclinic

Common.

## ALLANITE – (Ce)

$(Ce,Ca,Y)_2(Al,Fe^{+2},Fe^{+3})_3(SiO_4)_3(OH)$
Monoclinic

Forms an accessory component of the Beinn Bhreac pegmatite boulder, Tongue, Sutherland, and in the quartz-diorite near New Abbey, Kirkcudbright-shire. A common accessory mineral.

## ALLOPHANE

An amorphous hydrous aluminium silicate

Present as a white coating and intermixed with hydrohonessite and copper analogues of hydro-honessite on dump material at an old nickel mine 12km south-west of Inveraray, Argyllshire.

## ALMANDINE

$Fe_3^{+2}Al_2(SiO_4)_3$     Cubic

In the cordierite-garnet-biotite-plagioclase-ortho-clase-quartz hornfels of the contact aureole at Black Hill, Loch Muick, Grampian, and elsewhere.

## ALTAITE

PbTe     Cubic

Found as inclusions in galena in a quartz chalcopy-rite galena pyrite vein approximately 4.5km south-west of Aberfeldy, Perthshire.

## ALUMINO-KATOPHORITE

$Na_2Ca(Fe^{+2},Mg)_4AlSi_7AlO_{22}(OH)_2$     Monoclinic

Forms very small grains in rocks from the Falahill Formation (Ordovician) near Heriot, Galashiels area, Scottish Borders.

## ALUNOGEN

$Al_2(SO_4)_3.17H_2O$     Triclinic

As minute clear platy crystals and yellow-brown aggregates associated with glassy potassium alum, and tamarugite, Hurlet, Renfrewshire.

## *Amalgam*

Ag and Hg alloy     Cubic

Occurs as tiny stringers in safflorite, with native silver and erythrite in barite gangue from a mine dump, Silver Glen, Alva, Clackmannanshire.

## *Amblygonite-montebrasite* mineral

From Deannie Tunnel, Glen Strathfarrar, Ross-shire.

## ANALCIME

$NaAlSi_2O_6.H_2O$     Cubic

Found commonly in amygdaloidal basalts of Tertiary age in north Skye, Inverness-shire.

## ANATASE

$TiO_2$     Tetragonal

Occasionally as well-formed plates of steel-blue colour in heavy mineral concentrates prepared from sandstones of Carboniferous age.

## ANCYLITE-(Ce)

$SrCe(CO_3)_2(OH).H_2O$     Orthorhombic

Pale violet microcrystals associated with platy barite and drusy quartz with calcite, on gangue, Bellsgrove Lodge dump, Strontian, Argyllshire.

## ANDALUSITE

$Al_2SiO_5$     Orthorhombic

Found commonly in regionally metamorphosed rocks, especially in Aberdeenshire and Banffshire.

## ANDESINE

(An30-50)     Triclinic

Present as an essential component of numerous igneous and metamorphic rocks.

## ANDRADITE (Melanite)

$Ca_3Fe_2^{+3}(SiO_4)_3$     Cubic

An essential mineral in borolanite, associated with orthoclase, plagioclase, biotite and pyroxene, Cnoc-na-Sròine, Loch Borolan, Sutherland.

## ANGLESITE

$PbSO_4$     Orthorhombic

Formed by the oxidation of galena, especially at Leadhills, Lanarkshire, and Wanlockhead, Dum-friesshire.

## ANHYDRITE

$CaSO_4$     Orthorhombic

In the phonolite at Traprain Law, East Lothian, and within coarsely nodular carbonate beds of calcite and dolomite of the Lower Carboniferous cementstone group of the Tweed area, Burnmouth, Berwickshire.

## ANKERITE

$Ca(Fe^{+2},Mg,Mn)(CO_3)_2$     Trigonal

Widespread, and especially in the Leadhills, Lanarkshire and Wanlockhead, Dumfriesshire orefield.

## ANNABERGITE

$Ni_3(AsO_4)_2.8H_2O$                    Monoclinic

Within cavities in nickeline associated with galena, Pibble Mine, near Creetown, Kirkcudbrightshire.

## ANNITE

$KFe_3^{+2}AlSi_3O_{10}(OH,F)_2$            Monoclinic

Occurs in the Lewisian banded iron formation of amphibolite facies at Flowerdale, near Gairloch, Wester Ross.

## ANORTHITE

$CaAl_2Si_2O_8$ (An 90-100)              Triclinic

Found within a vein in a contact metamorphosed limestone at Dalnabo, Glen Gairn, Aberdeenshire.

## ANORTHOCLASE

$(Na,K)AlSi_3O_8$                        Triclinic

As phenocrysts in a monchiquite dyke, west side of Camas an Fhais, 0.8km north-east of Rudha Fionn Aird, Argyllshire.

## ANTHOPHYLLITE

$(Mg,Fe^{+2})_7Si_8O_{22}(OH)_2$            Orthorhombic

Found associated with actinolite and serpentine at Hillswick, Shetland.

## ANTIGORITE

$(Mg,Fe^{+2})_3Si_2O_5(OH)_4$            Monoclinic

Found within the serpentinite mass at Glen Urquhart, Inverness-shire, and serpentinite lenses in the Langavat Valley, South Harris, Outer Hebrides.

## *Apatite* group

$Ca_5(PO_4)_3F$                          Hexagonal

Apatite, when used alone, is generally synonymous with fluorapatite; see **fluorapatite**.

## *Apophyllite* group

$KCa_4Si_8O_{20}(F,OH).8H_2O$                —

See **hydroxyapophyllite**.

## ARAGONITE

$CaCO_3$                                Orthorhombic

Occurs as long (8cm) radiating transparent crystals, Leadhills, Lanarkshire.

## ARFVEDSONITE

$Na_3(Fe^{+2},Mg)_4Fe^{+3}Si_8O_{22}(OH)_2$      Monoclinic

Occurs as extremely irregular patches between idiomorphic feldspars or as spongy ophitic masses around feldspars in the Tertiary microgranite of Ailsa Craig, Firth of Clyde.

## *Argentite*

$Ag_2S$                                 Cubic

A silver sulphide found in late uraniferous mineralisation at the Main Mine, Tyndrum, Perthshire has the chemical composition of argentite but may be acanthite.

## ARGENTOPENTLANDITE

$Ag(Fe,Ni)_8S_8$                         Cubic

Occurs within the copper-nickel mineralisation near the base of a diorite intrusion at the Glen of the Bar, near Talnotry, Newton Stewart, Kirkcudbrightshire.

## ARMENITE

$BaCa_2Al_6Si_9O_{30}.2H_2O$   Orthorhombic ps. hexagonal

Found as porphyroblasts in quartzite at Coire Loch Kander near Braemar, Grampian.

## ARROJADITE

$KNa_4CaMn_4^{+2}Fe_{10}^{+2}Al(PO_4)_{12}(OH,F)_2$   Monoclinic

Honey-brown glassy grains in bluish-black coal-like areas up to 1cm in lepidolite tourmaline-bearing pegmatite, Glenbuchat, Aberdeenshire.

## *Arsenic*

As                                      Trigonal

This mineral requires confirmation.

## ARSENOPYRITE

FeAsS              Monoclinic ps. orthorhombic

From an old mine dump, with quartz, on the east bank of the Palnure Burn near the Glen of the Bar, near Talnotry, Newton Stewart, Kirkcudbrightshire.

## ATACAMITE

$Cu_2^{+2}Cl(OH)_3$                        Orthorhombic

Green staining in uranium-bearing veins cutting the aureole of the Criffel granodiorite near Dalbeattie, Kirkcudbrightshire.

## AUGITE

$(Ca,Na)(Mg,Fe,Al,Ti)(Si,Al)_2O_6$       Monoclinic

Widespread.

## AURICHALCITE

$(Zn,Cu^{+2})_5(CO_3)_2(OH)_6$              Orthorhombic

Investing carbonate gangue, with chalcopyrite and hydrozincite, Black Craig Mine, Newton Stewart, Kirkcudbrightshire.

## AUTUNITE

$Ca(UO_2)_2(PO_4)_2.10-12H_2O$            Tetragonal

Disseminated in a boulder of porphyritic granite, Caen Burn, Helmsdale, Sutherland.

## AXINITE (undifferentiated)

$(Ca,Fe^{+2},Mn^{+2})_3Al_2BSi_4O_{15}(OH)$    Triclinic

In thin quartz veins in hornblende-schist and associated with tourmaline and epidote, north-east of Loch Maree, Ross and Cromarty. (Used here as a species.)

## AZURITE

$Cu_3^{+2}(CO_3)_2(OH)_2$    Monoclinic

Associated with malachite and magnetite, Kilchrist, Skye, Inverness-shire, and investing sandstone in the quarry above Torduff Reservoir, near Currie, Midlothian.

## BADDELEYITE

$ZrO_2$    Monoclinic

As colourless, anhedral to subhedral grains $c.20\mu m$ associated with apatite, chlorite, amphibole and biotite in the mesostasis areas of gabbros of the layered pluton, Rhum, Inverness-shire.

## BAOTITE

$Ba_4(Ti,Nb)_8Si_4O_{28}Cl$    Tetragonal

Occurs as two inclusions (~15 x 10μm) in coarse pyrite crystals found in sulphide-bearing barite rock at Foss, near Aberfeldy, Perthshire.

## *Barbertonite*

$Mg_6Cr_2(CO_3)(OH)_{16}.4H_2O$    Hexagonal

The mineral described as stichtite (*q.v.*) which replaces chromite in serpentinite, Hoo Field, Cunningsburgh, Shetland, may be barbertonite.

## BARITE

$BaSO_4$    Orthorhombic

Common.

## BARROISITE

$NaCa(Mg,Fe^{+2})_3Al_2(Si_7Al)O_{22}(OH)_2$    Monoclinic

In dark-green amphibolites near Knockormal in the Girvan-Balantrae complex, South Ayrshire.

## BARYTOCALCITE

$BaCa(CO_3)_2$    Monoclinic

As an inclusion in pyrite in a sample of manganoan calcite and barite from the strata-bound barium-zinc mineralisation in Dalradian schists, near Aberfeldy, Perthshire.

## BAZIRITE

$BaZrSi_3O_9$    Hexagonal

Late-stage colourless interstitial grains associated with elpidite in the aegirine-riebeckite granite of Rockall, Inverness-shire.

## BENLEONARDITE

$Ag_8(Sb,As)Te_2S_3$    Tetragonal

Found as inclusions (<1-2μm) with hessite, altaite, coloradoite and rare electrum in galena from the veins at Calliachar-Urlar Burn, 4km south-west of Aberfeldy, Perthshire.

## *Benstonite*

$(Ba,Sr)_6(Ca,Mn)_6Mg(CO_3)_{13}$    Trigonal

Grains of pyrite in the Aberfeldy barium-zinc deposits, Perthshire, encapsulate barytocalcite, analysis of which shows considerable variation in Mg, Ca, Sr and Ba content. Some of the grains may be benstonite. This mineral requires confirmation.

## BERRYITE

$Pb_3(Ag,Cu)_5Bi_7S_{16}$    Monoclinic

In quartz veins hosted by the Loch Shin monzo-granite, 2km west of Lairg, Sutherland.

## BERTHIERINE

$(Fe^{+2},Fe^{+3},Mg,Al)_{2-3}(Si,Al)_2O_5(OH)_4$    Monoclinic

A chamosite previously described from Ayrshire is now classed as berthierine.

## BERTRANDITE

$Be_4Si_2O_7(OH)_2$    Orthorhombic

Pale-brown, small crystals (3mm) associated with genthelvite in miarolitic cavities in adamel-lite, Coire an Lochain, Cairngorm Mountains, Grampian.

## BERYL

$Be_3Al_2Si_6O_{18}$    Hexagonal

As apple-green crystals in drusy cavities in the Cairngorm granite, Aberdeenshire, and as euhedral crystals in the Loch Nevis mica prospect, Inverness-shire.

## BERZELIANITE

$Cu_2Se$    Cubic

As inclusions in two gold grains panned from the Smid Hope Burn tributary of the River Tweed near Tweedhope, Peeblesshire.

## *Betafite* see pyrochlore

## BETEKHTINITE

$Cu_{10}(Fe,Pb)S_6$    Orthorhombic

In galena-bearing veins cutting the Grudie granite, near Lairg, Sutherland.

## BEUDANTITE

$PbFe_3^{+3}(AsO_4)(SO_4)(OH)_6$    Trigonal

Occurs as a red-brown encrustation on quartz, High Pirn Mine, Wanlockhead, Dumfriesshire.

## BINDHEIMITE

$Pb_2Sb_2O_6(O,OH)$                    Cubic

Present as a yellow, small, earthy encrustation on stibnite, Hare Hill, Ayrshire.

## BIOTITE

$K(Mg,Fe^{+2})_3(Al,Fe^{+3})Si_3O_{10}(OH,F)_2$        Monoclinic

Common.

## BIRNESSITE

$Na_4Mn_{14}O_{27}.9H_2O$                Monoclinic

Small black grains within a manganese pan in a cutting through a fluvio-glacial gravel deposit at Birness, Aberdeenshire.

## BISMOCLITE

BiOCl                              Tetragonal

Associated with atacamite, connellite, and secondary uranium minerals in veins cutting the aureole of the Criffel granodiorite, near Dalbeattie, Kirkcudbrightshire.

## BISMUTH

Bi                                  Trigonal

Distributed through gangue minerals in veins cutting the aureole of the Criffel granodiorite, near Dalbeattie, Kirkcudbrightshire.

## BISMUTHINITE

$Bi_2S_3$                            Orthorhombic

Silver-grey massive material associated with pyrite in quartz veins, Lotus Hill, Beeswing, Kirkcudbrightshire.

## BISMUTITE

$Bi_2(CO_3)O_2$                        Tetragonal

Noted in two heavy mineral concentrates from the Helmsdale granite, Sutherland.

## Blende = SPHALERITE

## BÖHMITE

AlO(OH)                            Orthorhombic

As minute crystals in pisolitic bauxitic clay at Saltcoats and Kilwinning, Ayrshire.

## BOLTWOODITE

$HK(UO_2)SiO_4.1\frac{1}{2}H_2O$            Monoclinic

With zeunerite as alteration products of pitchblende in veins cutting the aureole of the Criffel granodiorite, near Dalbeattie, Kirkcudbrightshire.

## BORNITE

$Cu_5FeS_4$                  Orthorhombic ps. cubic

Anhedral to sub-hedral masses, up to 1cm, in a carbonate-quartz matrix; found as a loose boulder in glacial drift, Loganlee, Pentland Hills, Midlothian.

## BOTALLACKITE

$Cu_2^{+2}Cl(OH)_3$                      Monoclinic

Minute area of platy blue-green crystals associated with atacamite in radioactive rock with connellite, from Southwick cliffs, south of Dalbeattie, Kirkcudbrightshire.

## BOURNONITE

$PbCuSbS_3$                          Orthorhombic

Irregular grains (~1mm) associated with sphalerite, arsenopyrite and pyrite, in a specimen from the Glendinning Louisa antimony mine, Glenshanna Burn, Dumfriesshire.

## BRANNERITE

$(U,Ca,Y,Ce)(Ti,Fe)_2O_6$                Monoclinic

Metamict metallic grain (6 x 2mm) in massive quartz on heating gave a brannerite X-ray powder pattern, confirmatory chemistry obtained. Associated minerals include bismutite, cassiterite, fluorite and molybdenite, from Little Craig, Glen Gairn, Aberdeenshire.

## BRAUNITE

$Mn^{+2}Mn_6^{+3}SiO_{12}$                Tetragonal

Forms the fine-grained groundmass of the manganese ore at Dalroy Burn, Dalroy, Inverness-shire.

## BRAVOITE

$(Ni,Fe)S_2$                          Cubic

As minute crystals in dump material from Hilderston silver mine, Cairnpapple Hill, near Bathgate, West Lothian.

## BREDIGITE

$Ca_7Mg(SiO_4)_4$                      Orthorhombic

In the limestone contact-zone of the Camas Mòr gabbro, Muck, Inverness-shire.

## BREITHAUPTITE

NiSb                                Hexagonal

A composite platinum-group mineral grain in a chromite sample from Harold's Grave chromite quarry in the Baltasound, Haroldswick area, Unst, Shetland, contains an area of breithauptite.

## Breunnerite = ferroan MAGNESITE

## BREWSTERITE (s.l.)

$(Sr,Ba,Ca)Al_2Si_6O_{16}.5H_2O$      Monoclinic

Found associated with calcite, barite and harmotome at the mines at Strontian, Argyllshire.

## BROCHANTITE

$Cu_4^{+2}(SO_4)(OH)_6$      Monoclinic

Found as an alteration product of chalcopyrite.

## BROCKITE

$(Ca,Th,Ce)(PO_4).H_2O$      Hexagonal

A thorium-bearing grain from a panning concentrate taken on the Etive complex at the head of Loch Etive, Argyllshire, gave an X-ray powder pattern similar to brockite.

## Bronzite = ferroan ENSTATITE

Common.

## BROOKITE

$TiO_2$      Orthorhombic

Reddish-brown plates (<8 x 0.5mm) in calcite from Killin, Perthshire.

## BRUCITE

$Mg(OH)_2$      Trigonal

Forms plates in veins, at Swinna Ness, Unst, Shetland.

## BRUGNATELLITE

$Mg_6Fe^{+3}(CO_3)(OH)_{13}.4H_2O$      Hexagonal

Pearly-white to pale-brown scaly efflorescence on sheared brucite-bearing vein in serpentinite, Swinna Ness, Unst, Shetland.

## BUSTAMITE

$(Mn^{+2},Ca)_3Si_3O_9$      Triclinic

As a minor component, associated with hematite, of the manganese ore found at Dalroy Burn, Dalroy, Inverness-shire.

## BYTOWNITE

(An 70-90)      Triclinic

Present as an essential mineral in the allivalite rocks of Rhum, Inverness-shire.

## CALCITE

$CaCO_3$      Trigonal

Common.

## CALEDONITE

$Pb_5Cu_2(CO_3)(SO_4)_3(OH)_6$      Orthorhombic

Usually associated with leadhillite and other rare secondary lead minerals in the Leadhills, Lanarkshire and Wanlockhead, Dumfriesshire orefield.

## CASSITERITE

$SnO_2$      Tetragonal

Occurs, with magnetite, in the foliated granite-gneiss of Carn Chuinneag, Ross-shire.

## CELADONITE

$K(Mg,Fe^{+2})(Fe^{+3},Al)Si_4O_{10}(OH)_2$      Monoclinic

As a bright-green lining to amygdales, and frequently forms a thin skin on agates.

## CELESTINE

$SrSO_4$      Orthorhombic

On the shore opposite the Bass Rock, North Berwick, East Lothian, pale-blue, diverging crystals in altered igneous rock, also at Muirshiels barite mine, Renfrewshire.

## CELSIAN

$BaAl_2Si_2O_8$      Monoclinic

In Dalradian quartzites and schists from sulphide mineralized horizons south of Loch Tummel, Perthshire.

## CERUSSITE

$PbCO_3$      Orthorhombic

Formed by the oxidation of galena, especially at Leadhills, Lanarkshire and Wanlockhead, Dumfriesshire.

## CHABAZITE (s.l.)

$CaAl_2Si_4O_{12}.6H_2O$      Trigonal

Commonly found in the zeolitized basalts of Skye, Inverness-shire.

## CHABAZITE-K

$(K,Na,Ca)AlSi_2O_6.3H_2O$      Trigonal

Forms milky-white rhombs, up to 6mm, covering some $7cm^2$ on altered basalt from Lynedale, Loch Snizort, Skye, Inverness-shire.

## CHALCANTHITE

$Cu^{+2}SO_4.5H_2O$      Triclinic

Efflorescent pale-blue curved crystals associated with chalcopyrite, pyrite and quartz, Lauchentyre Mine, west of Gatehouse of Fleet, Kirkcudbrightshire.

## CHALCOCITE

$Cu_2S$      Monoclinic

Associated with malachite, at Elmsford, near Duns, Berwickshire and in lavas at Glenfarg, Perthshire.

## CHALCOPHANITE

$(Zn,Fe^{+2},Mn^{+2})Mn_3^{+4}O_7.3H_2O$          Trigonal

Patchy outcrops of iron oxides, with minor chalcophanite-bearing pod-like manganese oxides, in breccia, occur about 8km south-east of Tomintoul, Banffshire.

## CHALCOPYRITE

$CuFeS_2$          Tetragonal

From numerous localities, especially at copper, lead or zinc mines.

## CHAMOSITE

$(Fe^{+2},Mg,Fe^{+3})_5Al(Si_3Al)O_{10}(OH,O)_8$          Monoclinic

As ooliths in the Raasay Ironstone, Raasay, Inverness-shire.

## CHENITE

$Pb_4Cu^{+2}(SO_4)_2(OH)_6$          Triclinic

Forms minute, transparent to translucent sky-blue singly terminated crystals on oxidised galena, with chalcopyrite, at Leadhills, Lanarkshire.

## CHESTERITE

$(Mg,Fe^{+2})_{17}Si_{20}O_{54}(OH)_6$          Orthorhombic

Intergrown chesterite polytype with jimthompsonite and clinojimthompsonite on actinolite cores in Lewisian amphibolite facies ultramafic rocks near Achmelvich, Lochinver, Sutherland.

## CHEVKINITE

$(Ca,Ce,Th)_4(Fe^{+2},Mg)_2(Ti,Fe^{+3})_3Si_4O_{22}$   Monoclinic

An accessory mineral in the Tertiary granites of Glen Bay, Na h-Eagan and Conachair in the Isles of St Kilda.

## CHLORITOID

$(Fe^{+2},Mg,Mn)_2Al_4Si_2O_{10}(OH)_4$
Monoclinic and Triclinic

Large purplish-black crystals, and clots, with muscovite and quartz in schists on the west coast of Ness of Hillswick, Hillswick, Shetland.

## CHONDRODITE

$(Mg,Fe^{+2})_5(SiO_4)_2(F,OH)_2$          Monoclinic

In ore skarns at contacts of the Durness dolomite limestone with the Beinn an Dubhaich granite, Broadford area, Skye, Inverness-shire.

## CHROMITE

$Fe^{+2}Cr_2O_4$          Cubic

From podiform chromitites on the island of Unst, Shetland, and elsewhere.

## CHRYSOBERYL

$BeAl_2O_4$          Orthorhombic

Detected in panned heavy mineral concentrates from Glen Lui, Aberdeenshire.

## CHRYSOCOLLA

$(Cu^{+2},Al)_2H_2Si_2O_5(OH)_4.nH_2O$          Monoclinic

Found on gangue at Wanlockhead, Dumfriesshire.

## Chrysolite = FORSTERITE

Common.

## CHRYSOTILE

$Mg_3Si_2O_5(OH)_4$          —

Present in serpentinites.

## CINNABAR

$HgS$          Trigonal

Trace amounts in pan concentrate samples from a small stream at Hartside and from Woodend Burn, both draining from Devonshaw Hill, about 3km south-west of Lamington, Lanarkshire.

## CLAUSTHALITE

$PbSe$          Cubic

As inclusions (<1–5 μm) in two gold grains panned from the Smid Hope Burn tributary of the River Tweed near Tweedhope, Peeblesshire.

## CLINOATACAMITE

$Cu_2(OH)_3Cl$          Monoclinic

Intimately associated with botallackite on quartz matrix from Castletown Mine, Ballimore, Lochgilphead, Argyllshire.

## CLINOCHLORE

$(Mg,Fe^{+2})_5Al(Si_3Al)O_{10}(OH)_8$          Monoclinic

Common.

## CLINOCLASE

$Cu_3^{+2}(AsO_4)(OH)_3$          Monoclinic

As rosettes on a slab of quartz and associated with cornwallite and cornubite, possibly from Fitful Head or probably Quendale Bay, Shetland.

## CLINOHUMITE

$(Mg,Fe^{+2})_9(SiO_4)_4(F,OH)_2$          Monoclinic

In ore skarns at contacts of the Durness dolomite limestone with the Beinn an Dubhaich granite, Broadford area, Skye, Inverness-shire.

## CLINOJIMTHOMPSONITE

$(Mg,Fe^{+2})_5Si_6O_{16}(OH)_2$      Monoclinic

Forms euhedral wedge-shaped overgrowths, up to 1mm, in the *b*-axis direction on actinolite cores, and associated with jimthompsonite and chesterite polytype, in Lewisian amphibolite facies ultramafic rocks near Achmelvich, Lochinver, Sutherland.

## CLINOPTILOLITE *(s.l.)*

$(Na,K,Ca)_{2-3}Al_3(Al,Si)_2Si_{13}O_{36}.12H_2O$    Monoclinic

Found in Palaeozoic lavas of the Midland Valley at two localities, as vein material, about 3km west of Kilmacolm, Renfrewshire with heulandite-calcite - quartz, and some 4km north-east of Dumbarton, Dunbartonshire, with stilbite and laumontite.

## CLINOZOISITE

$Ca_2Al_3(SiO_4)_3(OH)$      Monoclinic

In symplectite-bearing nodules in the Ardgour marble, Coire Dubh, Argyllshire.

## CLINTONITE

$Ca(Mg,Al)_3(Al_3Si)O_{10}(OH)_2$      Monoclinic

One of the many metamorphic minerals developed in the Durness limestone horizons in the aureole of the Beinn an Dubhaich granite, Broadford area, Skye, Inverness-shire.

## COALINGITE

$Mg_{10}Fe_2^{+3}(CO_3)(OH)_{24}.2H_2O$      Trigonal

Golden-brown micaceous flakes and plates on chromitite from Hagdale, Unst, Shetland Islands are a mixture of coalingite and pyroaurite.

## COBALTITE

$CoAsS$      Orthorhombic ps.cubic

Occurs within sulphide mineralisation in the Fore Burn igneous complex some 24km east of Girvan, Ayrshire.

## COFFINITE

$U(SiO_4)_{1-x}(OH)_{4x}$      Tetragonal

Present as submicroscopic grains and up to 500µm, and along cracks in arkose at Ousdale, Caithness.

## Collophane

A massive fine-grained phosphate of the apatite group; shapeless clots in compact limestone, Clippens Limeworks, near Straiton, Midlothian.

## COLORADOITE

$HgTe$      Cubic

Rare inclusions of coloradoite with hessite and altaite are present in galena from material derived from a pyrite, galena, chalcopyrite quartz vein approximately 4.5km south-west of Aberfeldy, Perthshire.

## COLUMBITE

$(Fe,Mn)(Nb,Ta)_2O_6$      Orthorhombic

Found as crystals in the Chaipaval pegmatite, South Harris, Outer Hebrides.

## CONICHALCITE

$CaCu^{+2}(AsO_4)(OH)$      Orthorhombic

Forms an apple-green 'stain' thinly covering an area of approximately 28-30cm² on a pale, buff-coloured rock from Braid Hills, Edinburgh, Midlothian.

## CONNELLITE

$Cu_{19}^{+2}Cl_4(SO_4)(OH)_{32}.3H_2O$      Hexagonal

As a blue secondary mineral associated with ata-camite in uranium-bearing veins cutting the aureole of the Criffel granodiorite, near Dalbeattie, Kirkcudbrightshire.

## COPIAPITE

$Fe^{+2}Fe_4^{+3}(SO_4)_6(OH)_2.20H_2O$      Triclinic

As a powdery yellow efflorescence on altered pyrite 'from 160 level', Wanlockhead, Dumfriesshire.

## COPPER

$Cu$      Cubic

Either as crystals or small sheets associated with calcite and prehnite at Boyleston Quarry, near Barrhead, Renfrewshire.

## COQUIMBITE

$Fe_2^{+3}(SO_4)_3.9H_2O$      Trigonal

Small clusters of minute, clear glassy crystals associated with halotrichite, römerite, pyrite and voltaite, from an old railway cutting west of Stanely, near Paisley, Renfrewshire.

## CORDIERITE

$Mg_2Al_4Si_5O_{18}$      Orthorhombic

In argillaceous hornfels at Belhelvie, Aberdeenshire and Glen Clova, Angus.

## CORNUBITE

$Cu_5^{+2}(AsO_4)_2(OH)_4$      Triclinic

As botryoidal patches encrusting a slab of quartz, associated with cornwallite and clinoclase, possibly from Fitful Head or probably Quendale Bay, Shetland.

## CORNWALLITE

$Cu_5^{+2}(AsO_4)_2(OH)_4.H_2O$      Monoclinic

As botryoidal patches encrusting a slab of quartz, associated with cornubite and clinoclase, possibly from Fitful Head or probably Quendale Bay, Shetland.

## CORONADITE

$Pb(Mn^{+4},Mn^{+2})_8O_{16}$     Monoclinic ps. tetragonal

Predominantly as fine-grained veinlets and rare grains in a manganese vein, about 1km north-west of Arndilly House, Arndilly, Banffshire.

## CORRENSITE

?       –

Occurs as vein filling in a dolerite sill, Hillhouse Quarry, Ayrshire.

## CORUNDUM

$Al_2O_3$     Trigonal

In corundum-spinel xenoliths in norite at Haddo House, Aberdeenshire, and in aluminous xenoliths in sills, Mull, Argyllshire.

## COSALITE

$Pb_2Bi_2S_5$     Orthorhombic

In wall-rocks of a quartz-hübnerite vein near Gairnshiel Bridge, Glen Gairn, Aberdeenshire.

## COTUNNITE

$PbCl_2$     Orthorhombic

With leadhillite from the Hopeful Vein, Leadhills, Lanarkshire, and pyromorphite, mimetite and plumbogummite from the High Pirn Mine, Wanlockhead, Dumfriesshire.

## COVELLITE

$CuS$     Hexagonal

With malachite in a vein or fissure of about 4cm in sandstone, Kingsteps, Angus.

## COWLESITE

$CaAl_2Si_3O_{10}.5\text{-}6H_2O$     Orthorhombic

An amygdale mineral in basalt from Ouisgill Bay, Duirinish, Skye, and Kingsburgh, Loch Snizort, Skye, Inverness-shire.

## *Crichtonite* group

Titanium-rich minerals found in a sulphide-quartz-celsian rock, and sulphide-banded barite rock, at Aberfeldy, Perthshire.

## CRISTOBALITE

$SiO_2$     Tetragonal

In cavities in aegirine granite, Rockall, Inverness-shire.

## CROCOITE

$PbCrO_4$     Monoclinic

Associated with cerussite, pyromorphite and leadhillite in dump material from the Hopeful Vein, Leadhills, Lanarkshire.

## CROSSITE

$Na_2(Mg,Fe^{+2})_3(Al,Fe^{+3})_2Si_8O_{22}(OH)_2$     Monoclinic

Small sodic amphibole grains in rocks from the Falahill Formation (Ordovician) near Heriot, Galashiels area, Scottish Borders.

## CRYPTOMELANE

$K(Mn^{+4},Mn^{+2})_8O_{16}$     Monoclinic ps. tetragonal

Cryptomelane-lithiophorite mixtures occur as massive, black, fine-grained aggregates at the Lecht mines, Tomintoul, Banffshire.

## CUBANITE

$CuFe_2S_3$     Orthorhombic

As exsolutions in chalcopyrite occurring in small patches of sulphides in allivalite, Huntly, Aberdeenshire.

## CUMMINGTONITE

$(Mg,Fe^{+2})_7Si_8O_{22}(OH)_2$     Monoclinic

In cummingtonite-garnet schist occurring 0.7km south-west of the outlet of Loch Bad-na-Sgalaig, between Loch Maree and Gairloch, Ross and Cromarty.

## CUPRITE

$Cu_2O$     Cubic

Occurs with native copper and malachite in a mixture of analcime, calcite and prehnite at Boyleston Quarry, Barrhead, Renfrewshire.

## CUSPIDINE

$Ca_4Si_2O_7(F,OH)_2$     Monoclinic

As lanceolate crystals associated with magnetite, diopside, wollastonite, calcite and other minerals in dolomite contact skarns, Broadford, Skye, Inverness-shire.

## CYANOTRICHITE

$Cu_4^{+2}Al_2(SO_4)(OH)_{12}.2H_2O$     Orthorhombic

From Leadhills, Lanarkshire.

## CYMRITE

$BaAl_2Si_2(O,OH)_8.H_2O$     Monoclinic

Present as a minor constituent in schistose quartz-celsian rock in strata-bound barium-zinc mineralisation in Dalradian schist near Aberfeldy, Perthshire.

## DATOLITE

$Ca_2B_2Si_2O_8(OH)_2$     Monoclinic

Associated with calcite and prehnite in lavas at Glenfarg, Perthshire.

## DELLAITE

$Ca_6Si_3O_{11}(OH)_2$      Monoclinic(?)

Veinlets and bladed crystals associated with calcite in spurrite-rich rocks in Jurassic limestones, Kilchoan, Ardnamurchan, Argyllshire.

## DESCLOIZITE

$PbZn(VO_4)(OH)$      Orthorhombic

Found as brown spherules on pyromorphite from Wanlockhead, Dumfriesshire.

## DEVILLINE

$CaCu_4^{+2}(SO_4)_2(OH)_6.3H_2O$      Monoclinic

Common as blue-green pearly crystals in cavities in slag from Meadowfoot smelter dumps, Wanlockhead, Dumfriesshire.

## DIASPORE

$AlO(OH)$      Orthorhombic

As minute crystals (0.03mm) associated with böhmite in pisolitic bauxitic clay, Saltcoats and Kilwinning, Ayrshire.

## DICKITE

$Al_2Si_2O_5(OH)_4$      Monoclinic

White coatings on rocks from a pipeline excavation in highly altered volcanic rock, near Hillend Roadhouse, Edinburgh, Midlothian.

## DIGENITE

$Cu_9S_5$      Cubic

Irregular masses several centimetres in diameter, and crystals, associated with calcite-scapolite veins, Copper Geo, Fair Isle, Shetland.

## DIOPSIDE

$CaMgSi_2O_6$      Monoclinic

Common, especially in calc-silicate assemblages and marbles.

## Dipyre = MARIALITE

## DJURLEITE

$Cu_{31}S_{16}$      Monoclinic

Forms bluish-black areas, up to 3mm across, on barite gangue from Silver Glen, Alva, Clackmannanshire.

## DOLOMITE

$CaMg(CO_3)_2$      Trigonal

Common, especially in dolomite limestones and marbles.

## DRAVITE

$NaMg_3Al_6(BO_3)_3Si_6O_{18}(OH)_4$      Trigonal

Small crystals present in fragments of quartz-mica schist from Loch Assapol, Ross of Mull, Argyllshire.

## DUFTITE

$PbCu(AsO_4)(OH)$      Orthorhombic

Dull, black encrustation on botryoidal pyromorphite associated with white earthy apatite; specimen from an old dump at Tait's level, Whyte's Cleugh, Wanlockhead, Dumfriesshire.

## DUMORTIERITE

$Al_7(BO_3)(SiO_4)_3O_3$      Orthorhombic

In paragneiss from near Newburgh, Aberdeenshire; from pegmatite veins in schists near Colehill Farm, near Ellon, Aberdeenshire and from foliated granite of Bogierow near Portsoy, Banffshire.

## DUNDASITE

$PbAl_2(CO_3)_2(OH)_4.H_2O$      Orthorhombic

Snow-white sprays and hemispheres on anglesite and gangue, Susanna Vein, Leadhills, Lanarkshire.

## DYPINGITE

$Mg_5(CO_3)_4(OH)_2.5H_2O$      Monoclinic(?)

Milky-white crust covering 1.5-2cm² on chromite-bearing serpentinite from Hagdale, Unst, Shetland.

## EDENITE

$NaCa_2(Mg,Fe^{+2})_5Si_7AlO_{22}(OH)_2$      Monoclinic

In a tonalite of the Morven-Strontian complex, Inverness-shire.

## EDINGTONITE

$BaAl_2Si_3O_{10}.4H_2O$    Orthorhombic and tetragonal

From near Old Kilpatrick, Dunbartonshire.

## ELBAITE

$Na(Li,Al)_3Al_6(BO_3)_3Si_6O_{18}(OH)_4$      Trigonal

In transparent prismatic crystals of various shades of pink, green and blue (commonly up to 2cm x 2mm) associated with muscovite and sericite in loose boulders of lepidolite-bearing pegmatite, Glenbuchat, Aberdeenshire.

## *Electrum*

$(Au,Ag)$      Cubic

Part of the series gold-silver.

## ELPIDITE

$Na_2ZrSi_6O_{15}.3H_2O$      Orthorhombic

As a needle-shaped mineral occurring in miarolitic spots in the rockallite of Rockall, Inverness-shire.

## ELYITE

$Pb_4Cu^{+2}(SO_4)(OH)_8$ — Monoclinic

Sprays of lilac needles, up to 0.4mm, in cavities in a small specimen of litharge, massicot and cerussite found in a stream bed, Leadhills, Lanarkshire.

## EMPLECTITE

$CuBiS_2$ — Orthorhombic

In small fractures in galena, and associated with electrum, bismuth and schirmerite, from old mine dumps at Corrie Buie, Meal nan Oighreag, south of Loch Tay, Perthshire.

## ENARGITE

$Cu_3AsS_4$ — Orthorhombic

Trace amounts of tennantite and enargite form rims separating galena and chalcopyrite in the Calliachar-Urlar Burn veins, 4km south-west of Aberfeldy, Perthshire.

## ENSTATITE

$Mg_2Si_2O_6$ — Orthorhombic

Common, especially in gabbros of the Cullin Hills, Skye, and Rhum, Inverness-shire.

## EPIDOTE

$Ca_2(Fe^{+3},Al)_3(SiO_4)_3(OH)$ — Monoclinic

Common.

## EPISTILBITE

$CaAl_2Si_6O_{16}.5H_2O$ — Monoclinic

Small pale, flesh-coloured crystals found rarely in vesicular basalt at Talisker Bay, Skye, Inverness-shire, and associated with scolecite at Dearg Sgeir, Mull, Argyllshire.

## EPSOMITE

$MgSO_4.7H_2O$ — Orthorhombic

White silky crystals, in efflorescences on shale, Hurlet, Renfrewshire.

## ERIONITE (s.l.)

$(K_2,Ca,Na_2)_2Al_4Si_{14}O_{36}.15H_2O$ — Hexagonal

Creamy-coloured, platy-fibrous crystal aggregates with garronite, in a cavity in vesicular basalt, Storr, Skye, Inverness-shire.

## ERYTHRITE

$Co_3(AsO_4)_2.8H_2O$ — Monoclinic

Encrusting gangue from the silver mines at Silver Glen, Alva, Clackmannanshire.

## ETTRINGITE

$Ca_6Al_2(SO_4)_3(OH)_{12}.26H_2O$ — Hexagonal

Fibrous crystals in sharp fissure veins less than 0.5mm wide in metamorphosed limestones at Kilchoan, Ardnamurchan, Argyllshire.

### *Eucolite-eudialyte* mineral

Incompletely characterized colourless or pale-yellow mineral, in a cavity 1mm across in aegirine-granite, Rockall, Inverness-shire.

### Eulite = FERROSILITE

## EULYTITE

$Bi_4(SiO_4)_3$ — Cubic

Small pale-green, translucent to transparent spherulitic masses, *c.*0.25mm, on material from Southwick Cliffs, Solway Firth, south of Dalbeattie, Kirkcudbrightshire.

## EUXENITE - (Y)

$(Y,Ca,Ce,U,Th)(Nb,Ta,Ti)_2O_6$ — Orthorhombic

Associated with allanite and thorite in pegmatites from near Kinlochbervie, Sutherland.

## FAYALITE

$Fe_2^{+2}SiO_4$ — Orthorhombic

Found in the epigranites of the Marsco area, Skye, Inverness-shire.

## FERGUSONITE

$YNbO_4$ — Tetragonal

Occurs with cassiterite, columbite, ilmenorutile, monazite and xenotime in the heavy mineral suite of stream sediments from Glen Lui, Aberdeenshire. (The species is here undifferentiated.)

### *Feroxyhyte*

$\delta$-$Fe^{+3}O(OH)$ — Hexagonal

Possibly present in a thin iron-pan, 1-3mm thick, formed from basalt at Killiechronan, just southwest of Salen, Mull, Argyllshire.

## FERRI-BARROISITE

$CaNaMg_3Fe_2^{+3}Si_7AlO_{22}(OH)_2$ — Monoclinic

Forms very small grains in rocks from the Falahill Formation (Ordovician) near Heriot, Galashiels Area, Scottish Borders.

## FERRIERITE (s.l.)

$(Na,K)_2MgAl_3Si_{15}O_{36}(OH).9H_2O$ — Orthorhombic and Monoclinic

Occurs in veins in tuffs, within amygdales in basalt, and as a replacement mineral in volcanic rocks of Carboniferous age at numerous localities in the Midland Valley of Scotland.

## FERRIHYDRITE

$5Fe_2O_3.9H_2O$      Hexagonal

A thin iron-pan, 1–3mm thick, formed from basalt at Killiechronan just south-west of Salen, Mull, Argyllshire, contains ferrihydrite; feroxyhyte and akaganeite are also possibly present.

## FERRIMOLYBDITE

$Fe_2^{+3}(Mo^{+6}O_4)_3.8H_2O(?)$      Orthorhombic(?)

Yellow earthy coating, associated with molybdenite, in a quartz vein cutting granite, Screel Hill, Kirkmirran, Kirkcudbrightshire.

## Ferroaugite = AUGITE

## FERROBUSTAMITE

$Ca(Fe^{+2},Ca,Mn^{+2})Si_2O_6$      Triclinic

An 'iron rhodonite' from Skye, Inverness-shire, has been shown to possess the bustamite structure.

## FERRO-EDENITE

$NaCa_2(Fe^{+2},Mg)_5(Si_7,Al)O_{22}(OH)_2$      Monoclinic

Occurs in the western part of the Glen Bay Gabbro, Isles of St Kilda.

## FERROGLAUCOPHANE

$Na_2(Fe^{+2},Mg)_3Al_2Si_8O_{22}(OH)_2$      Monoclinic

Forms very small grains in rocks from the Falahill formation (Ordovician) near Heriot, Galashiels area, Midlothian.

## Ferrohedenbergite = AUGITE

## FERROHORNBLENDE

$Ca_2(Fe^{+2},Mg)_4Al(Si_7,Al)O_{22}(OH)_2$      Monoclinic

Occurs in a garnet-biotite-epidote-albite amphibolite, 0.3km N51°E from northern end of Loch-na-Craige, near Achahoish, southern Knapdale, Argyllshire.

## Ferrohypersthene = FERROSILITE

## FERROPARGASITE

$NaCa_2(Fe^{+2},Mg)_4Al(Si_6Al_2)O_{22}(OH)_2$      Monoclinic

Occurs in the Lewisian banded-iron formation of amphibolite facies at Flowerdale, near Gairloch, Wester Ross.

## Ferropigeonite = iron-rich PIGEONITE

## FERRORICHTERITE

$Na_2Ca(Fe^{+2},Mg)_5Si_8O_{22}(OH)_2$      Monoclinic

Sub-1mm anhedra intergrown with aenigmatite, together with Fe-Ti oxides, pyroxene and allanite in the Southern Prophyritic Epigranite, western Redhills, Skye, Inverness-shire.

## FERROSILITE

$(Fe^{+2},Mg)_2Si_2O_6$      Orthorhombic

In an amphibole-garnet rock from Druideag Lodge, Loch Duich, Ross-shire.

## FERROTSCHERMAKITE

$Ca_2(Fe^{+2},Mg)_3Al_2(Si_6Al_2)O_{22}(OH)_2$      Monoclinic

In prophyroblastic amphibole gneiss, near Beasdale rail bridge, South Morar, Inverness-shire.

## FLUOBORITE

$Mg_3(BO_3)(F,OH)_3$      Hexagonal

Associated with ludwigite, datolite, szaibelyite and harkerite in the dolomite contact skarns of the Broadford area, Skye, Inverness-shire.

## FLUORAPATITE

$Ca_5(PO_4)_3F$      Hexagonal

Common.

## FLUORITE

$CaF_2$      Cubic

Purple and green fluorite occurs in the granite at the Pass of Ballater, near Ballater, Aberdeenshire, and elsewhere.

## FORSTERITE

$Mg_2SiO_4$      Orthorhombic

Common.

## FOSHAGITE

$Ca_4Si_3O_9(OH)_2$      Triclinic

As colourless fibrous aggregates commonly perpendicular to the margin of thin (<2mm) veins in thermally metamorphosed Jurassic limestones at Kilchoan, Ardnamurchan, Argyllshire.

## FRANCEVILLITE

$(Ba,Pb)(UO_2)_2V_2O_8.5H_2O$      Orthorhombic

Trace amounts occur immediately beneath the Carboniferous unconformity near Dalveen, Dumfriesshire.

## FREIBERGITE

$(Ag,Cu,Fe)_{12}(Sb,As)_4S_{13}$      Cubic

As inclusions (<100μm) in fine-grained massive galena from the 'Hard Vein', Tyndrum, Perthshire.

## FRIEDRICHITE

$Pb_5Cu_5Bi_7S_{18}$      Orthorhombic

Compositions approaching friedrichite are reported from an assemblage which contains aikinite, hammarite, lindströmite, krupkaite, gladite and pekoite in quartz veins hosted by the Loch Shin monzogranite, 2km west of Lairg, Sutherland.

## GADOLINITE - (Y)

$Y_2Fe^{+2}Be_2Si_2O_{10}$      Monoclinic

Forms dark-green vitreous prismatic crystals (<2.5mm) associated with, and growing on, albite and orthoclase from miarolitic cavities within the northern portion of the Arran granite, Isle of Arran.

## GAHNITE

$ZnAl_2O_4$      Cubic

In the Chaipaval pegmatite vein, near Northton, South Harris, Outer Hebrides, Inverness-shire.

## GALENA

PbS      Cubic

Common.

## GALENOBISMUTITE

$PbBi_2S_4$      Orthorhombic

Present as plates (<0.5cm) in galena containing altered pyrrhotite, from old mine dumps at Corrie Buie, Meal nan Oighreag, south of Loch Tay, Perthshire.

## GARRONITE

$Na_2Ca_5Al_{12}Si_{20}O_{64}.27H_2O$      Tetragonal

Occurs as amygdale fillings in basalts of Tertiary age at Storr, Skye, Inverness-shire.

## GEDRITE

$(Mg,Fe^{+2})_5Al_2(Si_6Al_2)O_{22}(OH)_2$      Orthorhombic

In certain schists and granulites from near Strathy, Sutherland.

## GEHLENITE

$Ca_2Al(Al,Si)O_7$      Tetragonal

In the gabbro-limestone contact zone of Camas Mòr, Muck, Inverness-shire; associated minerals are wollastonite, monticellite, larnite and rankinite.

## GEIKIELITE

$MgTiO_3$      Trigonal

Extremely rare isolated euhedral crystals associated with rutile, zircon, baddeleyite and quandilite occur in siliceous dolomites of the Allt Guibhsachain area near Ballachulish, Argyllshire.

## GENKINITE

$(Pt,Pd)_4Sb_3$      Tetragonal

Found as two irregular crystals 4 x 2µm and 6 x 3µm in two chromite-rich rocks from the disused quarry at Harold's Grave in the Baltasound-Haroldswick area, Unst, Shetland.

## GENTHELVITE

$Zn_4Be_3(SiO_4)_3S$      Cubic

Green and brown tetrahedral crystals (<1.3mm) with bertrandite in miarolitic cavities in adamellite at Coire an Lochain, Cairngorm Mountains, Grampian.

## GERSDORFFITE

NiAsS      Cubic

Occurs in a cross-cut vein up to 5cms wide, in quartzose bands in schists, associated with chalcopyrite and pentlandite, at the mine between Coillebhraghad and Eas a Chosain about one mile west of Inveraray, Argyllshire.

## GEVERSITE

$Pt(Sb,Bi)_2$      Cubic

A grain of geversite was located in a chromite-rich sample from Jimmie's Quarry near Hagdale, Unst, Shetland.

## GIBBSITE

$Al(OH)_3$      Monoclinic

In soils and associated with aluminous vermiculite-chlorite and kaolinite, derived from a biotite-rich quartz-gabbro, Strathdon area, Aberdeenshire.

## GISMONDINE

$Ca_2Al_4Si_4O_{16}.9H_2O$      Monoclinic

Found in a lopolith at Clauchland's Point, 3km north-east of Lamlash, Isle of Arran.

## GLADITE

$PbCuBi_5S_9$      Orthorhombic

As coarse, fibrous crystals in quartz veins hosted by the Loch Shin monzogranite, 2km west of Lairg, Sutherland.

## GLAUCONITE

$(K,Na)(Fe^{+3},Al,Mg)_2(Si,Al)_4O_{10}(OH)_2$      Monoclinic

Dark-green flaky material on micaceous sandstone from Salisbury Crags, Edinburgh, Midlothian.

## GLAUCOPHANE

$Na_2(Mg,Fe^{+2})_3Al_2Si_8O_{22}(OH)_2$      Monoclinic

Occurs as a blue amphibole in an epidote-hornblende schist from the Girvan – Ballantrae complex, South Ayrshire.

## GLUSHINSKITE

$Mg(C_2O_4).2H_2O$      Monoclinic

Occurs at the lichen-rock interface on serpentinite colonised by *Lecanora atra* at Mill of Johnston, 7km south-west of Insch, Aberdeenshire.

## GMELINITE (*s.l.*)

$(Na_2,Ca)Al_2Si_4O_{12}.6H_2O$      Hexagonal

Milky-white and flesh-pink glassy crystals in amygdales in basalt at Talisker, Skye, Inverness-shire.

## GOBBINSITE

$Na_4(Ca,Mg,K_2)Al_6Si_{10}O_{32}.12H_2O$   Orthorhombic
ps. tetragonal

From a dyke on the foreshore near the Bunnahabhainn Distillery, 3.5km north of Port Askaig, Islay.

## GODLEVSKITE

$(Ni,Fe)_7S_6$      Orthorhombic

Associated with finely disseminated metals, alloys, arsenides and sulphides in the Unst ophiolite belt, Shetland.

## GOETHITE

$\propto\text{-}Fe^{+3}O(OH)$      Orthorhombic

Common.

## GOLD

Au      Cubic

Occurs as loose grains and scales in many river gravels, and within disseminated sulphides in a quartz vein at Cononish near Tyndrum, and Calliachar Burn near Aberfeldy, Perthshire.

## GOLDMANITE

$Ca_3(V,Al,Fe^{+3})_2(SiO_4)_3$      Cubic

A single grass-green grain from the detrital heavy mineral suite of a sandstone, Palaeocene in age, from the Montrose Group of the East Shetland Platform.

## GONNARDITE

$Na_2CaAl_4Si_6O_{20}.7H_2O$      Orthorhombic

Occurs with saponite in vesicular basalt, Allt Ribhein, Fiskavaig Bay, Skye, Inverness-shire.

## GORCEIXITE

$BaAl_3(PO_4)(PO_3OH)(OH)_6$   Monoclinic ps. trigonal

Microscopic, greenish, six-sided platy crystals with halloysite-(7Å) in a vein at Hospital Quarry, Elgin, Morayshire.

## GOSLARITE

$ZnSO_4.7H_2O$      Orthorhombic

Found as a feathery growth attached to the sides and timbers of some of the upper and dry old levels in the various lead-zinc mines at Leadhills, Lanarkshire and Wanlockhead, Dumfriesshire.

## GOYAZITE

$SrAl_3(PO_4)_2(OH)_5.H_2O$      Trigonal

In the clay fraction of a soil derived from Middle Old Red Sandstone rocks and collected from the side of the B9163 road near the junction with road B9169, north-west side of the Black Isle, Ross and Cromarty.

## *Goyazite* group

A cryptocrystalline, yellow-brown phosphate found in a clay rock occurring a short distance above the Pinkie Four Foot Coal in bores in the Musselburgh area, Edinburgh, Midlothian.

## GRAFTONITE

$(Fe^{+2},Mn^{+2},Ca)_3(PO_4)_2$      Monoclinic

Grey grains (<4mm) associated with other iron-manganese phosphates in small (15-30cm) pods within Moine striped gneisses exposed near the entry of Glen Chosaidh, Loch Quoich, Inverness-shire.

## GRANDIDIERITE

$(Mg,Fe^{+2})Al_3(BO_4)(SiO_4)O$      Orthorhombic

As pale-blue to colourless grains in a biotite-rich pelite xenolith in norite in the Haddo House complex and green-blue plates in the inner contact aureole of the Comrie granodiorite, Aberdeenshire.

## GRAPHITE

C      Hexagonal and trigonal

In veins, schists and marbles from numerous localities, especially schists from several locations in Perthshire.

## GREENALITE

$(Fe^{+2},Fe^{+3})_{2-3}Si_2O_5(OH)_4$      Monoclinic

As microscopic rounded or irregular greenish granules in Ordovician chert, Glenluce, Wigtownshire.

## GREENOCKITE

CdS      Hexagonal

Associated with prehnite, natrolite, analcime, thomsonite and calcite in material from Bishopton Railway Tunnel, Bishopton, Renfrewshire.

## GREIGITE

$Fe^{+2}Fe_2^{+3}S_4$      Cubic

Present as a soft, black corrosion product on a gold-garnet sword pommel unearthed from soil in East Fife.

## GROSSULAR

$Ca_3Al_2(SiO_4)_3$       Cubic

In contact metamorphosed limestones, especially at Dalnabo, Glen Gairn, Aberdeenshire, and numerous other localities.

## GROUTITE

$Mn^{+3}O(OH)$       Orthorhombic

Occurs in veins of manganese and iron oxides within the Jura Quartzite on the south-west coast of the Isle of Islay.

## GRUNERITE

$(Fe^{+2},Mg)_7Si_8O_{22}(OH)_2$       Monoclinic

Quartz-magnetite rocks, with grunerite, occur in the Lewisian Loch Maree Group in Flowerdale near Gairloch, Wester Ross.

## GUSTAVITE

$PbAgBi_3S_6$       Orthorhombic

Bladed aggregates up to 0.5cm and individual small blades are found in galena cubes occurring in massive quartz from the Corrie Buie veins, Meal nan Oighreag, Perthshire.

## GYPSUM

$CaSO_4.2H_2O$       Monoclinic

As satin spar traversing clays in the banks of the Whiteadder River, Berwickshire, and at numerous other localities.

## GYROLITE

$NaCa_{16}(Si_{23}Al)O_{60}(OH)_5.15H_2O$    Triclinic
                       ps. hexagonal

As a pearly-white to creamy-white amygdale mineral, associated with zeolites, in basalts of Tertiary age, Skye, Inverness-shire.

## HALITE

$NaCl$       Cubic

In cavities in oil shales, associated with barite and calcite, at Midcalder, Pumpherston, Midlothian.

## HALLOYSITE

$Al_2Si_2O_5(OH)_4$       Monoclinic

Found in soils derived from gabbro, around Insch, Aberdeenshire.

## HALOTRICHITE

$Fe^{+2}Al_2(SO_4)_4.22H_2O$       Monoclinic

Silky, fibrous growths derived from pyrite from a railway cutting, west of Stanely, Renfrewshire.

## HAMMARITE

$Pb_2Cu_2Bi_4S_9$       Orthorhombic

Forms coarse, fibrous crystals in quartz veins, hosted by the Loch Shin monzogranite, 2km west of Lairg, Sutherland.

## HARKERITE

$Ca_{24}Mg_8Al_2(SiO_4)_8(BO_3)_6(CO_3)_{10}.2H_2O$    Cubic

Associated with ludwigite, fluoborite, datolite and szaibelyite in dolomite contact skarns, Broadford area, Skye, Inverness-shire.

## HARMOTOME

$(Ba,K)_{1-2}(Si,Al)_8O_{16}.6H_2O$       Monoclinic

Colourless to milky-white crystals, up to 2.5cm long, associated with calcite, barite, and rarely brewsterite, found in the lead mines at Strontian, Argyllshire.

## HASTINGSITE
### potassian, magnesian

$NaCa_2(Fe^{+2},Mg)_4Fe^{+3}(Si_6Al_2)O_{22}(OH)_2$   Monoclinic

Widespread in the Loch Eil and Glenfinnan divisions of the Moine west of the Great Glen, Inverness-shire.

## HAUSMANNITE

$Mn^{+2}Mn_2^{+3}O_4$       Tetragonal

Lustrous, brownish-black crystals up to 2mm, with barite, in an area 3 x 1cm in massive manganese ore from Dalroy manganese mine, 10km east of Inverness, Inverness-shire.

## HEAZLEWOODITE

$Ni_3S_2$       Trigonal

Minute grains in chromitite, Hagdale Quarry, Unst, Shetland.

## HEDENBERGITE

$CaFe^{+2}Si_2O_6$       Monoclinic

Green porphyroblasts (<4mm) associated with aggregates of smaller crystals (<0.5mm) in hedenbergite-garnet-magnetite rocks, Loch Duich, Ross-shire.

## HEMATITE

$\alpha$-$Fe_2O_3$       Trigonal

Common.

## HEMIMORPHITE

$Zn_4Si_2O_7(OH)_2.H_2O$       Orthorhombic

Usually as glassy crystals, or botryoidal coatings on oxidised zone rocks, especially at Wanlockhead, Dumfriesshire.

## HERCYNITE

$Fe^{+2}Al_2O_4$                    Cubic

Deep-green, irregular microporphyroblastic grains (<0.4mm) often associated with iron-ore, in a silica-poor argillaceous hornfels, Sparcraigs, near Whitecairns, 13km north of Aberdeen, Aberdeenshire.

## HESSITE

$Ag_2Te$                    Monoclinic

Found as irregularly shaped grains, $c$.0.4mm, or as myrmekitic intergrowths in galena, in quartz-calcite-fluorite veins in the Loch Duich area, Ross and Cromarty.

## HEULANDITE (s.l.)

$(Na,Ca)_{2.3}Al_3(Al,Si)_2Si_{13}O_{36}.12H_2O$    Monoclinic

Red heulandite crystals, up to 4cm long, are associated with stilbite, calcite and quartz in lavas from Kilpatrick Hills, and Cochno, Dunbartonshire.

## HEXAHYDRITE

$MgSO_4.6H_2O$                    Monoclinic

As a pale-green secondary growth coating the surface of an exposed drill core (serpentinite) from North Ballaird bore no. 3, Balsalloch Farm, near Ballantrae, Ayrshire.

## HIBSCHITE

$Ca_3Al_2(SiO_4)_{3-x}(OH)_{4x}$            Cubic

In the gabbro-limestone contact zone of Camas Mòr, Muck, Inverness-shire.

## HINSDALITE

$(Pb,Sr)Al_3(PO_4)(SO_4)(OH)_6$        Trigonal

Forms small, ivory-coloured blebs and coatings on ferruginous 'gozzan-breccia' from the upper reaches of Penkiln Burn, 13km north-north-east of Newton Stewart, Kirkcudbrightshire.

## HOLLANDITE

$Ba(Mn^{+4},Mn^{+2})_8O_{16}$    Monoclinic ps. tetragonal

Found associated with coronadite, wad and pyrolusite in a manganese vein about 1km north-west of Arndilly House, Arndilly, Banffshire.

## HOLLINGWORTHITE

$(Rh,Pt,Pd)AsS$                    Cubic

Chromite samples from the disused quarries at Harold's Grave and Cliff, in the Baltasound – Haroldswick area, Unst, Shetland, contain hollingworthite in chloritised silicates.

## HONESSITE

$Ni_6Fe_2^{+3}(SO_4)(OH)_{16}.4H_2O$        Trigonal

Citron-yellow crusts, associated with green theophrastite on chromitite from Hagdale Quarry, Unst, Shetland, are a mixture of reevesite and honessite.

## HONGSHIITE

$PtCu(?)$                    Trigonal

Present as two grains or as a network of branches and enclosing a Ni-Cu alloy adjacent to a sperrylite crystal in chromitite from the Baltasound – Haroldswick area, Unst, Shetland.

## Hortonolite = magnesian manganoan FAYALITE

## HÜBNERITE

$Mn^{+2}WO_4$                    Monoclinic

As crystals up to 3cm long in a quartz-hübnerite vein in zinnwaldite granite near Gairnshiel Bridge, Glen Gairn, Aberdeenshire.

## HUMITE

$(Mg,Fe^{+2})_7(SiO_4)_3(F,OH)_2$        Orthorhombic

Associated with chondrodite, clinohumite, cuspidine, forsterite and monticellite in ore skarns at contacts of the Durness dolomite with the Beinn an Dubhaich granite, Broadford area, Skye, Inverness-shire.

## HYALOPHANE

$(K,Ba)Al(Si,Al)_3O_8$                Monoclinic

A minor constituent in strata-bound barium-zinc mineralisation in Dalradian schist, near Aberfeldy, Perthshire.

## HYDROCERUSSITE

$Pb_3(CO_3)_2(OH)_2$                Trigonal

Occurs as thin, scaly, pearly white coatings in cavities in galena, associated with cerussite, at Belton Grain Mine, Wanlockhead, Dumfriesshire.

## Hydrogrossular = HIBSCHITE

## HYDROHONESSITE

$Ni_6Fe_2^{+3}(SO_4)(OH)_{16}.7H_2O$        Hexagonal

With honessite and reevesite in citron-yellow crusts associated with theophrastite on chromitite from Hagdale Quarry, Unst, Shetland.

## HYDROMAGNESITE

$Mg_5(CO_3)_4(OH)_2.4H_2O$            Monoclinic

In the aureole beyond dolomite skarn zones of the Broadford area, Skye, Inverness-shire, and investing serpentinite, Hagdale, Nikka Vord, Unst, Shetland.

## HYDROTALCITE

$Mg_6Al_2(CO_3)(OH)_{16}.4H_2O$            Trigonal

Investing chromitite, from the south-west headwaters of Poundland Burn, Colmonell, Carrick, Ayrshire.

## HYDROTUNGSTITE

$H_2WO_4.H_2O$                        Monoclinic

As a minute bright-yellow, powdery coating on a hübnerite crystal in a quartz-hübnerite vein in zinnwaldite granite near Gairnshiel Bridge, Glen Gairn, Aberdeenshire.

## HYDROXYAPOPHYLLITE

$KCa_4Si_8O_{20}(OH,F).8H_2O$         Tetragonal

Occurs as linings to vugs in basalt from Quiraing, Dunvegan and Loch Bracadale, Skye, Inverness-shire.

## HYDROZINCITE

$Zn_5(CO_3)_2(OH)_6$                  Monoclinic

Forms a white encrustation on sphalerite and rock matrix from Wood of Cree Mine near Newton Stewart, Kirkcudbrightshire, and on the walls of some mine levels, and dump material at Leadhills, Lanarkshire and Wanlockhead, Dumfriesshire.

## Hypersthene = ENSTATITE

## IKAITE

$CaCO_3.6H_2O$                        Monoclinic

Elongate (19x2cm) pointed dark leathery-brown crystals (now calcite) with wrinkled surfaces and edges, from muds in the River Clyde at Cardross, Dunbartonshire, formed originally as ikaite.

## ILLITE

$(K,H_3O)(Al,Mg,Fe)_2(Si,Al)_4O_{10}(OH)_2.H_2O$
                                      Monoclinic

In a crush band in decomposed granite 1.6km north-east of Ballater, Aberdeenshire. (Used as a species since undifferentiated.)

## ILMENITE

$Fe^{+2}TiO_3$                        Trigonal

Common.

## ILMENORUTILE

$(Ti,Nb,Fe^{+3})_3O_6$                Tetragonal

Present in a heavy mineral suite of stream sediments from Glen Lui, Cairngorms, Aberdeenshire, associated with fergusonite, columbite, monazite, xenotime and cassiterite.

## IMOGOLITE

$Al_2SiO_3(OH)_4$                     —

Within Scottish soils, as gel coatings, in various horizons and at numerous localities.

## IRARSITE

$(Ir,Ru,Rh,Pt)AsS$                    Cubic

Occurs with other platinum group minerals as inclusions in chromite from the disused quarries at Harold's Grave and Cliff in the Baltasound – Haroldswick area, Unst, Shetland.

## Iridosmine – OSMIUM

## ISOMERTIEITE

$Pb_{11}Sb_2As_2$                     Cubic

Within an area of $20\mu m^2$ two isomertieite grains were found to be associated with sperrylite and Pb,Ag,Bi and Pd tellurides in a clinopyroxenite from the Loch Ailsh alkaline igneous complex, 25km north-east of Ullapool.

## Jacobsite

$(Mn^{+2},Fe^{+2},Mg)(Fe^{+3},Mn^{+3})_2O_4$   Cubic

An unidentified iron-manganese oxide thought to be jacobsite occurs in a mixture of oxides and silicates in the manganese mineralisation at Dalroy Burn, Dalroy, Inverness-shire.

## JAHNSITE

$CaMn(Mg,Fe^{+2})_2Fe_2^{+2}(PO_4)_4(OH)_2.8H_2O$
                                      Monoclinic

Rare small (<1mm) brown anhedral grains in small 15-30cm pods within the Moine striped kyanite-bearing gneiss, Loch Quoich, Inverness-shire.

## Jamesonite

$Pb_4FeSb_6S_{14}$                    Monoclinic

Reported in the Leadhills, Lanarkshire and Wanlockhead, Dumfriesshire orefield and in the Glendinning area, Dumfriesshire, but requires confirmation.

## JAROSITE

$K_2Fe_6^{+3}(SO_4)_4(OH)_{12}$       Trigonal

Yellowish earthy patches, or pale-yellow crystals in sulphide-rich highly altered gneisses, Loch Garbhaig, Letterewe Forest, Loch Maree, Ross and Cromarty.

## JIMTHOMPSONITE

$(Mg,Fe^{+2})_5Si_6O_{16}(OH)_2$      Orthorhombic

Occurs intergrown with clinojimthompsonite and chesterite polytype on actinolite cores in Lewisian amphibolite facies ultramafic rocks near Achmelvich, Lochinver, Sutherland.

## Johannsenite

$CaMn^{+2}Si_2O_6$                    Monoclinic

Reported to occur in the manganese deposit at Dalroy Burn, Dalroy, Inverness-shire. This mineral requires confirmation.

## JOHNSOMERVILLEITE

$Na_2Ca(Mg,Fe^{+2},Mn)_7(PO_4)_6$     Trigonal

Vitreous dark-brown grains (<1.5mm) in small 15-30cm pods within the Moine striped kyanite-bearing gneiss, Loch Quoich, Inverness-shire.

## JOSEITE-B

$Bi_4Te_2S$                                   Trigonal

Forms inclusions in bismuthinite found in quartz veins cutting greywacke peripheral to a porphyry complex at Cairngarroch Bay, south of Portpatrick, Wigtownshire.

## JULGOLDITE

$Ca_2Fe^{+2}(Fe^{+3},Al)_2(SiO_4)(Si_2O_7)(OH)_2.H_2O$

Monoclinic

Black aggregates (<1cm) or plumose clusters in, or on, pectolite from the dolerite quarries, Ratho, near Edinburgh, Midlothian; also similarly at Auchinstarry, Kilsyth, Stirlingshire.

## KAERSUTITE

$NaCa_2(Mg,Fe^{+2})_4Ti(Si_6Al_2)O_{22}(OH)_2$    Monoclinic

Forms large phenocrysts in a monchiquite, Wart Holm, Copinshay, Orkney.

## KAHLERITE

$Fe^{+2}(UO_2)_2(AsO_4)_2.10\text{-}12H_2O$          Tetragonal

Present as well-crystallized encrustations on joint and cavity surfaces in shaly to slaty country rock adjacent to U-Cu-As-Bi-Co mineralisation near Dalbeattie, Kirkcudbrightshire.

## KAOLINITE

$Al_2Si_2O_5(OH)_4$                          Triclinic

Forms thin beds at Lamb Hoga, Fetlar, Shetland.

## KASOLITE

$Pb(UO_2)SiO_4.H_2O$                     Monoclinic

With pitchblende at Tyndrum, Perthshire.

## *Kasolite*-like mineral

Predominantly a thorium-lead silicate surrounding uraninite in the Chaipaval granite pegmatite, South Harris, Inverness-shire.

## *Kermesite*

$Sb_2S_2O$                         Triclinic ps. monoclinic

Investing stibnite at Hare Hill, New Cumnock, Ayrshire. x-ray diffraction revealed it to be non-crystalline. This mineral requires confirmation.

## KESTERITE

$Cu_2(Zn,Fe)SnS_4$                           Tetragonal

Rare stannite-group minerals form symplectite-like intergrowths within chalcopyrite from silicified greissen are closer to kesterite in composition than stannite; from hübnerite vein, Gairnshiel Bridge, Glen Gairn, Aberdeenshire.

## KILCHOANITE

$Ca_3Si_2O_7$                                Orthorhombic

Crystals (<2mm) of kilchoanite have replaced rankinite in limestones thermally metamorphosed by gabbro at Kilchoan, Ardnamurchan, Argyllshire.

## Knebelite = manganoan FAYALITE

## KOECHLINITE

$Bi_2MoO_6$                                  Orthorhombic

Associated with cassiterite, acanthite, cosalite and molybdenite in wall rocks of a quartz-hübnerite vein near Gairnshiel Bridge, Glen Gairn, Aberdeenshire.

## KRUPKAITE

$PbCuBi_3S_6$                                Orthorhombic

As coarse, fibrous crystals in quartz veins hosted by the Loch Shin monzogranite, 2km west of Lairg, Sutherland.

## KYANITE

$Al_2SiO_5$                                  Triclinic

Common, especially in high-grade, regionally metamorphosed rocks.

## LABRADORITE

(An50-70)                                    Triclinic

Common, especially in gabbroic rocks of Rhum and Skye, Inverness-shire.

## LANARKITE

$Pb_2(SO_4)O$                                Monoclinic

One of the rarest lead secondary minerals in the Leadhills, Lanarkshire and Wanlockhead, Dumfriesshire orefield.

## LANGITE

$Cu_4^{+2}(SO_4)(OH)_6.2H_2O$                Monoclinic

Forms blue scales or tiny (0.3mm) clear-blue tabular crystals associated with posnjakite, brochantite, serpierite and wroewolfeite on sulphide ore, West Blackcraig Mine dumps, near Newton Stewart, Kirkcudbrightshire.

## LARNITE

$\beta\text{-}Ca_2SiO_4$                     Monoclinic

As drop-like grains in rankinite in a gehlenite-rankinite-larnite-spinel assemblage in the gabbro-limestone contact zone of Camas Mòr, Muck, Inverness-shire.

## LAUMONTITE

$CaAl_2Si_4O_{12}.4H_2O$ — Monoclinic

Occurs with tobermorite, gyrolite and mesolite on Mull, Argyllshire and on Skye, Inverness-shire, and at several localities in andesitic basalts near Glasgow.

## LAURITE

$RuS_2$ — Cubic

Found as two grains in chromite from the harzburgite west of Baltasound, Unst, Shetland.

## LAVENDULAN

$NaCaCu_5^{+2}(AsO_4)_4Cl.5H_2O$ — Orthorhombic

Bright-blue spherulitic aggregates (0.12-0.25mm) in vugs in altered radioactive rock, Southwick cliffs area, near Dalbeattie, Kirkcudbrightshire.

## LEADHILLITE

$Pb_4(SO_4)(CO_3)_2(OH)_2$ — Monoclinic

Has been found in all the Leadhills veins in Lanarkshire, especially the Susanna Vein, and also at Wanlockhead, Dumfriesshire.

## LEPIDOCROCITE

$\gamma$-$Fe^{+3}O(OH)$ — Orthorhombic

A lepidocrocite-rich fraction was obtained from clay material forming the iron-manganese pan of an imperfectly drained, brown forest soil from Blairmore Estate, near Huntly, Aberdeenshire.

## LEPIDOLITE

$K(Li,Al)_3(Si,Al)_4O_{10}(F,OH)_2$ — Monoclinic

Lilac-coloured, scaly aggregates (<3mm) in tourmaline-bearing pegmatite, Glenbuchat, Aberdeenshire.

## Lepidomelane = ferrian BIOTITE

## LEUCOPHOSPHITE

$KFe_2^{+3}(PO_4)_2(OH).2H_2O$ — Monoclinic

Waxy, orange coloured isotropic grains and pale-yellow fibrous crystals (<0.003mm) in granite, Rockall, Inverness-shire.

## LEVYNE (s.l.)

$(Ca,Na_2,K_2)Al_2Si_4O_{12}.6H_2O$ — Trigonal

Small (1-2mm) milky-white crystals occur in vesicles in basalts of Tertiary age at Storr, and with fibrous offretite overgrowths at Quiraing, Skye, Inverness-shire.

## LIEBIGITE

$Ca_2(UO_2)(CO_3)_3.11H_2O$ — Orthorhombic

Secondary uranium minerals associated with calcite and sulphides include zeunerite, kasolite, uranospinite and liebigite at Tyndrum, Perthshire.

## Lillianite homologues

Present as bladed aggregates, up to 0.5cm, in galena from the Corrie Buie veins, Meal nan Oighreag, Perthshire.

## Limonite = mostly GOETHITE

## LINARITE

$PbCu^{+2}(SO_4)(OH)_2$ — Monoclinic

Especially from Lady Anne Vein, Brown's Vein and the Susanna Vein at Leadhills, Lanarkshire and the High Pirn Mine, Belton Grain Vein, Wanlockhead, Dumfriesshire.

## LINDSTRÖMITE

$Pb_3Cu_3Bi_7S_{15}$ — Orthorhombic

Associated with aikinite, hammarite, krupkaite, gladite and pekoite in quartz veins hosted by the Loch Shin monzogranite, 2km west of Lairg, Sutherland.

## LINNAEITE

$Co^{+2}Co_2^{+3}S_4$ — Cubic

As a replacement of pyrite in cromaltite, near Bad na L'Achlaise, Ledmore, Assynt, Sutherland.

## LITHARGE

$PbO$ — Tetragonal

Associated with massicot, cerussite and elyite in a sample from a stream bed, Leadhills, Lanarkshire.

## LITHIOPHORITE

$(Al,Li)Mn^{+4}O_2(OH)_2$ — Trigonal

Lithiophorite-cryptomelane mixtures occur as massive black, fine-grained aggregates at the Lecht mines, Tomintoul, Banffshire.

## LIZARDITE

$Mg_3Si_2O_5(OH)_4$ — Trigonal and hexagonal

In the 'bastite'-serpentinite mass in Glen Urquhart, Inverness-shire.

## LOELLINGITE

$FeAs_2$ — Orthorhombic

As inclusions in dolomite in uranium-bearing veins cutting the aureole of the Criffel granodiorite, Southwick, near Dalbeattie, Kirkcudbrightshire.

## LUDWIGITE

$Mg_2Fe^{+3}BO_5$ — Orthorhombic

Associated with fluoborite, datolite, szaibelyite and harkerite at the contact of dolomite and granite, Beinn an Dubhaich, Skye, Inverness-shire.

## MACAULAYITE

$(Fe^{+3},Al)_{24}Si_4O_{43}(OH)_2$     Monoclinic

Forms a pigmentary material associated with kaolinite and illite and found as bright-red patches in an outcrop of deeply weathered granite some 7km west-north-west of Inverurie, Aberdeenshire.

## MACKINAWITE

$(Fe,Ni)_9S_8$     Tetragonal

Included in pentlandite as anhedral crystals (50-100µm); more rarely in chalcopyrite, in allivalite, Huntly, Aberdeenshire.

## MACPHERSONITE

$Pb_4(SO_4)(CO_3)_2(OH)_2$     Orthorhombic

Present as a dumb-bell-shaped composite crystal projecting from a cavity in quartz and associated with leadhillite, susannite, cerussite, caledonite, pyromorphite and mattheddleite, Leadhills, Lanarkshire.

## MAGHEMITE

$\gamma\text{-}Fe_2O_3$     Cubic

In altered basalts from the British Tertiary Province occurring on the islands of Rhum, Eigg, Canna and Muck, Inverness-shire.

## MAGNESIOHORNBLENDE

$Ca_2(Mg,Fe^{+2})_4Al(Si_7Al)O_{22}(OH,F)_2$     Monoclinic

Found in a coarse appinite in the Glen Tilt complex, Perthshire.

## MAGNESIORIEBECKITE

$Na_2(Mg,Fe^{+2})_3Fe_2^{+3}Si_8O_{22}(OH)_2$     Monoclinic

In blue felted aggregates, with aegirine and hematite, infilling joints and shears in the Inverness area, Inverness-shire.

## MAGNESIOTARAMITE

$Na_2Ca(Mg,Fe^{+2})_3Al_2(Si_6Al_2)O_{22}(OH)_2$     Monoclinic

In an orthopyroxene-clinopyroxene amphibole rock at the head of Glen Rodel, South Harris, Inverness-shire.

## MAGNESITE

$MgCO_3$     Trigonal

Pale-brown, pearly rhombohedra of magnesite with chlorite and talc occur at North Cross Geo, Haroldswick, Unst, Shetland.

## MAGNETITE

$Fe^{+2}Fe_2^{+3}O_4$     Cubic

Common, especially as an accessory in many rock types.

## MALACHITE

$Cu_2^{+2}(CO_3)(OH)_2$     Monoclinic

Coats cerussite from Brown's Vein, Glengonner, Leadhills, Lanarkshire, and common as a 'green stain'.

## MANGANITE

$Mn^{+3}O(OH)$     Monoclinic

Forms crystals (<2mm) in cavities throughout massive pyrolusite, Bridge of Don, near Aberdeen, Aberdeenshire.

## MANGANOTANTALITE

$Mn^{+2}Ta_2O_6$     Orthorhombic

As iridescent, very dark, brown grains (<1mm) in lepidolite-bearing pegmatite, Glenbuchat, Aberdeenshire.

## MARCASITE

$FeS_2$     Orthorhombic

Forms a cockscomb development on psammite from Netherglen Quarry, 9.5km south of Elgin, Moray.

## MARGARITE

$CaAl_2(Al_2Si_2)O_{10}(OH)_2$     Monoclinic

Occurs both as a primary phase, and as an alteration product of kyanite, in schists in Aberdeenshire, Banffshire and Perthshire.

## MARIALITE

$3NaAlSi_3O_8.NaCl$     Tetragonal

Forms a minor constituent of garnet-amphibolites, Beinn a Chapuill, Glenelg, Inverness-shire.

## Martite = HEMATITE

Pseudomorphous after **magnetite**.

## MASCAGNITE

$(NH_4)_2SO_4$     Orthorhombic

Fibrous and mealy encrustations in shale heaps, Emily coal pit, Arniston, Midlothian.

## MASSICOT

$PbO$     Orthorhombic

Soft, cream-yellow earthy material, associated with litharge, cerussite and elyite in a sample from a stream bed, Leadhills, Lanarkshire.

## MATILDITE

$AgBiS_2$     Hexagonal

Matildite and galena, formed by breakdown of a complex AgBiPbS phase in quartz veins in the monzonite of the Ratagan complex, Loch Duich area, Ross-shire.

## MATTHEDDLEITE

$Pb_5(Si_{1.5},S_{1.5})_3O_{12}(Cl,OH)$      Hexagonal

Tiny crystals (100μm x 10-20μm) line cavities in quartz and associated with macphersonite, leadhillite, susannite and other lead secondary minerals from Leadhills, Lanarkshire.

## MAUCHERITE

$Ni_{11}As_8$      Tetragonal

Occurs as grains up to a few tens of microns across with pentlandite, godlevskite and heazlewoodite in chromitite and chromite-bearing rocks from the Cliff region in the Baltasound - Haroldswick area, Unst, Shetland.

## MEIONITE

$3CaAl_2Si_2O_8.CaCO_3$      Tetragonal

Anhedral crystals enclosing other minerals, sometimes apatite, in symplectite-bearing nodules in the Ardgour marble, Coire Dubh, Argyllshire.

## Melanite = titanian ANDRADITE

## MELANTERITE

$Fe^{+2}SO_4.7H_2O$      Monoclinic

A yellowish-green stalactitic growth contains melanterite and rozenite, West Mains coal mine, West Calder, West Lothian.

## MERCURY

$Hg$      —

Tiny particles in close association with acanthite from dump material at the Hilderston silver mine, 3km north-east of Bathgate, West Lothian.

## MERTIEITE (or/stibiopalladinite)

Found in chromite as grains (9 x 4μm) in the disused chromite quarries at Cliff and Harold's Grave in the Baltasound - Haroldswick area, Unst, Shetland.

## MERWINITE

$Ca_3Mg(SiO_4)_2$      Monoclinic

Present as interpenetrant twinned laths (0.25mm) set in calcite, in a monticellite and gehlenite-bearing marble in the gabbro-limestone contact zone, Camas Mòr, Muck, Inverness-shire.

## MESOLITE

$Na_2Ca_2Al_6Si_9O_{30}.8H_2O$      Monoclinic

Thin, hair-like mesolite needles are common in amygdales in the basalts of Tertiary age, Skye, Inverness-shire.

## META-AUTUNITE

$Ca(UO_2)_2(PO_4)_2.2\text{-}6H_2O$      Tetragonal

Occurs with metatorbernite, quartz and barite, in boulders believed to be derived from an adjacent silicified fracture zone 2.4km north of Helmsdale, Sutherland.

## METAKAHLERITE

$Fe^{+2}(UO_2)_2(AsO_4)_2.8H_2O$      Tetragonal

As yellow scales closely associated with kahlerite and other secondary uranium minerals in veins cutting the aureole of the Criffel granodiorite, near Dalbeattie, Kirkcudbrightshire.

## METATORBERNITE

$Cu^{+2}(UO_2)_2(PO_4)_2.8H_2O$      Tetragonal

Occurs with meta-autunite, quartz and barite, in boulders believed to be derived from an adjacent silicified fracture zone 2.4km north of Helmsdale, Sutherland.

## METAZEUNERITE

$Cu^{+2}(UO_2)_2(AsO_4)_2.8H_2O$      Tetragonal

Found with meta-autunite, uranophane and schoepite in pitchblende in veins cutting the aureole of the Criffel granodiorite, Dalbeattie, Kirkcudbrightshire.

## MICROCLINE

$KAlSi_3O_8$      Triclinic

Common, especially in granite.

## MILLERITE

$NiS$      Trigonal

Hair-like crystals form a fan (7mm) in a cavity, associated with calcite and chalcopyrite, Docra Quarry, Beith, Ayrshire.

## MIMETITE

$Pb_5(AsO_4)_3Cl$      Hexagonal

Occurs as an orange-red encrustation on quartz, Belton Grain Vein, Wanlockhead, Dumfriesshire.

## MINIUM

$Pb_2^{+2}Pb^{+4}O_4$      Tetragonal

Found in the Susanna Vein, Leadhills, Lanarkshire.

## *Mirabilite*

$Na_2SO_4.10H_2O$      Monoclinic

In the Hurlet limestone, near Glasgow. This mineral requires confirmation.

## MITRIDATITE

$Ca_2Fe_3^{+3}(PO_4)_3O_2.3H_2O$      Monoclinic

As earthy dark yellow-green coatings on iron-manganese-phosphate minerals in small 15-30cm pods within the Moine striped kyanite-bearing gneiss, Loch Quoich, Inverness-shire.

## MIXITE

$BiCu_6^{+2}(AsO_4)_3(OH)_6 \cdot 3H_2O$     Hexagonal

Present as vitreous grass-green prismatic alterations from native bismuth, and as emerald-green mammillary growths within fractures, Needle's Eye, Southwick, near Dalbeattie, Kirkcudbrightshire.

## MOLYBDENITE

$MoS_2$     Hexagonal

In quartz or quartzo-feldspathic rocks and veins, especially at Tomnadashan, Perthshire and Hestan Island, Auchencairn Bay, Kirkcudbrightshire.

## MONAZITE - (CE)

$(Ce,La,Nd,Th)PO_4$     Monoclinic

As crystals (<6mm) in Lewisian acid pegmatites of South Harris, Inverness-shire.

## MONOHYDROCALCITE

$CaCO_3 \cdot H_2O$     Hexagonal

Forms clear, prismatic crystals, trigonal in section, with a glassy lustre in cavities in slag from the Meadowfoot smelter dumps, Wanlockhead, Dumfriesshire.

## MONTICELLITE

$CaMgSiO_4$     Orthorhombic

With periclase or brucite in marbles in the gabbro-limestone contact zone Camas Mòr, Muck, Inverness-shire.

## MONTMORILLONITE

$(Na,Ca)_{0.3}(Al,Mg)_2Si_4O_{10}(OH)_2 \cdot nH_2O$ Monoclinic

Occurs with kaolinite, quartz, feldspar, calcite and minor zeolites in brown and grey unctuous laminated argillaceous rocks from the Ardtun Leaf Beds, Slochd an Uruisge and Bremanoir, southwest Mull, Argyllshire.

176

## MORDENITE

$(Ca,Na_2,K_2)Al_2Si_{10}O_{24} \cdot 7H_2O$     Orthorhombic

Salmon-coloured acicular crystals, up to 2cm long, with calcite and quartz found near Tinkletop, Ballindean, Inchture, Perthshire.

### *Morenosite*

$NiSO_4 \cdot 7H_2O$     Orthorhombic

Possibly present as an alteration product of millerite. This mineral requires confirmation.

## MOTTRAMITE

$PbCu^{+2}(VO_4)(OH)$     Orthorhombic

Found in the Scar Vein, Leadhills Dod, Lanarkshire and Belton Grain Vein, Wanlockhead, Dumfriesshire.

## MULLITE

$Al_6Si_2O_{13}$     Orthorhombic

In fused phyllite occurring as inclusions in igneous intrusions, especially on the northern side of Loch Scridain, Mull, Argyllshire.

## MUSCOVITE

$KAl_2(Si_3Al)O_{10}(OH,F)_2$     Monoclinic ps. hexagonal

Common.

## NACRITE

$Al_2Si_2O_5(OH)_4$     Monoclinic

Occurs in orange and white clay gouges just northwest of the Southern Uplands Fault which was exposed during excavation work for the A74 road near Abington, Lanarkshire.

## NAKAURIITE

$Cu_8^{+2}(SO_4)_4(CO_3)(OH)_6 \cdot 48H_2O$     Orthorhombic

Occurs as small amounts of pale-blue fibrous material, interbedded with chrysotile, in Hagdale Quarry, Unst, Shetland.

## NATROLITE

$Na_2Al_2Si_3O_{10} \cdot 2H_2O$     Orthorhombic

Forms doubly terminated crystals up to 8cm long, with calcite, prehnite and laumontite from Boyleston Quarry, Renfrewshire.

## NEPHELINE

$(Na,K)AlSiO_4$     Hexagonal

A minor though essential component of some rocks, and associated with kaersutite, titanaugite and labradorite, in the Lugar sill, Ayrshire. Also in the sodalite-bearing phonolitic analcime - trachyte of Traprain Law, near Haddington, East Lothian.

## Niccolite = NICKELINE

## NICKELINE

$NiAs$     Hexagonal

Found in the New Glencrieff Vein, Glencreiff Mine, Wanlockhead, Dumfriesshire, and at Hilderston silver mine, Linlithgow, West Lothian.

## NONTRONITE

$Na_{0.3}Fe_2^{+3}(Si,Al)_4O_{10}(OH)_2 \cdot nH_2O$     Monoclinic

Occurs as a fibrous pale olive-green alteration product from actinolite in syenite, Noss Hill, South Mainland, Shetland.

## NORBERGITE

$Mg_3(SiO_4)(F,OH)_2$     Orthorhombic

Forms waxy yellow-brown crystals (<12mm) with arsenopyrite in crystalline limestone, Loch Ness, Inverness-shire.

## NORSETHITE

$BaMg(CO_3)_2$     Trigonal

Found in sulphide, barite and carbonate rocks from the Foss deposit, Aberfeldy, Perthshire.

## *Nosean*

Reported to occur in assyntite, 0.8km east of Ledbeg Cottage, Assynt, Sutherland. This mineral requires confirmation.

## NOVACEKITE

$Mg(UO_2)_2(AsO_4)_2.12H_2O$      Tetragonal

Present as greenish-yellow, botryoidal crusts and flakes, covering several square millimetres on pitchblende from Needle's Eye, Southwick, Dalbeattie area, Kirkcudbrightshire.

## OFFRETITE

$(K_2,Ca)_5Al_{10}Si_{26}O_{72}.30H_2O$      Hexagonal

Fibrous outgrowths on levyne crystals occurring in vesicles, up to 1.5cm, in basalts at Quiraing and other areas, Skye, Inverness-shire.

## OKENITE

$Ca_{10}Si_{18}O_{46}.18H_2O$      Triclinic

In amygdales, associated with pectolite, tobermorite, reyerite, tacharanite and xonotlite in basalts of Tertiary age of Mull and Morven, Argyllshire.

## OLIGOCLASE

(An10-30)      Triclinic

Common.

## OLIVENITE

$Cu_2^{+2}(AsO_4)(OH)$      Orthorhombic

Found as small olive-green tufted groups of crystals associated with chrysocolla and leadhillite at Brown's Vein, Glengonner, Leadhills, Lanarkshire.

## OMPHACITE

$(Ca,Na)(Mg,Fe,Al)Si_2O_6$      Monoclinic

Anhedral grains associated with garnet in eclogites from near Glenelg, Inverness-shire.

## OPAL

$SiO_2.nH_2O$      –

Found as a white skin on an agate from Mull, Argyllshire, and as botryoids on calcite in three specimens from near Totternish, Isle of Skye, Inverness-shire.

## ORCELITE

$Ni_{5-x}As_2$      Hexagonal

Occurs as minute grains in chlorite haloes in chromitite, with pentlandite, in chromite-bearing rocks from the Cliff region in the Baltasound - Haroldswick area, Unst, Shetland.

## ORTHOCLASE

$KAlSi_3O_8$      Monoclinic

Common.

## OSMIUM

(Os,Ir,Ru)      Hexagonal

Osmium occurs as tiny inclusions in laurite grains in the rims of chromite from the disused chromite quarries at Harold's Grave and Cliff in the Baltasound - Haroldswick area, Unst, Shetland.

## OURAYITE

$Ag_3Pb_4Bi_5S_{13}(?)$      Orthorhombic

Present as bladed aggregates in galena cubes found in quartz from the Corrie Buie veins, Meal nan Oighreag, Perthshire.

## PALLADIUM

Pd      Cubic

Found with berzelianite as inclusions in two gold grains from the Smid Hope Burn tributary of the River Tweed near Tweedhope, Peeblesshire.

## PALYGORSKITE

$(Mg,Al)_2Si_4O_{10}(OH).4H_2O$      Monoclinic and orthorhombic

Forms a large sheet, 45 x 33 x 2-3cm, from Burn of the Cairn, Cabrach, Aberdeenshire.

## PARAGONITE

$NaAl_2(Si_3Al)O_{10}(OH)_2$      Monoclinic

Found in red phyllites with quartz, muscovite, chlorite, hematite, rutile and apatite, near Dunoon, Argyllshire.

## PARALAURIONITE

PbCl(OH)      Monoclinic

Lath-like crystals up to several millimetres long by 0.05mm thickness and associated with lanarkite occur in cavities in slag from the Meadowfoot smelter dumps, Wanlockhead, Lanarkshire.

## PARATACAMITE

$Cu_2^{+2}Cl(OH)_3$      Trigonal

As a thin green coating on a small area (1 x 0.5mm) of calcite in a vein cutting the Portencorkrie diorite some 0.5km north-north-west of Barncorkrie Farm, Drummore, Wigtownshire.

## PARGASITE

$NaCa_2(Mg,Fe^{+2})_4Al(Si_6Al_2)O_{22}(OH)_2$      Monoclinic

In gneiss-like inclusions in the Tiree marble from the marble quarry, Balephetrish, Tiree, Argyllshire.

## PEARCEITE

$(Ag,Cu)_{16}As_2S_{11}$      Monoclinic

Occurs as inclusions (<100μm) in tennantite and galena in quartz veining in the monzonite of the Ratagan complex, Loch Duich area, Ross-shire.

## PECTOLITE

$NaCa_2Si_3O_8(OH)$       Triclinic

Forms a continuous greenish creamy-coloured lining to a cavity (15cm) in dolerite, comprised of 2cm pectolite fans with calcite, from a quarry north of the canal at Ratho, Midlothian.

## PEKOITE

$PbCuBi_{11}(S,Se)_{18}$       Orthorhombic

Associated with aikinite, hammarite, krupkaite, gladite and lindströmite in quartz veins hosted by the Loch Shin monzogranite, 2km west of Lairg, Sutherland.

## PENTLANDITE

$(Fe,Ni)_9S_8$       Cubic

With pyrite, pyrrhotite and chalcopyrite in chlorite schist near Eas a' Chosain Glen, 1.2km west-south-west of Inveraray, Argyllshire.

## PERICLASE

$MgO$       Cubic

Rounded yellow grains in the gabbro-limestone contact zone of Camas Mòr, Muck, Inverness-shire.

## PEROVSKITE

$CaTiO_3$       Orthorhombic ps. cubic

In the gabbro-limestone contact zone of Camas Mòr, Muck, Inverness-shire.

## PETZITE

$Ag_3AuTe_2$       Cubic

Forms rare inclusions (<20µm) with sylvanite in pyrite cemented by quartz in a brecciated host rock vein 50cm wide, near Tyndrum, Perthshire.

## PHENAKITE

$Be_2SiO_4$       Trigonal

Found as colourless crystals 1.5cm long, with microcline, albite, tourmaline and magnetite in pegmatitic veins in syenite, Sgor Chaonasaid, Ben Loyal, Sutherland.

## PHILLIPSITE (s.l.)

$(K,Na,Ca)_{1-2}(Si,Al)_8O_{16}.6H_2O$       Monoclinic

As an inclusion in apatite, and as a crust on a sapphire-bearing camptonite dyke from Loch Roag, Isle of Lewis, Ross and Cromarty.

## PHLOGOPITE

$KMg_3Si_3AlO_{10}(F,OH)_2$       Monoclinic

Common.

## PHOENICOCHROITE

$Pb_2(CrO_4)O$       Monoclinic

Found with pyromorphite, leadhillite and cerussite in the Hopeful Vein, Leadhills, Lanarkshire.

## PHOSGENITE

$Pb_2(CO_3)Cl_2$       Tetragonal

As colourless to yellow transparent crystals, up to 5 x 4mm, or as tabular crystals, with galena and sphalerite in cherty rock from Lossiemouth, Moray.

## PHOSPHOSIDERITE

$Fe^{+3}PO_4.2H_2O$       Monoclinic

Found as a minute patch in an iron-manganese-phosphate assemblage in small 15-30cm pods within the Moine striped kyanite-bearing gneiss, Loch Quoich, Inverness-shire.

## PICKERINGITE

$MgAl_2(SO_4)_4.22H_2O$       Monoclinic

Aggregates of pale-yellow fibrous crystals on slate, Dunkeld, Perthshire.

## PICROPHARMACOLITE

$H_2Ca_4Mg(AsO_4)_4.11H_2O$       Triclinic

Small, white spheres on pink barite matrix with erythrite from old dumps, Silver Glen, Alva, Clackmannanshire.

## PIGEONITE

$(Mg,Fe^{+2},Ca)(Mg,Fe^{+2})Si_2O_6$       Monoclinic

Occurs as unzoned phenocrysts in andesite, Mull, and in the pitchstone of Ardnamurchan, Argyllshire, and Rhum, Inverness-shire.

Pitchblende = massive URANINITE

## PLATINUM

$Pt$       Cubic

Platinum occurs with irarsite and an unnamed Rh-Sb-S phase in a composite platinum group mineral grain found in a chromite-rich sample from Jimmie's Quarry, near Hagdale, Unst, Shetland.

## PLATTNERITE

$PbO_2$       Tetragonal

Black botryoidal masses associated with yellow and green pryomorphite and cerussite in a 1.5 x 1.5cm cavity in quartz, Belton Grain Vein, Wanlockhead, Dumfriesshire.

## PLOMBIERITE

$Ca_5H_2Si_6O_{18}.6H_2O(?)$       —

Replaces larnite or merwinite in small patches in metamorphosed limestone at Kilchoan, Ardnamurchan, Argyllshire.

## PLUMBOGUMMITE

$PbAl_3(PO_4)_2(OH)_5.H_2O$      Trigonal

With plattnerite from the dumps of the Raik Vein, Leadhills, Lanarkshire, and with pyromorphite, mimetite and cotunnite, High Pirn Mine, Wanlockhead, Dumfriesshire.

## POLYBASITE

$(Ag,Cu)_{16}Sb_2S_{11}$      Monoclinic ps. hexagonal

Present with freibergite and gersdorffite in the gold-bearing Calliachar-Urlar veins 4km south-west of Aberfeldy, Perthshire.

## POSNJAKITE

$Cu_4^{+2}(SO_4)(OH)_6.H_2O$      Monoclinic

Dark-blue crusts associated with serpierite or as minute crystals (<0.3mm) in dump material from West Blackcraig Mine, Blackcraig, Kirkcudbrightshire.

## POTARITE

PdHg      Tetragonal

Found as a single grain within a chromite grain rim from the disused chromite quarry at Cliff, near the Baltasound - Haroldswick area, Unst, Shetland.

## POTASSIUM ALUM

$KAl(SO_4)_2.12H_2O$      Cubic

Occurs as an efflorescence on shales at Hurlet, Renfrewshire and Moffat, Dumfriesshire.

## POWELLITE

$CaMoO_4$      Tetragonal

A single greenish-black crystal (*c.*1cm) associated with apophyllite and analcime, in a geode in phonolite, Traprain Law, near Haddington, East Lothian.

## PREHNITE

$Ca_2Al_2Si_3O_{10}(OH)_2$      Orthorhombic

In altered lavas and gabbros, especially as large pale- green mamillations in druses at Boyleston Quarry, Barrhead, Renfrewshire.

## PSEUDOBROOKITE

$(Fe^{+3},Fe^{+2})_2(Ti,Fe^{+3})O_5$      Orthorhombic

Associated with corundum, magnetite, spinel and mullite in emery-like rocks adjacent to a dolerite plug, Sithean Sluaigh, Strachur, Loch Fyne, Argyllshire.

## PSEUDOMALACHITE

$Cu_5^{+2}(PO_4)_2(OH)_4$      Monoclinic

Forms an olive-green to blackish-green, partially vitreous lining to a 1cm-long cavity in quartz from the barite mine dump at Auchencairn, Rerrick, Kirkcudbrightshire.

## Psilomelane = ROMANECHITE

## Ptilolite = MORDENITE

## PUMPELLYITE - (MG)

$Ca_2MgAl_2(SiO_4)(Si_2O_7)(OH)_2.H_2O$      Monoclinic

Needle-like clusters with a stellate habit in quartz veins occupying sheer planes in Ordovician spilitic lavas near Knockormal Farm, 1.6km east of Lendalfoot, Ayrshire.

## PYRARGYRITE

$Ag_3SbS_3$      Trigonal

As grains in galena associated with sphalerite, chalcopyrite, pyrite and barite in a quartz gangue, Tyndrum, Perthshire.

## PYRITE

$FeS_2$      Cubic

Common.

## PYROAURITE

$Mg_6Fe_2^{+3}(CO_3)(OH)_{16}.4H_2O$      Trigonal

Found as cream to pale-brown soft coatings on serpentine from the serpentinite mass of Unst, Shetland.

## *Pyrochlore*

The mineral occurring in the Chaipaval pegmatite vein, South Harris, Inverness-shire, reported as 'a betafite-like mineral' is probably pyrochlore.

## PYROLUSITE

$Mn^{+4}O_2$      Tetragonal

Massive, with manganite and barite, found loose in till, Bridge of Don, Aberdeenshire. Also as a compact mass (9 x 7 x 4cm) of radiating crystals (<5cm) with minor quartz, Hoy, Orkney Islands.

## PYROMORPHITE

$Pb_5(PO_4)_3Cl$      Hexagonal

Common as a secondary mineral in the Leadhills, Lanarkshire and Wanlockhead, Dumfriesshire orefield.

## PYROPE

$Mg_3Al_2(SiO_4)_3$      Cubic

Occurs as isolated xenocrysts in the matrix of a volcanic agglomerate, which is rich in glassy alkali basalt fragments, from a vent at Elie, Fife.

## PYROPHANITE

$Mn^{+2}TiO_3$                    Trigonal

As grains in braunite, with manganoan-ilmenite, in the manganese ore at Dalroy Burn, Dalroy, Inverness-shire.

## PYROPHYLLITE

$Al_2Si_4O_{10}(OH)_2$           Monoclinic and triclinic

In kyanite-bearing rocks which normally contain small amounts of chloritoid, within regionally metamorphosed Dunrossness phyllites, Dunrossness, South Mainland, Shetland.

## PYROXMANGITE

$Mn^{+2}SiO_3$                    Triclinic

Pink grains (<5mm) in amphibole-garnet schist, Glen Beag, Glenelg, Inverness-shire.

## PYRRHOTITE

$Fe_{1-x}S$                       Monoclinic and hexagonal

Found associated with chalcopyrite and nickeline, in diorite at the old mine at Talnotry, 7km northeast of Newton Stewart, Kirkcudbrightshire.

## QUANDILITE

$(Mg,Fe^{+2})_2(Ti,Fe^{+2},Al)O_4$   Cubic

Extremely rare isolated euhedral crystals associated with geikielite, rutile, zircon and baddeleyite occur in metamorphosed siliceous dolomite of the Allt Guibhsachain area near Ballachulish, Argyllshire.

## QUARTZ

$SiO_2$                           Trigonal

Common.

## QUEITITE

$Pb_4Zn_2(SiO_4)(Si_2O_7)(SO_4)$   Monoclinic

As dark greyish-brown hemispheres up to 1mm, showing a fibrous, radial structure in quartz, from dump material at Horner's Vein, Leadhills, Lanarkshire.

## QUENSELITE

$PbMn^{+3}O_2(OH)$                Monoclinic

Quaternary gravels in a raised beach deposit at Luce Bay, Wigtownshire, are cemented mainly by wad which contains grains of birnessite, lithiophorite and quenselite.

## RAMMELSBERGITE

$NiS_2$                           Orthorhombic

Occurs as thin borders on nickeline from the Glencrieff Mine, Wanlockhead, Dumfriesshire, and from Menimuir Burn, Cassencarie, Kirkcudbrightshire.

## RAMSDELLITE

$Mn^{+4}O_2$                      Orthorhombic

Found as a minor component of small pod-like manganese oxides associated with outcrops of iron oxides in breccia about 8km south-east of Tomintoul, Banffshire.

## RANKINITE

$Ca_3Si_2O_7$                     Monoclinic

Forms rounded grains, often including laths of gehlenite, in the gabbro-limestone contact zone of Camas Mòr, Muck, Inverness-shire.

## REEVESITE

$Ni_6Fe_2^{+3}(CO_3)(OH)_{16}\cdot4H_2O$   Trigonal

Yellow, very fine-grained encrustations, associated with theophrastite, on chromitite, Hagdale Quarry, Unst, Shetland.

## RETGERSITE

$NiSO_4\cdot6H_2O$                Tetragonal

Pale blue to blue-green crusts and minute curved or twisted crystals associated with annabergite and nickeline, Menimuir Burn, Cassencarie, Kirkcudbrightshire.

## REYERITE

$(Na,K)_2Ca_{14}(Si,Al)_{24}O_{58}(OH)_8\cdot6H_2O$   Trigonal

Coarse pale-green radiating aggregates of plates filling amygdales in basalts, associated with analcime, thomsonite and natrolite, 'S Airde Beinn, 5km west of Tobermory, Mull, Argyllshire.

## RHODOCHROSITE

$Mn^{+2}CO_3$                     Trigonal

Found loose, associated with hematite and barite, in an old manganese mine dump, Dalroy Burn, near Culloden Moor, Inverness-shire.

## RHODONITE

$(Mn^{+2},Fe^{+2},Mg,Ca)SiO_3$   Triclinic

As veins (<2.5cm) in an amphibole-garnet schist associated with pyroxmangite, Glen Beag, Glenelg, Inverness-shire.

## RICHTERITE

$Na_2Ca(Mg,Fe^{+2})_5Si_8O_{22}(OH)_2$   Monoclinic

Occurs with aegirine-augite, alkali feldspar, biotite, apatite and sphene in fenitised quartzites of Glas Choille, adjacent to the Borrolan complex, Assynt, Sutherland.

## RIEBECKITE

$Na_2(Fe^{+2},Mg)_3Fe_2^{+3}Si_8O_{22}(OH)_2$   Monoclinic

In granophyric riebeckite-felsite, at the east end of the Beorgs of Uyea, North Roe, Shetland.

## ROCKBRIDGEITE

$(Fe^{+2},Mn^{+2})Fe_4^{+3}(PO_4)_3(OH)_5$     Orthorhombic

Dark olive-green anhedral grains (<1mm) in an iron-manganese-phosphate assemblage in small 15-30cm pods within the Moine striped kyanite-bearing gneiss, Loch Quoich, Inverness-shire.

## ROMANECHITE

$(Ba,H_2O)(Mn^{+4},Mn^{+3})_5O_{10}$     Monoclinic

Black cryptocrystalline bands associated with manganite, in highly crystalline manganiferous blocks in a supraglacial till, Bridge of Don, Aberdeenshire.

## RÖMERITE

$Fe^{+2}Fe_2^{+3}(SO_4)_{4}.14H_2O$     Triclinic

Small clusters of minute amber-coloured crystals associated with halotrichite, pyrite, coquimbite and voltaite, old railway cutting west of Stanely, near Paisley, Renfrewshire.

## ROSCOELITE

$K(V^{+3},Al,Mg)_2(AlSi_3)O_{10}(OH)_2$     Monoclinic

Forms a cement in sandstone and replaces plagioclase in a lithic clast of granitic composition within Lower Old Red Sandstone at Gamrie Bay, 11km east of Banff, Banffshire.

## ROZENITE

$Fe^{+2}SO_4.4H_2O$     Monoclinic

Forms porcelain-white patches replacing a greenish melanterite stalactite, West Mains coal mine, West Calder, Midlothian.

## RUSSELLITE

$Bi_2WO_6$     Orthorhombic

Irregular patches of russellite have been identified from silicified wallrocks adjacent to hübnerite veins near Gairnshiel Bridge, Glen Gairn, Aberdeenshire.

## RUSTUMITE

$Ca_{10}(Si_2O_7)_2(SiO_4)Cl_2(OH)_2$     Monoclinic

Occurs in a zone between spurrite-dominant and kilchoanite-dominant assemblages in thermally metamorphosed Jurassic limestones at Kilchoan, Ardnamurchan, Argyllshire.

## RUTHENIRIDOSMINE

$(Ir,Os,Ru)$     Hexagonal

Three light-grey metallic grains found in an alluvial gold sample from two localities in the Ochil Hills, Central Lowlands.

## RUTILE

$TiO_2$     Tetragonal

Red-brown crystals (<7cm) in quartz with chlorite, Loch na Lairige, near Killin, Perthshire. Also as a common accessory mineral.

## SAFFLORITE

$(Co,Fe)As_2$     Orthorhombic

Tin-white patches associated with native silver crystals (2mm) and erythrite in barite gangue from an old mine dump, Silver Glen, Alva, Clackmannanshire.

## SAL-AMMONIAC

$NH_4Cl$     Cubic

Forms a grey-white scaly efflorescence on fused shale/coal fragments, Emily coal pit, Arniston, Midlothian.

## *Samarskite*

Metamict thorium, rare-earth niobium oxides, which may well be samarskite, occur in a corundum-bearing xenolithic dyke in the Loch Roag area of Lewis, Inverness-shire.

## SANIDINE

$KAlSi_3O_8$     Monoclinic

Occurs as large (~6cm) glassy phenocrysts or xenocrysts in a trachybasalt dyke cutting lavas at Bangley Quarry near Haddington, East Lothian. Also as megacrysts, particularly in alkali olivine basalts of East Fife.

## SAPONITE

$(Ca/2,Na)_{0.3}(Mg,Fe^{+2})_3(Si,Al)_4O_{10}(OH)_2.4H_2O$     Monoclinic

As brown unctuous material from Cathkin Quarry, Carmunnock, 3.2km south of Rutherglen, Lanarkshire, and as white infilling to amygdales in basalt at Storr, Skye, Inverness-shire.

## SAPPHIRINE

$(Mg,Al)_8(Al,Si)_6O_{20}$     Monoclinic

Associated with spinel and orthopyroxene in a hornfels occurring adjacent to an ultrabasic intrusion, Coire nam Muc, Ardgour, Argyllshire.

## SCAWTITE

$Ca_7Si_6O_{18}(CO_3).H_2O$     Monoclinic

Scaly aggregates replacing spurrite in late-stage veins in thermally metamorphosed Jurassic limestones, Kilchoan, Ardnamurchan, Argyllshire.

## SCHEELITE

$CaWO_4$     Tetragonal

Pale blue to green massive material in granite at Craignair Quarry, Dalbeattie, Kirkcudbrightshire.

## SCHIRMERITE

$Ag_3Pb_3Bi_9S_{18}$ to $Ag_3Pb_6Bi_7S_{18}$     Orthorhombic

In small fractures in galena, and with electrum, bismuth and emplectite, from old mine dumps at Corrie Buie, Meal nan Oighreag, Perthshire.

## SCHMIEDERITE

$Pb_2Cu_2^{+2}(Se^{+4}O_3)(Se^{+6}O_4)(OH)_4$  Monoclinic

Lath-like, blue-mauve crystals associated with barite, serpierite, caledonite and anglesite in cavities in slag from the Meadowfoot smelter dumps, Wanlockhead, Dumfriesshire.

## SCHOEPITE

$UO_3.2H_2O$  Orthorhombic

Lemon-yellow crystals (~0.1mm) in aggregates or massive fissure fillings and encrustations with pitchblende, in rock from Southwick cliffs area, Dalbeattie, Kirkcudbrightshire.

## SCHORL

$NaFe_3^{+2}Al_6(BO_3)_3Si_6O_{18}(OH)_4$  Trigonal

Deep blue-black crystals, commonly 3 x 1cm, associated with elbaite in quartz feldspar pegmatite, Glenbuchat, Aberdeenshire.

## SCHORLOMITE

$Ca_3Ti_2^{+4}(Fe_2^{+3}Si)O_{12}$  Cubic

With ferroan wollastonite and titanaugite in a pegmatite near Camphouse, Ardnamurchan, Argyllshire.

## SCOLECITE

$CaAl_2Si_3O_{10}.3H_2O$  Monoclinic

Scolecite frequently intergrown with mesolite occurs in the basalts of Tertiary age on Skye, Inverness-shire and as large radiating masses up to 8-10cm in the metamorphosed basalts at Maol nan Damh, Ben More, Mull, Argyllshire.

## SCORODITE

$Fe^{+3}AsO_4.2H_2O$  Orthorhombic

As an alteration of arsenopyrite in a tourmaline-arsenopyrite-pyrite-chalcopyrite vein associated with the Fore Burn igneous complex, Glenthraig, 9km south-west of Dalmellington, Ayrshire.

## SCOTLANDITE

$PbSO_3$  Monoclinic

Occurs as pale yellow to greyish white and colourless chisel-shaped crystals, up to 1mm long, in vuggy anglesite in a large cavity in barite from Leadhills, Lanarkshire.

## SEMSEYITE

$Pb_9Sb_8S_{21}$  Monoclinic

Intergrown crystals (0.5mm) in cavities in stibnite, from the antimony mine at Glendinning, Glenshanna Burn, Eskdale, Dumfriesshire.

## SENARMONTITE

$Sb_2O_3$  Cubic

Rare octahedral crystals (<0.5mm) in cavities in stibnite, Hare Hill, 5km south-east of New Cumnock, Ayrshire.

## SEPIOLITE

$Mg_4Si_6O_{15}(OH)_2.6H_2O$  Orthorhombic

Soft white powdery material in talc-carbonate rock, Swinna Ness, Unst, Shetland.

## SERPIERITE

$Ca(Cu^{+2},Zn)_4(SO_4)_2(OH)_6.3H_2O$  Monoclinic

Sky-blue, pearly, fibrous to lath-like aggregates partly investing specimens from mine dumps, Blackcraig, Kirkcudbrightshire.

## SIDERITE

$Fe^{+2}CO_3$  Trigonal

Found as a vein in the Lower Old Red Sandstone at Sandlodge Mine, Sandwick, Mainland, Shetland.

## SIEGENITE

$(Ni,Co)_3S_4$  Cubic

Fine-grained bronze-like patches in mine dump material from Blackcraig, Kirkcudbrightshire.

## SILLIMANITE

$Al_2SiO_5$  Orthorhombic

Common in gneisses and migmatites in the Grampian Highlands and north-eastern Scotland.

## SILVER

$Ag$  Cubic and hexagonal

As dendrites on barite and calcite, with hydrocarbon, erythrite, copper minerals and safflorite, Silver Glen, Alva, Clackmannanshire.

## *Smaltite*

An arsenic-deficient variety of skutterudite, included here because only the variety has been found rather than the species.

## SMITHSONITE

$ZnCO_3$  Trigonal

In a mineralised fault breccia, associated with calcite, siderite and sulphides, including sphalerite, about 1.6km north of Loch Calder, Caithness.

## SODALITE

$Na_8Al_6Si_6O_{24}Cl_2$ Cubic

Associated with analcime in small irregular patches in the groundmass or as roundish patches with ophitic relations to the feldspars, in the phonolite at Traprain Law, near Haddington, East Lothian.

## SPERRYLITE

$PtAs_2$ Cubic

Anhedral grains in chromite-bearing samples from the disused chromite quarry near the Balta-sound – Haroldswick area, Unst, Shetland.

## SPESSARTINE

$Mn_3^{+2}Al_2(SiO_4)_3$ Cubic

In the Lewisian granite pegmatites of South Harris, Inverness-shire.

## SPHALERITE

$(Zn,Fe)S$ Cubic

Common, especially in the Leadhills, Lanarkshire and Wanlockhead, Dumfriesshire orefield.

## Sphene = TITANITE

## SPINEL

$MgAl_2O_4$ Cubic

Forms octahedra and clumps in the marbles of Glenelg, Inverness-shire and elsewhere as an accessory mineral.

## SPURRITE

$Ca_5(SiO_4)_2(CO_3)$ Monoclinic

As isolated grains in calcite or intimately intergrown with gehlenite in the gabbro-limestone contact zone of Camas Mòr, Muck, Inverness-shire.

## Stannite see KESTERITE

## STANNOIDITE

$Cu_8(Fe,Zn)_3Sn_2S_{12}$ Orthorhombic

In the hübnerite veins-silicified wall rock assemblages of a zinnwaldite-bearing granite, Gairnshiel Bridge, Glen Gairn, Aberdeenshire.

## STAUROLITE

$(Fe^{+2},Mg,Zn)_2Al_9(Si,Al)_4O_{22}(OH)_2$
Monoclinic ps. orthorhombic

Common in medium-grade schists derived from argillaceous sediments, found in Glen Urquhart, Inverness-shire, Glen Clova and Glen Isla, Angus.

## STELLERITE

$CaAl_2Si_7O_{18}.7H_2O$ Orthorhombic

Occurs with mordenite in brick-red amygdales in an olivine basalt at Todhead Point, 9km south of Stonehaven, Kincardineshire.

## STEVENSITE

$(Ca/?)_{0.3}Mg_3Si_4O_{10}(OH)_2$ Monoclinic

As a brown gel-like coating on pectolite, from Corstorphine Hill, Edinburgh, Midlothian.

## STIBICONITE

$Sb^{+3}Sb_2^{+5}O_6(OH)$ Cubic

Soft earthy yellow to white alteration product of stibnite, old mine dumps, Hare Hill, 5km south-east of New Cumnock, Ayrshire.

## Stibiopalladinite see MERTIEITE

## STIBNITE

$Sb_2S_3$ Orthorhombic

Plumose groups of stibnite occur in quartz, with dolomite and stibiconite, old mine dumps, Hare Hill, 5km south-east of New Cumnock, Ayrshire.

## Stichtite see BARBERTONITE

## STILBITE

$NaCa_2Al_5Si_{13}O_{36}.14H_2O$ Monoclinic and triclinic

Found in the Lower Carboniferous altered andesites and basalts around Glasgow and in the basalts of Tertiary age on Skye, Inverness-shire.

## STILPNOMELANE

$K(Fe^{+2},Mg,Fe^{+3})_8(Si,Al)_{12}(O,OH)_{27}$
Monoclinic and triclinic

Golden-brown in thin-section and associated with a muddy-brown biotite in a low-grade epidiorite from Cnoc-na-Faire, near Carsaig, north of Tayvallich, Argyllshire.

## STOLZITE

$PbWO_4$ Tetragonal

Occurs as a pseudomorphous replacement after hübnerite in a sericite greisen within a zinnwaldite-bearing granite, Gairnshiel Bridge, Glen Gairn, Aberdeenshire. Also as two grains from Wanlockhead, Dumfriesshire.

## STRONTIANITE

$SrCO_3$ Orthorhombic

With calcite, barite, harmotome and brewsterite at the mines at Strontian, Argyllshire.

## SULPHUR

S                                    Orthorhombic

As an efflorescence on fused shale, Emily coal pit, Arniston, Midlothian and in a basalt vesicle near the Old Pier, at Loch Begg, Craignish, Argyllshire.

## SUSANNITE

$Pb_4(SO_4)(CO_3)_2(OH)_2$          Trigonal

Associated with leadhillite, lanarkite, caledonite, cerussite and pyromorphite, Susanna Mine, Leadhills, Lanarkshire, and elsewhere (rare) throughout the orefield.

## SYLVANITE

$(Au,Ag)_2Te_4$                     Monoclinic

Forms rare inclusions (<20μm) in pyrite associated with petzite, hessite, native silver, electrum and galena in a brecciated host rock vein 50cms wide near Tyndrum, Perthshire.

## SYLVITE

KCl                                  Cubic

Within saline fluid inclusions in quartz found in the Lagalochan Late Caledonian intrusion complex near Loch Melfort, Argyllshire.

## SZAIBELYITE

$MgBO_2(OH)$                         Monoclinic

In the ore skarns developed in the Durness limestone horizons in the aureole of the Beinn an Dubhaich granite near Broadford, Skye, Inverness-shire.

## TACHARANITE

$Ca_{12}Al_2Si_{18}O_{51}.18H_2O$   Monoclinic

Compact and white, altering to tobermorite and gyrolite, in vesicles of an olivine-dolerite, 0.8km south of Portree on the Staffin road, Skye, Inverness-shire.

184

## TALC

$Mg_3Si_4O_{10}(OH)_2$              Monoclinic and triclinic

Found in chlorite-talc-quartz rocks at Hillswick, and as pale-green veins (6-7cms) in talc-carbonate rocks of Queyhouse Quarry, Quoys, Unst, Shetland.

## TAMARUGITE

$NaAl(SO_4)_2.6H_2O$                Monoclinic

White fluffy encrustations associated with alunogen and potassium alum, Hurlet, Paisley, Renfrewshire.

## TELLUROBISMUTHITE

$Bi_2Te_3$                           Trigonal

Occurs with nickeline, gersdorffite, pyrrhotite, pyrite and other sulphides within the copper-nickel mineralisation near the base of a diorite intrusion at the Glen of the Bar, near Talnotry, Newton Stewart, Kirkcudbrightshire.

## TENNANTITE

$(Cu,Ag,Fe,Zn)_{12}As_4S_{13}$      Cubic

Associated with pyrite, chalcopyrite, covellite, bornite, sphalerite and galena in sulphide nodules (<20mm) in the Devonian Upper Melby Fish Bed, Metta Taing, Walls and Sandness, West Mainland, Shetland.

## *Tenorite*

$Cu^{+2}O$                           Monoclinic

This mineral requires confirmation.

## TETRADYMITE

$Bi_2Te_2S$                          Trigonal

Coarse, fibrous crystals of gladite and krupkaite with inclusions of tetradymite occur in quartz veins hosted by the Loch Shin monzogranite, 2km west of Lairg, Sutherland.

## TETRAHEDRITE

$(Cu,Fe,Ag,Zn)_{12}Sb_4S_{13}$      Cubic

As free-standing crystals (<4mm) with pyrite, or masses up to 5 x 3cm, intimately associated with pyrite and calcite, Tomnadashan, Perthshire.

## THAUMASITE

$Ca_6Si_2(CO_3)_2(SO_4)_2(OH)_{12}.24H_2O$   Hexagonal

Found with xonotlite, pectolite and gyrolite in veins in serpentine from a quarry on the north side of Colla Firth, North Mainland, Shetland.

## THEOPHRASTITE

$Ni(OH)_2$                           Trigonal

Bright-green encrustations on chromitite from Hagdale Quarry, Unst, Shetland.

## THOMSONITE

$NaCa_2Al_5Si_5O_{20}.6H_2O$        Orthorhombic

Common in the Carboniferous andesitic-basaltic lavas around Glasgow and in the basalts of Tertiary age on Skye, Inverness-shire.

## THORIANITE

$ThO_2$                              Cubic

Occurs with thorite associated with apatite in syenite, Cnoc nan Cullean, Ben Loyal, Sutherland.

## THORITE

$(Th,U)SiO_4$                        Tetragonal

As an accessory mineral, in an amazonite-quartz-biotite pegmatite boulder, Beinn Bhreac, Tongue, Sutherland.

## THOROGUMMITE

$Th(SiO_4)_{1-x}(OH)_{4x}$          Tetragonal

As inclusions of glassy red or earthy red-brown grains (3-6mm) with thorite in biotite crystals in the Sletteval pegmatite, South Harris, Inverness-shire.

## TIEMANNITE

HgSe                                           Cubic

Sub-rounded inclusions, up to 1μm, in pyrobitumen found in spoil heap material at the silver mine, Silver Glen, Alva, Clackmannanshire.

## TILLEYITE

$Ca_5Si_2O_7(CO_3)_2$                          Monoclinic

Typically developed in sponge-like single crystals (<1.5mm) with gehlenite enclosures in the gabbro-limestone contact zone of Camas Mòr, Muck, Inverness-shire.

## TITANITE

$CaTiSiO_5$                                    Monoclinic

Clove-brown crystals (<3cm) in granitic pegmatite-like rock, from a roadside quarry, North Ford, North Uist, Inverness-shire. Also as a common accessory mineral.

## TOBERMORITE

$Ca_5Si_6(O,OH)_{18}.5H_2O$                    Orthorhombic

In amygdales, usually associated with mesolite or reyerite in basalts of Tertiary age, Mull, Argyllshire and Skye, Inverness-shire.

## TODOROKITE

$(Mn^{+2},Ca,Mg)Mn_3^{+4}O_7.H_2O$             Monoclinic

Associated with birnessite, hematite and goethite in fine-grained, reddish-blue to black exhalative ferromanganese sediments at Noblehouse, Peeblesshire.

## TOPAZ

$Al_2SiO_4(F,OH)_2$                            Orthorhombic

As colourless to pale-blue crystals or rolled fragments, Beinn a' Bhuird, Cairngorms, Aberdeenshire.

## TORBERNITE

$Cu^{+2}(UO_2)_2(PO_4)_2.8-12H_2O$             Tetragonal

In large buried blocks of uraniferous arkose near Ousdale Burn, southern Caithness.

## TREMOLITE

$Ca_2(Mg,Fe^{+2})_5Si_8O_{22}(OH)_2$           Monoclinic

Common in impure metamorphosed limestones, especially in Glen Urquhart, Inverness-shire, Glen Tilt, Perthshire, and Shetland.

## TRIDYMITE

$SiO_2$                                        Monoclinic, ps. hexagonal

As a component of opal CT that occurs as a botryoidal coating on calcite, from a roadside cutting in a dolerite sill between Mealt and Lealt, Totternish, Skye, Inverness-shire.

## TRIPLITE

$(Mn^{+2},Fe^{+2},Mg,Ca)_2(PO_4)(F,OH)$        Monoclinic

Brown glassy anhedral grains found in bluish-black, wad-like areas up to 0.5cm across, in lepidolite-tourmaline-bearing pegmatite, Glenbuchat, Aberdeenshire.

## TRÖGERITE

$(UO_2)_3(AsO_4)_2.12H_2O(?)$                  Tetragonal(?)

Present in a quartz vein consisting of pitchblende, hydrocarbons, sulphides, native bismuth, atacamite, connellite, bismoclite and boltwoodite, on the coast at Marbruie, near Dalbeattie, Kirkcudbright-shire.

## TROILITE

FeS                                            Hexagonal

Tiny bronze-coloured flecks and grains (<2mm) in an impure marble from Sgiath Bheinn, Glenelg, Inverness-shire proved to be troilite with admixed pyrrhotite.

## TSCHERMAKITE

$Ca_2(Mg,Fe^{+2})_3Al_2(Si_6Al_2)O_{22}(OH)_2$   Monoclinic

In garnet-amphibolite from an eclogite, about 0.8km east of north of Beinn a Chapuill, Glenelg, Inverness-shire.

## TUNGSTITE

$WO_3.H_2O$                                    Orthorhombic

As a bright yellow infill of a rectangular cavity, 3 x 2mm, in quartz-hübnerite vein in zinnwaldite granite near Gairnshiel Bridge, Glen Gairn, Aberdeenshire.

## TYROLITE

$CaCu_5^{+2}(AsO_4)_2(CO_3)(OH)_4.6H_2O$

Orthorhombic

Blue-green radiating pearly blades forming encrustation on country rock found loose in a stream, Silver Glen, Alva, Clackmannanshire.

## ULLMANNITE

NiSbS                                          Cubic

As rare (20μm) euhedral grains in galena in a mineralised breccia boulder recovered from the Crom Allt River near Tyndrum, Perthshire.

## *Ulvöspinel*

$TiFe_2^{+2}O_4$                               Cubic

Occurs within magnetite of the Cambir dolerite, Isles of St. Kilda. This mineral requires confirmation.

## URANINITE

$UO_2$                                         Cubic

Occurs in the Chaipaval pegmatite, South Harris, Inverness-shire, and at Tyndrum, Perthshire.

## URANOPHANE

$Ca(UO_2)_2[SiO_3(OH)]_2.5H_2O$      Monoclinic

In a pegmatite north-east of Sletteval, near Loch a' Sgurr, South Harris, Inverness-shire.

## URANOSPINITE

$Ca(UO_2)_2(AsO_4)_2.10H_2O$      Tetragonal

Associated with pink calcite, sulphides and uranium secondary minerals in late uraniferous mineralisation at Tyndrum, Perthshire.

## VALENTINITE

$Sb_2O_3$      Orthorhombic

As tiny crystals found with stibiconite in stibnite-quartz samples from the antimony mine at Glendinning, Dumfriesshire and at Hare Hill antimony mine, The Knipe, near New Cumnock, Ayrshire.

## VALLERIITE

$4(Fe,Cu)S.3(Mg,Al)(OH)_2$      Hexagonal

Lamellae associated with small patches (<0.5mm) of sulphides composed of pentlandite, pyrrhotite, chalcopyrite, and carbonate in an allivalite, Huntly, Aberdeenshire.

## VANADINITE

$Pb_5(VO_4)_3Cl$      Hexagonal

Waxy-yellow to reddish orangy-brown globules on grey to white plumboan apatite, from the Belton Grain Vein, High Pirn Mine, Wanlockhead, Dumfriesshire.

## VANDENDRIESSCHEITE

$PbU_7^{+6}O_{22}.12H_2O$      Orthorhombic

Minute orange patches on rock matrix, Southwick cliffs area, south of Dalbeattie, Kirkcudbrightshire.

## VERMICULITE

$(Mg,Fe^{+2},Al)_3(Si,Al)_4O_{10}(OH)_2.4H_2O$    Monoclinic

As a decomposition product of biotite in soils, Glenbuchat, Aberdeenshire, and in an altered ultrabasic pod, Rodel, South Harris, Inverness-shire.

## VERNADITE

$\delta\text{-}MnO_2$      Hexagonal
$(Mn^{+4},Fe^{+3},Ca,Na)(O,OH)_2.nH_2O$

A Mn-enriched fraction of clay material from the iron-manganese pan of an imperfectly drained, brown forest soil from Blairmore Estate, near Huntly, Aberdeenshire.

## VESUVIANITE

$Ca_{10}Mg_2Al_4(SiO_4)_5(Si_2O_7)_2(OH)_4$    Tetragonal

In metamorphosed impure limestones especially at Dalnabo, Glen Gairn, Aberdeenshire.

## VESZELYITE

$(Cu^{+2},Zn)_3(PO_4)(OH)_3.2H_2O$    Monoclinic

Transparent to translucent, lustrous dark blue crystals, from cavities in quartz, associated with hemimorphite, malachite, aurichalcite and brochantite, Straitstep Vein dump, Wanlockhead, Dumfriesshire.

## VIOLARITE

$Fe^{+2}Ni_2^{+3}S_4$      Cubic

Occurs as a later replacement within copper-nickel mineralisation near the base of a diorite intrusion at the Glen of the Bar, near Talnotry, Newton Stewart, Kirkcudbrightshire.

## VISHNEVITE

$(Na,Ca,K)_6(Si,Al)_{12}O_{24}[(SO_4)(CO_3)Cl_2]_{2-4}.nH_2O$      Hexagonal

Large cleavable masses associated with orthoclase, melanite and dark mica, in pegmatites and veins in borolanite from a road quarry at Allt a' Mhuillin, Loch Borolan, Assynt, Sutherland.

## VIVIANITE

$Fe_3^{+2}(PO_4)_2.8H_2O$      Monoclinic

As crystal groups in the centre of bones, in a crannog, Loch Lee, Ayrshire.

## VOLTAITE

$K_2Fe_5^{+2}Fe_4^{+3}(SO_4)_{12}.18H_2O$      Cubic

Small clusters of minute black crystals, pale green in thin section, associated with halotrichite, pyrite, coquimbite and römerite, from an old railway cutting, west of Stanely, near Paisley, Renfrewshire.

## WAIRAKITE

$CaAl_2Si_4O_{12}.2H_2O$      Monoclinic

Found associated with laumontite in basalts at two localities in the Midland Valley.

## WALPURGITE

$Bi_4(UO_2)(AsO_4)_2O_4.2H_2O$      Triclinic

Yellow crystals (<1mm) in uranium-bearing veins cutting the aureole of the Criffel granodiorite, near Dalbeattie, Kirkcudbrightshire.

## WHEWELLITE

$CaC_2O_4.H_2O$      Monoclinic

Found in a creamy-white layer intermingled with the hyphae of lichen fungus at the lichen/rock interface on various rock types on the west coast of Mull, Argyllshire.

## WITHERITE

$BaCO_3$        Orthorhombic

Pseudomorphous after barite or as globular masses (<20cm) in a large cavity, associated with quartz, calcite, barite and galena, Glencrieff Mine, Wanlockhead, Dumfriesshire.

## WITTICHENITE

$Cu_3BiS_3$        Orthorhombic

Forms an inclusion in a gold grain from the Ochil Hills, Perthshire.

## WOLFRAMITE

$(Fe,Mn)(W,Nb,Ta)O_4$        Monoclinic

As crystals, up to 3cm, associated with cassiterite and scheelite in a quartz-wolframite vein in zinn-waldite-granite near Gairnshiel Bridge, Glen Gairn, Aberdeenshire.

## WOLLASTONITE

$CaSiO_3$        Monoclinic and triclinic

In thermally metamorphosed limestones, especially at Dalnabo, Glen Gairn and Crathie, Aberdeenshire.

## WOODRUFFITE

$(Zn,Mn^{+2})Mn_3{}^{+4}O_7.1-2H_2O$        Monoclinic

Minor pod-like manganese oxides with goethite, hematite, cryptomelane, ramsdellite and other manganese oxides are found with patchy outcrops of iron oxides in breccia, about 8km south-east of Tomintoul, Aberdeenshire.

## WROEWOLFEITE

$Cu_4{}^{+2}(SO_4)(OH)_6.2H_2O$        Monoclinic

Tiny dark blue grains or bladed aggregates associated with langite, posnjakite, brochantite and serpierite on sulphide ore from the dumps at West Blackcraig Mine near Newton Stewart, Kirkcudbrightshire.

## WULFENITE

$PbMoO_4$        Tetragonal

Occurs with galena, chalcopyrite and pyromorphite in quartz at Lauchentyre, Kirkcudbright-shire. Also found rarely in other lead deposits and in granite at Pass of Ballater, Aberdeenshire.

## WURTZITE

$(Zn,Fe)S$        Hexagonal and trigonal

Medium to dark-brown lath-like crystals (<4mm) are found with pyrite, sphalerite, apatite and dolomite in septarian nodules in fish-beds in Lower Carboniferous shales, Wardie, Edinburgh, Midlothian.

## Xanthophyllite = CLINTONITE

## XENOTIME-(Y)

$YPO_4$        Tetragonal

In heavy mineral residues from granites of Aberdeenshire and Banffshire.

## XONOTLITE

$Ca_6Si_6O_{17}(OH)_2$        Monoclinic and triclinic

As pure-white silky veins (~1cm) in gabbro from the Binhill Quarry near Huntly, Aberdeenshire.

## ZARATITE

$Ni_3(CO_3)(OH)_4.4H_2O$        Cubic

As a bright-green encrustation, with theophrastite, on chromitite from Hagdale Quarry, Unst, Shetland.

## ZEUNERITE

$Cu^{+2}(UO_2)_2(AsO_4)_2.10-16H_2O$        Tetragonal

As yellowish-green plates in vughs with sulphides and uranium secondary minerals, and as an alteration product of pitchblende, in pitchblende-bearing veins cutting the aureole of the Criffel granodiorite, near Dalbeattie, Kirkcudbrightshire.

## ZINKENITE

$Pb_9Sb_{22}S_{42}$        Hexagonal

Iridescent finely felted crystal aggregates associated with galena, sphalerite, semseyite and stibnite from the antimony mine at Glendinning, Glenshanna Burn, Eskdale, Dumfriesshire.

## ZINNWALDITE

$KLiFe^{+2}Al(Al,Si_3)O_{10}(F,OH)_2$        Monoclinic

In granite on the north side of the Balmoral-Tomintoul road immediately north of the bridge over the River Gairn, Gairnshiel Lodge, Aberdeenshire.

## ZIRCON

$ZrSiO_4$        Tetragonal

As a common accessory mineral and found loose with pyrope in the beach sand at Elie Ness, Fife.

## Zirconolite = ZIRKELITE

## ZIRKELITE

$(Ca,Th,Ce)Zr(Ti,Nb)_2O_7$        Monoclinic ps. cubic

Anhedral grains (<60µm) dark reddish-brown in thin-section found with baddeleyite, apatite, chlorite, amphibole and biotite in mesostasis areas of the Rhum layered pluton, Rhum, Inverness-shire.

## ZOISITE

$Ca_2Al_3(SiO_4)_3(OH)$        Orthorhombic

In coarse-grained marbles, with zoisite prisms up to several centimetres, with quartz, tremolite and diopside, at Millton, Glen Urquhart, Inverness-shire.

# APPENDIX 2

## MINERALS ORIGINALLY DISCOVERED IN SCOTLAND

Towards the turn of the 18th century Scottish geology and mineralogy came to the forefront to rival, and compliment, European developments in the sciences. New minerals were discovered and chemically analysed. The earlier references detailed below relate to works in which mineral *names* were first used for minerals deemed to be sufficiently well characterised. Synonymic names, especially chemical names, existed prior to the adoption of the currently accepted names. Thus, it is the publication of the mineral *name*, not the chemical composition or other characteristics, which determines the discovery date. The type locality is detailed for all 29 new species first discovered in Scotland.

### 1791

**STRONTIANITE**          $SrCO_3$

Named after the type locality, Strontian, Argyllshire (Sulzer, 1791).

### 1820

**THOMSONITE**          $NaCa_2Al_5Si_5O_{20}.6H_2O$

From Old Kilpatrick, Dunbartonshire. Named after Dr Thomas Thomson (1773-1852), Professor of Chemistry, Glasgow University, who analysed the mineral (Brooke, 1820a).

### 1822

**BREWSTERITE**          $(Sr,Ba,Ca)Al_2Si_6O_{16}.5H_2O$

The type locality is Strontian, Argyllshire. Named after Sir David Brewster (1781-1868), Scottish physicist who studied the optical properties of minerals (Brooke, 1822).

### 1825

**EDINGTONITE**          $BaAl_2Si_3O_{10}.4H_2O$

From Old Kilpatrick, Dunbartonshire. Named after Mr Edington, of Glasgow, who discovered the mineral (Haidinger, 1825a).

### 1832

**CALEDONITE**          $Pb_5Cu_2(CO_3)(SO_4)_3(OH)_6$

The type locality is Wanlockhead, Dumfriesshire. Named after Caledonia (Scotland). (Beudant, 1832)

**LANARKITE**          $Pb_2(SO_4)O$

Found in the Susanna Mine, Leadhills, Lanarkshire. Named after the county (Beudant, 1832).

**LEADHILLITE**          $Pb_4(SO_4)(CO_3)_2(OH)_2$

Named after the village of Leadhills, Lanarkshire (Beudant, 1832).

### 1840

**GREENOCKITE**          $CdS$

Discovered in excavated material during construction of Bishopton railway tunnel, Bishopton, Renfrewshire. Named after Lord Greenock (1783-1859). (Jameson, 1840)

### 1845

**SUSANNITE**          $Pb_4(SO_4)(CO_3)_2(OH)_2$

Named after the Susanna Mine, Leadhills, Lanarkshire (Haidinger, 1845).

**PLATTNERITE**          $PbO_2$

Originally described on specimens probably from Leadhills, Lanarkshire. Named after K.F. Plattner (1800-58), Professor of Metallurgy and Assaying, Freiberg, Germany (Haidinger, 1845).

### 1851

**GYROLITE**          $NaCa_{16}(Si_{23}Al)O_{60}(OH)_5.15H_2O$

The type locality is Storr, Skye, Inverness-shire. Named from the Greek γυρὸς for 'round', alluding to its form; spherical concretions (Anderson, 1851).

### 1856

**PENTLANDITE**          $(Fe,Ni)_9S_8$

The type locality is Craignure, Inveraray, Argyllshire. Named after Joseph Barclay Pentland (1797-1873), Irish natural scientist who first noted the mineral (Dufrénoy, 1856).

### 1880

**TOBERMORITE**          $Ca_5Si_6(O,OH)_{18}.5H_2O$

Found near, and named after Tobermory, Mull, Argyllshire (Heddle, 1880).

### 1924

**MULLITE**          $Al_6Si_2O_{13}$

Found at Seabank Villa and Nuns Pass, Mull, Argyllshire and named after the island (Bowen *et al.*, 1924).

### 1948

**HARKERITE**

$Ca_{24}Mg_8Al_2(SiO_4)_8(BO_3)_6(CO_3)_{10}.2H_2O$

From the Broadford area, Skye, Inverness-shire. Named after Alfred Harker (1859-1939), British petrologist of Cambridge University renowned for his work on Skye and the Inner Hebrides (Tilley, 1948).

## 1954

**LIZARDITE - 6T1** $Mg_3Si_2O_5(OH)_4$

A new lizardite polymorph, from Nikka Vord, Baltasound, Unst, Shetland. (Originally named after The Lizard, Cornwall, England.) (Brindley and Knorring, 1954.)

## 1956

**BIRNESSITE** $Na_4Mn_{14}O_{27}.9H_2O$

Named after the type locality, Birness, Aberdeenshire (Jones and Milne, 1956).

## 1961

**TACHARANITE** $Ca_{12}Al_2Si_{18}O_{51}.18H_2O$

Found just north of Portree, Skye, Inverness-shire. Readily alters to a mixture of tobermorite and gyrolite, hence named from the Gaelic *Tacharan* 'a changeling' (Sweet, 1961).

**KILCHOANITE** $Ca_3Si_2O_7$

The type locality is Kilchoan, Ardnamurchan, Argyllshire (Agrell and Gay, 1961).

## 1965

**DELLAITE** $Ca_6Si_3O_{11}(OH)_2$

The type locality is Kilchoan, Ardnamurchan, Argyllshire. Named after Della Martin Roy (1926- ) of Pennsylvania State University, U.S.A., who worked on synthetic calcium silicates (Agrell, 1965).

**RUSTUMITE** $Ca_{10}(Si_2O_7)_2(SiO_4)Cl_2(OH)_2$

The type locality is Kilchoan, Ardnamurchan, Argyllshire. Named after Rustum Roy (1924- ), materials chemist, Pennsylvania State University, U.S.A. (Agrell, 1965).

## 1973

**FERROBUSTAMITE** $Ca(Fe^{+2},Ca,Mn^{+2})Si_2O_6$

The type locality is Camas Malag, Skye, Inverness-shire. Originally described as iron wollastonite (iron rhodonite; Tilley, 1948a), renamed ferrobustamite. (Rapoport and Burnham, 1973).

## 1975

**BAZIRITE** $BaZrSi_3O_9$

The type locality is Rockall, Inverness-shire, in the North Atlantic, off Scotland. Named after the barium and zirconium constituents (Hawkes *et al.*, 1975; Young *et al.*, 1978).

## 1980

**JOHNSOMERVILLEITE**
$Na_2Ca(Mg,Fe^{+2},Mn)_7(PO_4)_6$

Found at Glen Cosaidh, Loch Quoich, Inverness-shire. Named after the late Mr John M Somerville who found the specimen in which the mineral occurred (Livingstone, 1980).

## 1984

**MACAULAYITE** $(Fe^{+3},Al)_{24}Si_4O_{43}(OH)_2$

The type locality is Inverurie, Aberdeenshire. Named after the Macaulay Institute for Soil Research, Aberdeen (Wilson *et al.*, 1984).

**MACPHERSONITE** $Pb_4(SO_4)(CO_3)_2(OH)_2$

From Leadhills, Lanarkshire. Named after H.G. Macpherson (1925-2001), Keeper of Minerals, Royal Museum, Edinburgh (Livingstone and Sarp, 1984).

**SCOTLANDITE** $PbSO_3$

Found on a sample from the Susanna Vein, Leadhills, Lanarkshire. Named after the country (Paar *et al.*, 1984).

## 1986

**CHENITE** $Pb_4Cu^{+2}(SO_4)_2(OH)_6$

The type specimen came from the Susanna Mine, Leadhills, Lanarkshire. Named after the mineralogist T.T. Chen (Paar *et al.*, 1986).

## 1987

**MATTHEDDLEITE** $Pb_5(Si_{1.5},S_{1.5})_3O_{12}(Cl,OH)$

From Leadhills, Lanarkshire. Named after Matthew Forster Heddle (1828-97), Professor of Chemistry, University of St Andrews, famous Scottish mineralogist and a founder member, and past President, of the Mineralogical Society (Livingstone *et al.*, 1987).

# APPENDIX 3

## SCOTTISH MINERALS 1991-96

Appendix 1 contains 33 entries (32 species) additional to those found in the *Glossary of Scottish Mineral Species 1981* (Macpherson and Livingstone, 1982) and *Glossary of Scottish Mineral Species – an update* (Livingstone, 1993a) which cover the period 1991-96. These minerals have come to light either from a literature survey or via personal contacts. For these 33 entries, mineral names, references, or other data, are collated below for ease of identifying the additional entries.

| | |
|---|---|
| BAOTITE | N.R. Moles, *pers. comm.*, 1993 |
| BENLEONARDITE | (Ixer *et al.*, 1996) |
| BERZELIANITE | (Leake *et al.*, 1993) |
| BRANNERITE | J. Faithfull, *pers. comm.*, 1995 |
| CHESTERITE | (Droop, 1994) |
| CLINOATACAMITE | Specimen NMS G 1991.13.1 |
| CLINOJIMTHOMPSONITE | see **chesterite** |
| DUNDASITE | Donated by Dr Max. Wirth (confirmed by XRD) |
| ENARGITE | see **benleonardite** |
| FRIEDRICHITE | (Lowry *et al.*, 1994) |

| | |
|---|---|
| GADOLINITE - (Y) | R.J. Gillanders, *pers. comm.*, 1996 |
| GEIKIELITE | (Ferry, 1996) |
| GISMONDINE | (Dyer *et al.*, 1996) |
| GOBBINSITE | see **gismondine** |
| HYDROTALCITE | J. Faithfull, *pers. comm.*, 1995 |
| ISOMERTIEITE | (Shaw *et al.*, 1994) |
| JIMTHOMPSONITE | see **chesterite** |
| KESTERITE | (Webb *et al.*, 1992) |
| LINDSTRÖMITE | see **friedrichite** |
| MERCURY | (Meikle, 1994) |
| PALLADIUM | see **berzelianite** |
| PEKOITE | see **friedrichite** |

| | |
|---|---|
| POLYBASITE | see **benleonardite** |
| QUANDILITE | see **geikielite** |
| ROSCOELITE | (Panhuys-Sigler *et al.*, 1996) |
| RUSSELLITE | see **kesterite** |
| RUTHENIRIDOSMINE | (Davidson, 1995a) |
| Stannite | see **kesterite** |
| STANNOIDITE | see **kesterite** |
| STOLZITE | (Livingstone, 1992) |
| SYLVITE | (Lowry *et al.*, 1995) |
| TROILITE | (Davidson, 1996) |
| WITTICHENITE | (Leake *et al.*, 1993) |

# REFERENCES AND BIBLIOGRAPHY

ABDUL-SAMAD, F. A., THOMAS, J. H., WILLIAMS, P. A. and SYMES, R. F. 1982. Chemistry of formation of lanarkite, $Pb_2OSO_4$. *Mineralogical Magazine*, vol. 46, pp. 499-501.

ADAMSON, G. F. S. 1988. *At the End of a Rainbow, the occurrence of gold in Scotland*. Albyn Press, Haddington, 94pp.

AGRELL, S. O. 1965. Polythermal metamorphism of limestones at Kilchoan, Ardnamurchan. *Mineralogical Magazine*, vol. 34, pp. 1-15.

AGRELL, S. O. and GAY, P. 1961. Kilchoanite, a polymorph of rankinite. *Nature*, vol. 189, p. 743.

AGRICOLA, G. 1546. *De Natura Fossilium*. Basel. (M. C. Bandy and J. A. Bandy, 1955, The Geological Society of America, Special Paper 63, 240pp.).

AIKIN, A. 1815. *A Manual of Mineralogy*. 2nd edn, Longman & Co., London, 164pp.

AKIZUKI, M., KUDOH, Y. and KURIBAYASHI, T. 1996. Crystal structures of the {011}, {610}, and {010} growth sectors in brewsterite. *American Mineralogist*, vol. 81, pp. 1501-6.

ALDRIDGE, D. 1988. *A plan of interpretation for the Wanlockhead and Leadhills Orefield*. Scottish Development Agency.

ALLAN, R. 1834. *A Manual of Mineralogy*. Adam and Charles Black, Edinburgh, 351pp.

ALLAN, R. 1834a. Abstract of a paper accompanying a suite of volcanic rocks from the Lipari Islands, presented to the Royal Society. *Transactions Royal Society Edinburgh*, vol. 12, pp. 531-37.

ALLAN, R. 1850. Daturin, in dem harn mit Stramonium vergifteter personen nachgewiesen. *Liebig Annals*, vol. 74, pp. 223-25.

ALLAN, R. 1855. On the condition of the Haukedair geysers of Iceland, July 1855. *British Association Report* 1855, part 2, pp. 75-78.

ALLAN, T. (READ 1808) 1812. Remarks on a mineral from Greenland, supposed to be crystallised gadolinite. *Transactions Royal Society Edinburgh*, vol. 6, pp. 345-51.

ALLAN, T. 1814. *Mineralogical Nomenclature, alphabetically arranged; with synoptic tables of the chemical analyses of minerals*. Archibald Constable and Company, Edinburgh; and for Longman, Hurst, Rees, Orme, and Brown, London.

ALLAN, T. (READ 1813) 1815. An account of the mineralogy of the Faroe Islands. *Transactions Royal Society Edinburgh*, vol. 7, pp. 229-67.

ANDERSON, T. 1851. Description and analysis of gurolite, (gyrolite) a new mineral species. *London, Edinburgh, and Dublin Philosophical Magazine*, 4th series, vol. 1, pp. 111-13.

ARMSTRONG, W. 1901. *Sir Henry Raeburn*. William Heinemann, London, pp. 104-5.

ATKINSON, S. 1619. The Discoverie and Historie of the Gold Mynes in Scotland, written by Stephen Atkinson in the year MDCXIX, pp. 47-50. From the original, prefaced by Gilbert Laing Meason, and presented to the Bannatyne Club, Edinburgh. James Ballantyne and Co., Edinburgh, 1825, 119pp.

BAILEY, E. B., CLOUGH, C. T., WRIGHT, W. B., RICHEY, J. E. and WILSON, G. V. 1924. *Tertiary and post-Tertiary Geology of Mull, Loch Aline, and Oban*. Memoirs of the Geological Survey, Scotland, Her Majesty's Stationery Office, 445pp.

BAINBRIDGE, J. 1980. Lord Breadalbane's Mines. *Scots Magazine*, new series, vol. 114, no. 1, pp. 38-45.

BANKS, J. 1772. 'Account of Staffa' in T. Pennant's *A Tour in Scotland and Voyage to the Hebrides*. (1776). 2nd edn, Benji White, London, vol. 2, part 1, pp. 299-309.

BARROW, G. 1893. On an intrusion of muscovite–biotite gneiss in the south-eastern Highlands of Scotland, and its accompanying metamorphism. *Quarterly Journal Geological Society London*, vol. 49, pp. 330-58.

BARROW, G. 1912. On the geology of lower Dee-side and the southern Highland border. *Proceedings Geologists' Association*, vol. 23, pp. 268-90.

BERG, G. 1901. Ueber einen neuen fundort des caledonites in Chile. *Mineralogische und Petrographische Mitteilungen*, vol. 20, pp. 390-98.

BEUDANT, F. S. 1832. *Traité de Minéralogie*. Paris, vol. 1, 752pp., vol. 2, 797pp.

BEVERIDGE, R., BROWN, S., GALLAGHER, M. J. and MERRITT, J. W. 1991. Economic Geology. *Geology of Scotland*. 3rd edn, ed. G. Y. Craig. The Geological Society, London, 612pp.

BIDEAUX, R. A. 1972. The Collector. *Mineralogical Record*, vol. 3, pp. 12 and 92.

BIOT, J. B. 1806. Relation d'un voyage fait dans le département de l'Orne, pour constater la réalité d'un météore observé à l'Aigle (26 Avril 1803). *Memoire Institute de France*, vol. 7 (Histoire), pp. 224-66.

BLACK, G. F. and BISSET, J. 1894. Catalogue of Dr Grierson's Museum, Thornhill, Edinburgh, 119pp.

BLUCK, B. J. 1985. The Scottish paratectonic Caledonides. *Scottish Journal of Geology*, vol. 21, pp. 437-64.

BLUCK, B. J. 1992. In Lawson, J. D. and Weedon, D. S. 1992. *Geological Excursions around Glasgow and Girvan*. Geological Society of Glasgow, 495pp.

BOAST, A. M., HARRIS, M. and STEFFE, D. 1990. Intrusive-hosted gold mineralization at Hare Hill, Southern Uplands, Scotland. *Transactions Institution Mining Metallurgy*, vol. 99, pp. B106-12.

BOGGILD, O. B. 1922. Re-examination of some zeolites (okenite, ptilolite, etc). *Det kgl Danske Videnskabernes Selskab, Mathematisk-Fysiske*, vol. 4, 42pp.

BOSWORTH, T. O. 1912. The heavy mineral grains in the sands of the Scottish Carboniferous. *Geological Magazine*, vol. 9, pp. 515-16.

BOUÉ, A. 1820. *Essai géologique sur L'Écosse*. Paris, 519pp.

BOURNON, J. L. 1808. *Traité de Minéralogie*. William Phillips, London, 3 vols.

BOWEN, N. L. 1940. Progressive metamorphism of siliceous limestone and dolomite. *Journal of Geology*, vol. 48, pp. 225-74.

BOWEN, N. L., GREIG, J. W. and ZIES, E. G. 1924. Mullite, a silicate of alumina. *Journal Washington Academy of Sciences*, vol. 14, pp. 183-91.

BOWIE, S. H. U. 1962. *Summary of Progress, British Geological Survey Great Britain*, for 1961, p. 67.

BREWSTER, D. 1820. Notice respecting some new species of lead-ore from Wanlockhead and Lead Hills. *Edinburgh Philosophical Journal*, vol. 3, pp. 138-40.

BREWSTER, D. 1821. On the connexion between the optical structure and chemical composition of minerals. *Edinburgh Philosophical Journal*, vol. 5, pp. 1-8.

BREWSTER, D. 1821a. Account of comptonite, a new mineral from Vesuvius. *Edinburgh Philosophical Journal*, vol. 4, pp. 131-33.

BREWSTER, D. 1825. Description of withamite, a new mineral species found in Glenco. *Edinburgh Journal of Science*, vol. 2, pp. 218-21.

BREWSTER, D. 1825a. Description of gmelinite, a new mineral species. *Edinburgh Journal of Science*, vol. 2, pp. 262-67.

BREWSTER, D. 1826. Farther observations on Levyne, a new mineral species. *Edinburgh Journal of Science*, vol. 4, pp. 316-17.

BREWSTER, D. 1826a. Description of hopeite, a new mineral, from Altenberg near Aix-la-Chapelle. *Transactions Royal Society Edinburgh*, vol. 10, pp. 107-11.

BREWSTER, D. and JAMESON, R. 1822. Notice of mineralogical journeys, and of a mineralogical system, by the late Rev. Dr. John Walker, Professor of Natural History in the University of Edinburgh. *Edinburgh Philosophical Journal*, vol. 6, pp. 88-94.

BRINDLEY, G. W. and KNORRING, O. von 1954. A new variety of antigorite (ortho-antigorite) from Unst, Shetland Islands. *American Mineralogist*, vol. 39, pp. 794-804.

BROCHANT, A. J. M. 1803. *Traité élémentaire de Minéralogie Suivant les Principles du Professor Werner*. Paris, 2 tom: tom 1, 644pp; tom 2, 674pp.

BROCK, C. H. 1983. *William Hunter 1718-1783. A Memoir*. University of Glasgow Press, 81pp.

BRONGNIART, A. 1807. *Traité élémentaire de Minéralogie*. Paris, vol. 1, 564pp., vol. 2, 445pp.

BROOKE, H. J. 1820. Account of three new species of lead-ore found at Leadhills. *Edinburgh Philosophical Journal*, vol. 3, pp. 117-20.

BROOKE, H. J. 1820a. On mesotype, needlestone, and thomsonite. *Annals Philosophical Journal*, vol. 16, pp. 193-94.

BROOKE, H. J. 1822. On the comptonite of Vesuvius, the brewsterite of Scotland, the stilbite and the heulandite. *Edinburgh Philosophical Journal*, vol. 6, pp. 112-15.

BROOKE, H. J. 1823. *Familiar Introduction to Crystallography*. London, p. 458.

BROWN, R. 1919. The Mines and Minerals of Leadhills. *Transactions Dumfriesshire and Galloway Natural History and Antiquarian Society*, 3rd series, vol. 6, pp. 124-37.

BROWN, R. 1925. More about the Mines and Minerals of Wanlockhead and Leadhills. *Transactions Dumfriesshire and Galloway Natural History and Antiquarian Society*, vol. 13, pp. 58-79.

BRYCE-WRIGHT, 1880. *Catalogue of Mineralogical, Geological, Conchological, and Archaeological Specimens, & c*. London, 56pp.

BURTON, C. J. and TODD, J. G. 1992. Trearne Quarry – Lower Carboniferous fossil faunas and their palaeoecology, in *Geological Excursions around Glasgow and Girvan*, eds J. D. Lawson and D. S. Weedon, Geological Society of Glasgow, 495pp.

BURTON, K. W., COHEN, A. S., O'NIONS, R. K. and O'HARA, M. J. 1994. Archaean crustal development in the Lewisian complex of northwest Scotland. *Nature*, vol. 370, pp. 552-55.

BUTLER, B. C. M. 1962. Biotite- and sphene-rich rocks in the Moine Series of Ardnamurchan, Argyllshire. *Geological Magazine*, vol. 99, pp. 173-82.

CADELL, H. M. 1925. The Hilderston Silver Mine, in *The Rocks of West Lothian*. Oliver and Boyd, Edinburgh, pp. 359-78.

CALVERT, S. E. and PRICE, N. B. 1970. Composition of manganese nodules and manganese carbonates from Loch Fyne, Scotland. *Contributions Mineralogy Petrology*, vol. 29, pp. 215-33.

CALLENDER, R. M. 1990. *Gold in Britain*. Goldspear (UK) Limited, Beaconsfield, 64pp.

CAMERON, W. E. 1976. Coexisting sillimanite and mullite. *Geological Magazine*, vol. 113, pp. 497-592.

CAMPBELL, R. and HOLMES, A. 1948. Record of papers and exhibits, 17 October 1945. *Transactions Edinburgh Geological Society*, vol. 14, part 11, pp. 283-84.

CARSWELL, J. 1950. *The Prospector, being the life and times of Rudolf Erich Raspe (1737-1794)*. The Cresset Press, London, 278pp.

CARTWRIGHT, I., FITCHES, W. R., O'HARA, M. J., BARNICOAT, A. C. and O'HARA, S. 1985. Archaean supracrustal rocks from the Lewisian near Stoer, Sutherland. *Scottish Journal of Geology*, vol. 21, pp. 187-96.

CASSIRER, F. W. and MARTIN, A. 1979. Memoirs of a mineral collector. *Mineralogical Record*, vol. 10, pp. 223-29.

CHITNIS, A. C. 1970. The University of Edinburgh's Natural History Museum and the Huttonian-Wernerian debate. *Annals of Science*, vol. 26, no. 2, pp. 85-94.

CLARINGBULL, G. F. and HEY, M. H. 1952. A re-examination of tobermorite. *Mineralogical Magazine*, vol. 29, pp. 960-62.

CLARK, A. M. 1993. *Hey's Mineral Index*. 3rd edn, Chapman and Hall, London, 852pp.

CLIFF, G., GARD, J. A., LORIMER, G. W. and TAYLOR, H. F. W. 1975. Tacharanite. *Mineralogical Magazine*, vol. 40, pp. 113-26.

COATS, J. S., FORTEY, N. J., GALLAGHER, M. J. and GROUT, A. 1984. Stratiform barium enrichment in the Dalradian of Scotland. *Economic Geology*, vol. 79, pp. 1585-95.

COATS, J. S., SHAW, M. H., GALLAGHER, M. J., ARMSTRONG, M., GREENWOOD, P. G., CHACKSFIELD, B. C., WILLIAMSON, J. P. and FORTEY, N. J. 1991. Gold in the Ochil Hills, Scotland. *Mineral Reconnaissance Programme Report 116*, British Geological Survey, 93pp.

COATS, J. S., SMITH, C. G., FORTEY, N. J., GALLAGHER, M. J., MAY, F. and McCOURT, W. J. 1980. Strata-bound barium-zinc mineralization in Dalradian schist near Aberfeldy, Scotland. *Transactions Institution Mining Metallurgy*, vol. 89, pp. B110-22.

COCHRAN-PATRICK, R. W. 1878. *Early records relating to mining in Scotland*. David Douglas, Edinburgh, 205pp.

COLLIE, N. 1889. On some Leadhills minerals. *Journal Chemical Society*, vol. 55, pp. 90-96.

CONNELL, A. 1830. On the chemical constitution of brewsterite. *Edinburgh New Philosophical Journal*, vol. 8, pp. 355-57.

CONNELL, A. 1840. Chemical Examination of Greenockite, or Sulphuret of Cadmium. *Edinburgh New Philosophical Journal*, vol. 28, pp. 392-95.

CONOLLY, M. F. 1866. *Biographical Dictionary of Eminent Men of Fife*. John C. Orr, Edinburgh, 492pp.

COOMBS, D. S. 1998. (Chairman) Recommended nomenclature for zeolite minerals: report of the subcommittee on zeolites of the International Mineralogical Association, Commission on New Minerals and Mineral Names. *Mineralogical Magazine*, vol. 62, pp. 533-71.

COOPER, I. 1976. Letter from a loner. *Gem Craft*, December, pp. 811-12.

CRAIG, W. S. 1976. *History of the Royal College of Physicians of Edinburgh*. Blackwell Scientific Publications, Oxford, 1125pp.

CRAIG, G. Y. 1991. *Geology of Scotland*. 3rd edn, The Geological Society, London, 612pp.

CRAWFORD, A. 1790. On the medicinal properties of the Muriated Barytes. *Medical Communications*, vol. 2, pp. 301-59.

CUNNINGHAM, R. J. H. 1843. Geognostical account of Banffshire. *Prize-essays and Transactions of the Highland and Agricultural Society of Scotland*, new series, vol. 8, pp. 447-502.

CURRIE, J. 1905. The Mineralogy of the Faeröes, arranged topographically. *Transactions Edinburgh Geological Society*, vol. 9, pp. 1-68.

CURRIE, J. 1907. Note on some new localities for gyrolite and tobermorite. *Mineralogical Magazine*, vol. 14, pp. 93-95.

CUVIER, G. 1815. *Essay on the Theory of the Earth*. Translated by R. Kerr, with mineralogical notes by Professor Jameson. 2nd edn, William Blackwood, Edinburgh, 332pp.

DANA, J. D. 1837. *System of Mineralogy*. 1st edn, New Haven, 444pp.

DANA, J. D. 1892. *System of Mineralogy*. 6th edn, London, 1134pp.

DAUBENTON 1784. *Tableaux méthodiques des Minéraux*. Paris (A classified catalogue only, from Tschernich, 1992).

DAVIDSON, P. J. 1989. Stevensite, a smectite-group mineral from Corstorphine Hill, Edinburgh. *Scottish Journal of Geology*, vol. 25, part 1, pp. 63-67.

DAVIDSON, P. J. 1995. Troilite from Glenelg, Highland Region, Scotland: The first British Isles occurrence. *Journal Russell Society*, vol. 6, pp. 52-53.

DAVIDSON, P. J. 1995a. Rutheniridosmine from the Ochil Hills, Scotland: The first British Isles occurrence. *Journal Russell Society*, vol. 6, pp. 54-55.

DAVY, H. 1808. Electro-chemical researches, on the decomposition of the earths; with observations on the metals obtained from the alkaline earths, and on the amalgam procured from ammonia. *Transactions Royal Philosophical Society London*, vol. 98, pp. 333-70.

DAWSON, W. R. 1958. *The Banks letters. A calendar of the manuscript correspondence of Sir Joseph Banks*, ed. W. R. Dawson. The British Museum, London, pp. 214-15.

DEAN, D. R. 1992. *James Hutton and the History of Geology*. Cornell University Press, Ithaca and London, 303pp.

DEER, W. A. 1935. The Cairnsmore of Carsphairn igneous complex. *Quarterly Journal Geological Society*, London, vol. 91, pp. 47-76.

DEFOE, D. 1748. *A tour thro' the whole island of Great Britain by a gentleman*. 4th edn, vol. 4, Birt, Osborne, Browne, Hodges, Millar & Robinson, London, pp. 79-80.

DELLOW, E. L. 1973. *Svedenstierna's Tour Great Britain 1802-3. The Travel Diary of an Industrial Spy*. David and Charles, Newton Abbot, 192pp.

DEWEY, H. 1920. Special reports on the mineral resources of Great Britain, vol. 15, Arsenic and Antimony Ores. *Memoirs of the Geological Survey*, p. 56.

DEWEY, J. F. 1971. A model for the Lower Palaeozoic evolution of the southern margin of the early Caledonides of Scotland and Ireland. *Scottish Journal of Geology*, vol. 7, pp. 219-40.

DREVER, H. I. 1940. The geology of Ardgour, Argyllshire. *Transactions Royal Society Edinburgh*, vol. 60, pp. 141-70.

DROOP, G. T. R. 1994. Triple-chain pyriboles in Lewisian ultramafic rocks. *Mineralogical Magazine*, vol. 58, pp. 1-20.

DUDGEON, P. 1867. Notes on some rare minerals occurring in the district. *Transactions and Journal of the Proceedings of the Dumfriesshire and Galloway Natural History and Antiquarian Society*, for the session 1864-1865, pp. 25-26.

DUDGEON, P. 1871. Minerals lately found in Dumfriesshire and Galloway, not hitherto noticed as occurring in these localities. *Transactions and Journal of the Proceedings of the Dumfriesshire and Galloway Natural History and Antiquarian Society*, for the session 1867-68, pp. 38-39.

DUDGEON, P. 1877. Historical notes on the occurrence of gold in the south of Scotland. *Mineralogical Magazine*, vol. 1, pp. 21-28.

DUDGEON, P. 1877a. Specimen of auriferous quartz. *Proceedings Royal Society Edinburgh*, vol. 9, p. 338.

DUDGEON, P. 1884. On the occurrence of linarite in slag. *Mineralogical Magazine*, vol. 5, p. 33.

DUDGEON, P. 1890. Notes on the minerals of Dumfries and Galloway. *Transactions and Journal of the Proceedings of the Dumfriesshire and Galloway Natural History and Antiquarian Society*, no 6. pp. 175-82.

DUDGEON, P. 1897. Occurrence of mispickel in the Stewartry of Kirkcudbright. *Mineralogical Magazine*, vol. 11, p. 15.

DUFF, D. 1980. *Queen Victoria's Highland Journals*. Webb and Bower, Exeter, p. 71.

DUFRÉNOY, A. 1856. *Traité de Minéralogie*. 1st edn, vol. 2, p. 549.

DUNNING, F. W., MERCER, I. F., OWEN, M. P., ROBERTS, R. H. and LAMBERT, J. L. M. 1978. *Britain Before Man*. Her Majesty's Stationery Office, 36pp.

DURANT, G. P. and ROLFE, W. D. I. 1984. William Hunter (1718-1783) as natural historian: his 'geological' interests. *Earth Sciences History*, vol. 3, pp. 9-24.

DYER, A. and WILSON, O. 1988. Zeolites. *UK Journal of Mines and Minerals*, no. 4, pp. 17-21.

DYER, A., WILSON, O. M., ENAMY, H. and WILLIAMS, C. D. 1993. Stellerite from Todhead Point, Grampian Region, Scotland. *Mineralogical Magazine*, vol. 57, pp. 540-42.

DYER, A., WILSON, O., WILLIAMS, C. D. and SMETHURST, J. 1996. Zeolite occurrences in basaltic dykes on the Western Isles of Scotland. *Journal Russell Society*, vol. 6, pp. 89-92.

EDWARDS, W. N. 1951. William Nicol and Henry Clifton Sorby. *Nature*, vol. 168, pp. 566-67.

EMBREY, P. G. 1977. *Manual of the Mineralogy of Great Britain and Ireland by Greg and Lettsom 1858*. A facsimile reprint with Supplementary Lists of British Minerals by L. J. Spencer, F.R.S. and a Fourth Supplementary List (1977) together with a forward by P.G. Embrey. Lapidary Publications, Kent, England, 483pp.

EMBREY, P. G. and SYMES, R. F. 1987. Minerals of Cornwall and Devon. *British Museum (Natural History) London*, 154pp.

EVANS, L. J. 1987. Low-grade regional metamorphism of Palaeozoic rocks in the Midland Valley of Scotland. Unpublished Ph.D. thesis, University of St Andrews.

EYLES, J. M. 1951. William Nicol and Henry Clifton Sorby: two centenaries. *Nature*, vol. 168, pp. 98-99.

EYLES, V. A. 1954. Robert Jameson and the Royal Scottish Museum. *Discovery*, April, London, pp. 155-62.

FAIRLEY, J. A. 1925. *Lauriston Castle the Estate and its owners*. Oliver and Boyd, Edinburgh, pp. 173-77.

FARRAR, W. V. and FARRAR, K. R. 1968. Thomas Allan, Mineralogist: An autobiographical fragment. *Annals of Science*, vol. 24, no. 2, pp. 115-20.

FAUJAS SAINT-FOND, B. 1799. *Travels in England, Scotland and the Hebrides; undertaken for the purpose of examining the state of the Arts, the Sciences, Natural History and Manners, in Great Britain*. James Ridgway, London, vol. 1, 361pp., vol. 2, 352pp.

FERRY, J. M. 1996. Three novel isograds in metamorphosed siliceous dolomites from the Ballachulish aureole, Scotland. *American Mineralogist*, vol. 81, pp. 485-94.

FETTES, D. J., MENDUM, J. R., SMITH, D. I. and WATSON, J. V. 1992. Geology of the Outer Hebrides. *Memoir of the British Geological Survey*, sheets (solid edition) Lewis and Harris, Uist and Barra (Scotland), 197pp.

FLEISCHER, M. and MANDARINO, J. A. 1991. *Glossary of Mineral Species*. The Mineralogical Record Inc., Tucson, 256pp.

FLEISCHER, M. and MANDARINO, J. A. 1995. *Glossary of Mineral Species 1995*. The Mineralogical Record Inc., Tucson, 280pp.

FLEMMING, J. 1807. *Economical Mineralogy of the Orkney and Zetland Islands*.

FLETCHER, L. 1889. The Renaissance of British Mineralogy. *Mineralogical Magazine*, vol. 8, pp. 138-45.

FLETT, J. 1936. The first geological map of Scotland. *Transactions Edinburgh Geological Society*, vol. 13, pp. 291-303.

FORBES, J. D. 1840. On the Optical Characters of Greenockite (Sulphuret of Cadmium). *London, Edinburgh, and Dublin Philosophical Magazine and Journal of Science*, 3rd series, vol. 17, p. 8.

FORDYCE, G. and ALCHORNE, S. 1779. An Examination of various Ores in the Museum of Dr. William Hunter. *Philosophical Transactions Royal Society London*, vol. 69, pp. 527-36.

FOWLER, M. B. 1992. Elemental and O-Sr-Nd isotope geochemistry of the Glen Dessary syenite, N. W. Scotland. *Journal Geological Society London*, vol. 149, pp. 209-20.

FRANCIS, E. H., FORSYTH, I. H., READ, W. A. and ARMSTRONG, M. 1970. The Geology of the Stirling District. *Memoirs of the Geological Survey of Great Britain, Scotland*. Her Majesty's Stationery Office, pp. 297-99.

FRASER, A. G. 1989. *The Building of Old College, Adam, Playfair and The University of Edinburgh*. Edinburgh University Press, 384pp.

FRENZEL, A. 1881. Vanadinit und Tritochorit. *Mineralogische und Petrographische Mitteilungen*, vol. 3, pp. 504-16.

FRIEND, J. N. and ALLCHIN, J. P. 1940. Colloidal gold as a colouring principle in minerals. *Mineralogical Magazine*, vol. 25, pp. 584-96.

FRONDEL, C. 1972. Jacob Forster (1739-1806) and his connections with forsterite and palladium. *Mineralogical Magazine*, vol. 38, pp. 545-50.

GALLAGHER, M. J. 1964. Rock alteration in some mineralized basic dykes in Britain. *Transactions Institution of Mining and Metallurgy*, vol. 73, pp. 825-40.

GARD, J. A. and TAYLOR, H. F. W. 1958. A further investigation of tobermorite from Loch Eynort, Scotland. *Mineralogical Magazine*, vol. 31, pp. 361-70.

GEIKIE, A. 1887. *The Scenery of Scotland*. 2nd edn, Macmillan and Co., London, 481pp.

GEIKIE, A. 1904. *Scottish Reminiscences*. James Maclehose and Sons, Glasgow, 447pp.

GEMMELL, E. W. 1903. Note in Proceedings. *Transactions Geological Society Glasgow*, vol. 12, p. 400.

GILLIES, W. A. 1938. *In Famed Breadalbane*. The Munro Press, Perth, 439pp.

GIUSEPPETTI, G., MAZZI, F. and TADINI, C. 1990. The crystal structure of leadhillite: $Pb_4(SO_4)(CO_3)_2(OH)_2$. *Neues Jahrbuch für Mineralogie, Monatshefte*, vol. 6, pp. 255-68.

GLEN, D. C. and YOUNG, J. 1876. List of minerals and rock specimens found in the Central, Southern, and Western Districts of Scotland. *British Association for the Advancement of Science, Glasgow*, pp. 156-64.

GLOCKER, E. F. 1839. *Grundriss der Mineralogie*. Nürnberg, p. 618.

GOODCHILD, J. G. 1897. Dr. Heddle and his geological work. *Transactions Edinburgh Geological Society*, vol. 7, pp. 317-27.

GOODCHILD, J. G. 1898. Dr Heddle, M.D., F.R.S.E., Emeritus Professor of Chemistry at St Andrews. Born 1828; died 19th November 1897. *Proceedings Royal Physical Society Edinburgh*. vol. 14, pp. 69-77.

GOODCHILD, J. G. 1901. A revised list of the minerals known to occur in Scotland. *British Association for Advancement of Science Report*, pp. 648-49.

GOODCHILD, J. G. 1903. The natural history of Scottish zeolites and their allies. *Transactions Geological Society Glasgow*, vol. 12 (supplement), pp. 1-65.

GOODMAN, S. and LAPPIN, M. A. 1996. The thermal aureole of the Lochnagar Complex: mineral reactions and implications from thermal modelling. *Scottish Journal of Geology*, vol. 32, pp. 159-72.

GORDON, M. M. 1869. *The Home Life of Sir David Brewster*. Edmonston and Douglas, Edinburgh, 440pp.

GOTTARDI, G. and GALLI, E. 1985. *Natural Zeolites*. Springer-Verlag, Berlin, Heidelberg, 409pp.

GRANT, J. 1882. *Old and New Edinburgh*. Cassell, Petter, Galpin and Co., London, vol. 1, 384pp., vol. 2, 384pp., vol. 3, 392pp.

GREEN, D. I. 1986. UK wulfenite. *UK Journal of Mines and Minerals*, no. 1, pp. 28-29.

GREEN, D. I. 1987. The minerals of Meadowfoot Smelter. *UK Journal of Mines and Minerals*, no. 2, pp. 3-9.

GREEN, D. I. 1990. Veszelyite a mineral new to Britain, from Wanlockhead, Scotland. *UK Journal of Mines and Minerals*, no. 8, pp. 6-7.

GREEN, D. I. and TODD, J. G. 1996. Zeolites and related minerals from Moonen Bay, Isle of Skye, Scotland. *UK Journal of Mines and Minerals*, no. 16, pp. 21-27.

GREEN, D. I. and WOOD, M. 1996. Epistilbite from the Isle of Skye, Scotland. *Journal Russell Society*, vol. 6, pp. 85-87.

GREG, R. P. and LETTSOM, W. G. 1858. *Manual of the Mineralogy of Great Britain and Ireland* (facsimile reprint), Lapidary Publications, Broadstairs, Kent, 483pp.

GREIG, D. C. 1988. Geology of the Eyemouth district. *Memoir British Geological Survey*, sheet 34 (Scotland), 78pp.

HAIDINGER, W. 1825. *Treatise on Mineralogy, on the Natural History of the Mineral Kingdom*, by Frederick Mohs. Translated from the German, with considerable additions. Archibald Constable and Co., Edinburgh; and Hurst, Robinson, and Co., London, vol. 1, 458pp., vol. 2, 472pp., and vol. 3, 319pp.

HAIDINGER, W. 1825a. Description of edingtonite, a new mineral species. *Edinburgh Journal of Science*, vol. 3, pp. 316-20.

HAIDINGER, W. 1826. Description of Fergusonite, a New Mineral Species. *Transactions Royal Society Edinburgh*, vol. 10, pp. 271-78.

HAIDINGER, W. 1826a. On the forms of crystallisation of the mineral called the sulphato-tri-carbonate of lead. *Transactions Royal Society Edinburgh*, vol. 10, pp. 217-30.

HAIDINGER, W. 1845. *Handbuch der bestimmenden Mineralogie*. Vienna, p. 505.

HALL, M. B. 1992. The library and archives of the Royal Society, 1660-1990. *Royal Society*, London, p. 60.

HALL, A. J., BANKS, D., FALLICK, A. E. and HAMILTON, P. J. 1989. An hydrothermal origin for copper-impregnated prehnite and analcime from Boylestone Quarry, Barrhead, Scotland. *Journal of the Geological Society*, London, vol. 146, pp. 701-13.

HALL, A. R. and TILLING, L. 1976. *The Correspondence of Isaac Newton*. Cambridge University Press, vol. 6, 1713-1718. p. 316.

HARDING, R. R., MERRIMAN, R. J. and NANCARROW, P. H. A. 1984. St. Kilda: an illustrated account of the geology. *British Geological Survey Report*, vol. 16, no. 7, 46pp.

HARRIS, P. G. and BRINDLEY, G. W. 1954. Mordenite as an alteration product of a pitchstone glass. *American Mineralogist*, vol. 39, pp. 819-24.

HARRISON, R. K., STONE, P., CAMERON, I. B., ELLIOT, R. W. and HARDING, R. R. 1987. Geology, petrology and geochemistry of Ailsa Craig, Ayrshire. *British Geological Survey Report*, vol. 16, no. 9, 29pp.

HARVEY, H. R. 1972. An inquiry into the contribution to science and the advancement of science of Sir David Brewster, K.H., D.C.L. (1781-1868). MA thesis, University of Exeter, 122pp.

HARWOOD, H. F. 1951. The greenockite locality at Bishopton, Scotland. *American Mineralogist*, vol. 36, p. 630.

HAUSMANN, J. F. L. 1809. *Versuch eines Entwurfs zu einer Einleitung in die Oryktognosie*. Cassel.

HAUSMANN, J. F. L. 1813. *Handbuch der Mineralogie*. 3 vols., Göttingen.

HAÜY, R. J. 1801. *Traité de Minéralogie*. Paris, vol. 1, 494pp., vol. 2, 617pp., vol. 3, 588pp., vol. 4, 592pp., vol. 5, 86 plates.

HAÜY, R. J. 1809. *Tableau comparatif des résultats de la cristallographie et de l'analyse chimique relativement à la classification des Minéraux*. Paris.

HAÜY, R. J. 1813. *Catalogue de la Collection Minéralogique du Comte De Bournon*. London, pp. 343-44.

HAÜY, R. J. 1822. *Traité de Minéralogie and Atlas*. Paris, vol. 1, 594pp., vol. 2, 613pp., vol. 3, 592pp., vol. 4, 604pp., atlas 120 plates.

HAWKES, J. R., MERRIMAN, R. J., HARDING, R. R. and DARBYSHIRE, D. P. F. 1975. Expeditions to Rockall 1971-72. In R. K. Harrison, *Institute of Geological Sciences Report*, no. 75/1, pp. 11-51.

HAYNES, V. 1983. *In Scotland – A New Study*, ed. C. M. Clapperton. David and Charles, London, 327pp.

HEANEY, P. J., PREWITT, C. T. and GIBBS, G. V. 1994. Silica physical behaviour, geochemistry and materials applications. *Reviews in Mineralogy*, vol. 29, Mineralogical Society of America, 606pp.

HEDDLE, M. F. 1855. Analysis of the mineral 'Edingtonite'. *London, Edinburgh, and Dublin Philosophical Magazine*, vol. 9, pp. 179-81.

HEDDLE, M. F. 1878. The County Geognosy and Mineralogy of Scotland. *Mineralogical Magazine*, vol. 2, pp. 9-35.

HEDDLE, M. F. 1878a. The County Geognosy and Mineralogy of Scotland. Reprinted from *Mineralogical Magazine*. Lake and Lake, Truro. (A large collection of Heddle papers, 1878-84.)

HEDDLE, M. F. 1880. Preliminary notice of substances which may prove to be new minerals. Part second: balvraidite, hydrated labradorite, tobermorite, walkerite. *Mineralogical Magazine*, vol. 4 (for 1882), pp. 117-23.

HEDDLE, M. F. 1882. Minerals new to Britain, and the geognosy and mineralogy of Scotland. *Mineralogical Magazine*, vol. 5 (for 1884) pp. 1-25 and 71-106.

HEDDLE, M. F. 1883. The geognosy and mineralogy of Scotland. Sutherland, including Parts IV, V & VI. *Mineralogical Magazine*, vol. 5 (for 1884) pp. 133-89; 217-63; 271-324.

HEDDLE, M. F. 1887. On the occurrence of greenockite at a new locality. *Mineralogical Magazine*, vol. 7, pp. 133-37.

HEDDLE, M. F. 1892. On the occurrence of sapphire in Scotland. *Mineralogical Magazine*, vol. 9, pp. 389-90.

HEDDLE, M. F. 1893. On new localities for zeolites. *Transactions Geological Society Glasgow*, vol. 9, pp. 72-79.

HEDDLE, M. F. 1893a. On pectolite and okenite from new localities: the former with new appearances. *Transactions Geological Society Glasgow*, vol. 9, pp. 241-55.

HEDDLE, M. F. 1897. Obituary. Patrick Dudgeon, F.R.S.E. *Mineralogical Magazine*, vol. 11, pp. 30-31.

HEDDLE, M. F. 1901. *The Mineralogy of Scotland*, ed. J. G. Goodchild. David Douglas, Edinburgh, vol. 1, 148pp., vol. 2, 247pp.

HEDDLE, M. F. and GREG, R. P. 1855. On British pectolites. *London, Edinburgh and Dublin Philosophical Magazine*, vol. 9, pp. 248-53.

HENRY, W. C. (undated). *A biographical notice of the late Robert Allan, Esq., F.R.S.E. and F.G.S; with extracts from his journals.* (For Private Circulation), 24pp.

HEY, M. H. 1934. Studies on the zeolites. Part VI. Edingtonite. *Mineralogical Magazine*, vol. 23, pp. 483-94.

HOPE, T. C. 1794. Postscript to the History. *Transactions Royal Society Edinburgh*, vol. 3, pp. 141-48.

HOPE, T. C. 1798. Account of a mineral from Strontian, and of a peculiar species of earth which it contains. *Transactions Royal Society Edinburgh*, vol. 4, pp. 3-39.

HOSACK, D. 1822. Extract of a letter from Dr. Hosack to Prof. Green. *American Journal of Science*, vol. 4, pp. 397-98.

HUMPHRIES, F. J. and CLIFF, R. A. 1982. Sm-Nd dating and cooling history of Scourian granulites, Sutherland. *Nature*, vol. 295, pp. 515-17.

HUTCHINSON, H. 1951. Gold from Scottish Hills. *Scottish Field*, July, pp. 24-26.

HUTTON, J. 1788. Theory of the Earth: or an Investigation of the Laws observable in the Composition, Dissolution and Restoration of the Land upon the Globe. *Transactions Royal Society Edinburgh*, vol. 1, part 2, pp. 209-304.

IMRIE, N. 1798. A short mineralogical description of the Mountain of Gibraltar. *Transactions Royal Society Edinburgh*, vol. 4, pp. 191-202.

IMRIE, N. 1810. Some remarks upon the Pudding or Conglomerate Rock, which stretches along the whole of the South Front of the Grampian Mountains in Scotland, from where they commence in the West, to where they finish their course towards the East in the German Ocean. *Memoirs of the Wernerian Natural History Society*, vol. 1, pp. 453-60.

IMRIE, N. 1812. A Geological Account of the Southern District of Stirlingshire, commonly called the Campsie Hills, with a few remarks relative to the two prevailing Theories as to Geology, and some examples given illustrative of these remarks. *Memoirs of the Wernerian Natural History Society*, vol. 2, pp. 24-50.

IMRIE, N. 1812a. A Description of the Strata which occur in ascending from the Plains of Kincardineshire to the Summit of Mount Battoc, one of the most elevated points in the Eastern District of the Grampian Mountains. *Transactions Royal Society Edinburgh*, vol. 6, pp. 3-19.

IMRIE, N. 1817. *A catalogue of specimens, illustrative of the geology of Greece, and part of Macedonia*, 15pp.

INGRAM, S. M. 1994. Agates from Ardownie Quarry, Monifieth, Tayside, Scotland. *UK Journal of Mines and Minerals*, no 14, pp. 17-20.

INGRAM, S. M., ANDERSON, D. G. and TODD, J. G. 1993. Molybdenite from Coire Buidhe, Glen Creran, Argyll. *UK Journal of Mines and Minerals*, no. 12, pp. 4-8.

INGRAM, S. M., TODD, J. G. and ANDERSON, D. G. 1992. Pyrite from Goat Quarry – An exceptional Scottish occurrence. *UK Journal of Mines and Minerals*, no. 11, pp. 8-10.

IXER, R. A. F., PATTRICK, R. A. D. and STANLEY, C. J. 1996. *The geology and genesis of gold mineralisation at Calliachar-Urlar Burn, Scotland*, in Smith (Chairman) Mineralisation in the Caledonides, the Mike Gallagher Memorial Meeting, Abstracts Volume. Institution of Mining and Metallurgy, Edinburgh Geological Society and Irish Association for Economic Geology.

JACKSON, B. 1984. Sapphire from Loch Roag, Isle of Lewis, Scotland. *Journal of Gemmology*, vol. 19, pp. 336-42.

JACKSON, B. 1990. Queitite, a first Scottish occurrence. *Scottish Journal of Geology*, vol. 26, pp. 57-58.

JAMESON, L. 1854. Biographical Memoir of the late Professor Jameson. *Edinburgh New Philosophical Journal*, vol. 57, pp. 1-49.

JAMESON, R. 1798. *An outline of the mineralogy of the Shetland Islands, and of the Isle of Arran. With an Appendix containing observations on Peat, Kelp and Coal.* William Creech, Edinburgh, 202pp.

JAMESON, R. 1800. *Outline of the mineralogy of the Scottish Isles; with mineralogical observations made in different parts of the mainland of*

*Scotland, and dissertations upon Peat and Kelp.* C. Stewart and Co., Edinburgh, vol. 1, 243pp., vol. 2, 289pp.

JAMESON, R. 1804-08. *System of Mineralogy.* William Blackwood, Edinburgh, vol. 1 (1804), 607pp., vol. 2 (1805), 625pp., vol. 3 (1808), 368pp.

JAMESON, R. 1805a. *Treatise on the external characters of minerals,* 84pp, *with Tabular view of the different generic and subordinate special external characters of minerals,* 32pp. Edinburgh.

JAMESON, R. 1805b. *A mineralogical description of the county of Dumfries.* Bell & Bradfute and E. Blackwood, Edinburgh, 185pp.

JAMESON, R. 1809. On cryolite. *Memoirs of the Wernerian Natural History Society,* vol. 1, pp. 465-68.

JAMESON, R. 1809a. On the topaz of Scotland. *Memoirs of the Wernerian Natural History Society,* vol. 1, pp. 447-52, and p. 628.

JAMESON, R. 1813. *Mineralogical travels through the Hebrides, Orkney and Shetland Islands, and Mainland of Scotland, with dissertations upon peat and kelp.* Edinburgh, vol. 1, 243pp., vol. 2, 289pp.

JAMESON, R. 1816. *A System of Mineralogy.* 2nd edn, Archibald Constable & Co., Edinburgh, vol. 1, 537pp., vol. 2, 489pp., vol. 3, 599pp.

JAMESON, R. 1816a. *A Treatise on the external, chemical, and physical characters of minerals.* 2nd edn, Archibald Constable & Co., Edinburgh, 304pp.

JAMESON, R. 1817. *A Treatise on the external, chemical, and physical characters of minerals.* 3rd edn, Archibald Constable & Co., Edinburgh, 314pp.

JAMESON, R. 1817a. Instructions to Collectors, in Sweet, 1972.

JAMESON, R. 1820. *A System of Mineralogy.* 3rd edn, Archibald Constable & Co., Edinburgh, vol. 1, 405pp., vol. 2, 632pp., vol. 3, 596pp.

JAMESON, R. 1821. *Manual of Mineralogy: containing An Account of Simple minerals, and also A Description and Arrangement of Mountain Rocks.* Archibald Constable & Co., Edinburgh, 501pp.

JAMESON, R. 1822. In Brewster and Jameson, editors; Notice of mineralogical journeys, and of a mineralogical system, by the late Rev. Dr. John Walker, Professor of Natural History in the University of Edinburgh. *Edinburgh Philosophical Journal,* vol. 6, pp. 88-94.

JAMESON, R. 1829. List of Geological and Mineralogical Collections in Great Britain and Ireland. *Edinburgh New Philosophical Journal,* vol. 7, pp. 113-15.

JAMESON, R. 1833. Chemical analysis of stratified rocks altered by plutonean agency; and analysis of Largo Law basaltic rock and wollastonite from Corstorphine Hill. *Edinburgh New Philosophical Journal,* vol. 15, pp. 386-88.

JAMESON, R. 1837. Mineralogy. *Encyclopaedia Britannica,* and published separately in 1837, Adam and Charles Black, Edinburgh, 283pp.

JAMESON, R. 1840. Notice of Greenockite, a new Mineral Species of the Order Blende. *Edinburgh New Philosophical Journal,* vol. 28, pp. 390-92.

JAMESON, R. 1843. Scientific Intelligence – Mineralogy and Chemistry. *Edinburgh New Philosophical Journal,* vol. 34, p. 180.

JARDINE, W. 1842. The Naturalist's Library. *Ornithology,* vol. 12, W. H. Lizars, Edinburgh, 349pp.

JEHU, T. J. 1932. Obituary Notices. *Proceedings Royal Society Edinburgh,* vol. 51, pp. 202-04.

JOHNSTON, J. D. 1995. Pseudomorphs after ikaite in a glaciomarine sequence in the Dalradian of Donegal, Ireland. *Scottish Journal of Geology,* vol. 31, pp. 3-9.

JOHNSTON, J. F. W. 1831. On the discovery of vanadium in Scotland, and on the vanadiate of lead, a new mineral species. *Edinburgh Journal of Science,* new series, vol. 5, pp. 166-68.

JONES, J. 1984. The geological collection of James Hutton. *Annals of Science,* vol. 41, pp. 223-44.

JONES, L. H. P. and MILNE, A. A. 1956. Birnessite, a new manganese oxide mineral from Aberdeenshire, Scotland. *Mineralogical Magazine,* vol. 31, pp. 283-88.

KAY, J. 1842. *Kay's Portraits, A series of original portraits and caricature etchings.* Hugh Paton, Edinburgh, vol. 2, pp. 178-82.

KEITH, A., KEITH, J. and BREWSTER, D. 1820. Account of the establishment of a scientific prize by the late Alexander Keith, Esq. of Dunottar. *Transactions Royal Society Edinburgh,* vol. 9, pp. 259-60.

KENNGOTT, A. 1868. Notiz über die Krystallgestalten des susannit und Leadhillit. *Neues Jahrbuch für Mineralogie,* pp. 319-20.

KHOMJAKOV, A. P., POLEZHAEVA, L. L. and SOKOLOVA, E. V. 1994. Crawfordite $Na_3Sr(PO_4)(CO_3)$ - a new mineral from the bradleyite family. *Proceedings Russian Mineralogical Society,* vol. 123, pp. 107-11.

KING, P. M. 1976. The secondary minerals of the Tertiary lavas of northern and central Skye – zeolite zonation patterns their origins and formation. Unpublished Ph.D. thesis, University of Aberdeen.

KIRWAN, R. 1794-96. *Elements of Mineralogy.* P. Elmsley, London, vol. 1, 1794, 510pp., vol. 2, 1796, 529pp.

KLAPROTH, M. H. 1795-1815. *Beiträge zur chemischen Kenntniss der Mineralkörpers.* Berlin and Posen, vol. 1, 1795; vol. 2, 1797; vol. 3, 1802; vol. 4, 1807; vol. 5, 1810; vol. 6, 1815.

KNELLER, B. C. 1985. In Conditions of Dalradian metamorphism. Regional corundum in the NE Dalradian. *Quarterly Journal Geological Society,* London, vol. 142, p. 4.

KOBELL, F. von 1838. *Grundzüge Mineralogie,* p. 283.

LACROIX, A. 1885. Sur la plumbocalcite de Wanlockhead (Écosse). *Bulletin Societé Minéralogie de France,* vol. 8, pp. 36-38.

LANDLESS, J. G. 1985. *A Gazetteer to the Metal Mines of Scotland.* The Wanlockhead Museum Trust, Occasional paper no. 1, 56pp.

LANG, V. von 1859. Versuch einer Monographic des Bleivitriols. *Sitzungsberichie-Kaiseruche Akademie Wissen.* Wien, vol. 36, pp. 241-92.

LAUDER LINDSAY, W. 1867. The goldfields of Scotland. *Journal of the Royal Geological Society of Ireland,* vol. 2, pp. 176-88.

LAUDER LINDSAY, W. 1880. Museum specimens of native Scottish gold. *Transactions Edinburgh Geological Society,* vol. 3, pp. 153-68.

LEAKE, B. E. 1978. Nomenclature of Amphiboles. *Mineralogical Magazine*, vol. 42, pp. 533-63.

LEAKE, B. E. 1997. (Chairman) Nomenclature of amphiboles: Report on the subcommittee on amphiboles of the International Mineralogical Association Commission on New Minerals and Mineral Names. *Mineralogical Magazine*, vol. 61, pp. 295-321.

LEAKE, R. C., BLAND, D. J. and COOPER, C. 1993. Source characterization of alluvial gold from mineral inclusions and internal compositional variation. *Transactions Institution Mining and Metallurgy*, vol. 102, pp. B65-82.

LEONHARD, C. C., MERZ, K. F. and KOPP, J. H. 1806. *Systematisch-tabellarische Übersicht und Charakteristik der Mineralkörper*. Frankfurt am Maine.

LIVINGSTONE, A. 1974. An occurrence of tacharanite and scawtite in the Huntly gabbro, Aberdeenshire. *Mineralogical Magazine*, vol. 39, pp. 820-21.

LIVINGSTONE, A. 1976. Julgoldite, new data and occurrences; a second recording. *Mineralogical Magazine*, vol. 40, pp. 761-63.

LIVINGSTONE, A. 1980. Johnsomervilleite, a new transition – metal phosphate mineral from the Loch Quoich area, Scotland. *Mineralogical Magazine*, vol. 43, pp. 833-36.

LIVINGSTONE, A. 1984. Fluorine in sarcolite: additional history and new chemical data. *Mineralogical Magazine*, vol. 48, pp. 107-12.

LIVINGSTONE, A. 1988. Reyerite, tobermorite, calcian analcime and bytownite from amygdales in a Skye basalt. *Mineralogical Magazine*, vol. 52, pp. 711-13.

LIVINGSTONE, A. 1989. Low-temperature, hydrothermal garnet associated with zeolites, from basalt lavas near Beith, Ayrshire. *Mineralogical Magazine*, vol. 53, pp. 125-29.

LIVINGSTONE, A. 1989a. A calcian analcime-bytownite intergrowth in basalt, from Skye, Scotland, and calcian analcime relationships. *Mineralogical Magazine*, vol. 53, pp. 382-85.

LIVINGSTONE, A. 1990. Matthew Forster Heddle (1828-1897), famous Scottish mineralogist. *Journal Russell Society*, vol. 3, pp. 61-65.

LIVINGSTONE, A. 1992. The composition of stolzite from Wanlockhead and wulfenite from various Scottish localities. *Journal Russell Society*, vol. 4, pp. 55-57.

LIVINGSTONE, A. 1993. Origin of the leadhillite polymorphs. *Journal Russell Society*, vol. 5, pp. 11-14.

LIVINGSTONE, A. 1993a. Glossary of Scottish mineral species – an update. *Scottish Journal of Geology*, vol. 29, pp. 87-101.

LIVINGSTONE, A. 1993b. Epilog: Rescue of historic early Giesecke specimens at Nairn museum, Scotland. *Mineralogical Record*, vol. 24(2), pp. 66-67.

LIVINGSTONE, A. 1994. Analyses of calcian phosphatian vanadinite, and apatite high in lead, from Wanlockhead, Scotland. *Journal Russell Society*, vol. 5, pp. 124-26.

LIVINGSTONE, A. and MacPHERSON, H. G. 1983. Fifth supplementary list of British minerals (Scottish). *Mineralogical Magazine*, vol. 47, pp. 99-105.

LIVINGSTONE, A. and RUSSELL, J. D. 1985. X-ray powder data for susannite and its distinction from leadhillite. *Mineralogical Magazine*, vol. 49, pp. 759-61.

LIVINGSTONE, A., RYBACK, G., FEJER, E. E. and STANLEY, C. J. 1987. Mattheddleite, a new mineral of the apatite group from Leadhills, Strathclyde Region. *Scottish Journal of Geology*, vol. 23, pp. 1-8.

LIVINGSTONE, A. and SARP. H. 1984. Macphersonite, a new mineral from Leadhills, Scotland, and Saint-Prix, France – a polymorph of leadhillite and susannite. *Mineralogical Magazine*, vol. 48, pp. 277-82.

LLOYD, B. and LLOYD, M. 2000. The Journals of Robert Ferguson (1767-1840). *Mineralogical Record*, vol. 31, pp. 425-42.

LOWRY, D., BOYCE, A. J., FALLICK, A. E. and STEPHENS, W. E. 1995. Genesis of porphyry and plutonic mineralisation systems in metaluminous granitoids of the Grampian Terrane, Scotland. *Transactions Royal Society Edinburgh*, vol. 85, pp. 221-37.

LOWRY, D., STEPHENS, W. E., HERD, D. A. and STANLEY, C. J. 1994. Bismuth sulphosalts within quartz veining hosted by the Loch Shin monzogranite, Scotland. *Mineralogical Magazine*, vol. 58, pp. 39-47.

LUDWIG, C. F. 1803, 1804. *Handbuch der Mineralogie nach A. G. Werner*. Leipzig, vol. 1, 1803, 369pp., vol. 2, 1804, 226pp.

LUMSDEN, G. I., TULLOCH, W., HOWELLS, M. F. and DAVIES, A. 1967. The Geology of the neighbourhood of Langholm. *Memoirs Geological Survey Scotland*, sheet 11, 255pp.

LYELL, C. 1881. *Life, letters and journals of Sir Charles Lyell*, ed. Mrs Lyell. J. Murray, vol. 1, pp. 156-57.

MACGREGOR, M., HERRIOT, A. and KING, B. C. 1972. *Excursion Guide to the Geology of Arran*. Geological Society of Glasgow, 199pp.

MacGREGOR, A. G., MacGREGOR, M. and ROBERTSON, T. 1944. Barytes in Central Scotland. *Wartime Pamphlet No. 38*, Geological Survey Great Britain, 34pp.

MACKENZIE, G. S. 1800. Experiments on the combustion of the diamond, the formation of steel by its combination with iron, and the pretended transmission of carbon through the vessels. *Journal of Natural Philosophy*, vol. 4, pp. 103-10.

MACKENZIE, G. S. 1815. An account of some geological facts observed in the Faroe Islands. *Transactions Royal Society Edinburgh*, vol. 7, pp. 213-27.

MACKENZIE, G. S. 1826. On the formation of chalcedony. *Transactions Royal Society Edinburgh*, vol. 10, pp. 82-104.

MACKENZIE, R. C. 1957. Saponite from Allt Ribhein, Fiskavaig Bay, Skye. *Mineralogical Magazine*, vol. 31, pp. 672-80.

MACMILLAN, D. 1990. *Scottish Art 1460-1990*. Mainstream Publishing Co., Edinburgh, 432pp.

MACNAIR, P. 1904. On Pseudogaylussite dredged from the Clyde at Cardross, and other recent additions to the Mineral Collections in the Kelvingrove Museum. *Proceedings Royal Philosophical Society Glasgow*, vol. 35, pp. 3-11.

MACPHERSON, H. G. 1989. *Agates*. British Museum (Natural History) London, and National Museums of Scotland, Edinburgh, 72pp.

MACPHERSON, H. G. and LIVINGSTONE, A. 1982. Glossary of Scottish mineral species 1981. *Scottish Journal of Geology*, vol. 18, pp. 1-47.

MALLET, J. W. 1859. On Brewsterite. *London, Edinburgh and Dublin Philosophical Magazine and Journal of Science*, vol. 18, pp. 218-20.

MARLAND, G. 1975. Stability of calcium carbonate hexahydrate (ikaite). *Geochimica et Cosmochimica Acta*, vol. 39, pp. 83-91.

McCALLIEN, W. J. 1937. *Scottish Gem Stones*. Blackie and Son, Ltd., London, 120pp.

McCONNELL, J. D. C. 1954. The hydrated calcium silicates riversideite, tobermorite, and plombierite. *Mineralogical Magazine*, vol. 30, pp. 293-305.

McCRACKEN, A. 1965. The Glendinning Antimony mine (Louisa Mine). *Transactions Dumfriesshire and Galloway Natural History and Antiquarian Society*, vol. 42, pp. 140-48.

McKAY, M. M. 1980. *The Rev. Dr. John Walker's Report on the Hebrides of 1764 and 1771*. John Donald, Edinburgh, 256pp.

McLACHLAN, G. R. 1951. The aegirine-granulites of Glen Lui, Braemar, Aberdeenshire. *Mineralogical Magazine*, vol. 29, pp. 476-95.

McMILLAN, A. A. 1997. *Quarries of Scotland*. Technical, Conservation, Research and Education Division, Historic Scotland, Edinburgh, 84pp.

McMULLEN, M. J. and TODD, J. G. 1990. Mineralisation of the Kinharvie Burn, South West Scotland. *UK Journal of Mines and Minerals*, no. 8, pp. 43-45.

McNEISH, C. 1996. *The Munros, Scotland's Highest Mountains*. Lomond Books, Edinburgh, 228pp.

MEIKLE, T. J. K. 1970. Mineral recognition for beginners, part 1. *Gems*, vol. 2, no. 2, pp. 7 and 8, and p. 28.

MEIKLE, T. J. K. 1970a. Mineral recognition for beginners, part 2: How to use geological maps. *Gems*, vol. 2, no. 3, pp. 11 and 12, and p. 23.

MEIKLE, T. J. K. 1989. The secondary mineralogy of the Clyde Plateau lavas, Scotland, part 1: Boyleston quarry. *Journal Russell Society*, vol. 2, pp. 11-14.

MEIKLE, T. J. K. 1989a. The secondary mineralogy of the Clyde Plateau lavas, Scotland, part 2: Loanhead quarry. *Journal Russell Society*, vol. 2, pp. 15-21.

MEIKLE, T. J. K. 1990. The secondary mineralogy of the Clyde Plateau lavas, Scotland, part 3: Hartfield Moss. *Journal Russell Society*, vol. 3, pp. 43-47.

MEIKLE, T. J. K. 1992. Growth of acanthite on native silver from the Clyde Plateau lavas, Scotland. *Journal Russell Society*, vol. 4, pp. 67-69.

MEIKLE, T. J. K. 1992a. Greenockite from Bishopton. The type locality – a review. *UK Journal of Mines and Minerals*, no. 11, pp. 4-6.

MEIKLE, T. J. K. 1994. Native silver from Hilderston Mine, West Lothian, Scotland. *Journal Russell Society*, vol. 5, pp. 83-90.

MEIKLE, T. J. K. 1994a. Loanhead Quarry – an update. *British Micromount Society Newsletter*, no. 38, pp. 2-3.

MEIKLE, T. K. and TODD, J. G. 1995. Silica-rich edingtonite and associated minerals from Loanhead quarry, Beith, Strathclyde – a new Scottish locality. *Journal Russell Society*, vol. 6, pp. 27-30.

MELLOR, J. W. 1929. *A comprehensive treatise on inorganic and theoretical chemistry*, vol. 9, Longmans Green, London, p. 714.

MENDES DA COSTA, E. 1812. Notices and Anecdotes of Literati, Collectors, etc. from a MS. by the late Mendes da Costa, and collected between 1747 and 1788. *Gentleman's Magazine*, vol. 82, pp. 205-7, 513 and 515.

MILLER, H. 1842. *The Old Red Sandstone*. 2nd edn, John Johnstone, Edinburgh, 288pp.

MILLER, J. M. and TAYLOR, K. 1966. Uranium mineralization near Dalbeattie, Kirkcudbrightshire. *Bulletin of the Geological Survey of Great Britain*, no. 25, pp. 1-18.

MILLER, J. W. 1973. A visit to Rabbie Burns country and return to Leadhills. *The South African Lapidary Magazine*, vol. 7, pp. 9-10.

MITCHELL, G. H., WALTON, E. K. and GRANT, D. (eds) 1960. *Edinburgh Geology, an excursion guide*. Oliver and Boyd, Edinburgh and London, p. 68.

MITCHELL, R. S. 1979. *Mineral Names What Do They Mean?* Van Nostrand Reinhold Co., London, p. 117.

MITCHELL, R. S. 1988. Who's who in Mineral Names. *Rocks and Minerals*, vol. 63, pp. 304-8.

MOHS, F. 1804. *Des Herrn von der Null mineralien kabinet nach einem, durchaus auf aussere Kennzeichnen gegründeten Systeme geordenet*, 3 vols, 3rd edn, Vienna.

MOHS, F. 1820. *The Characters of the Classes, Orders, Genera, and Species; Or, The Characteristic of the Natural History System of Mineralogy*. W. and C. Tait, Edinburgh, 109pp.

MOORBATH, S. 1962. Lead isotope abundance studies on mineral occurrences in the British Isles and their geological significance. *Philosophical Transactions Royal Society London*. Series a, vol. 254, pp. 295-360.

MORETON, S. 1996. The Alva Silver Mine Silver Glen, Alva Scotland. *Mineralogical Record*, vol. 27, pp. 405-14.

MORETON, S., ASPEN, P., GREEN, D. I. and INGRAM, S. M. 1998. The silver and cobalt mineralisation near Alva, Central Region, Scotland. *Journal Russell Society*, vol. 7, pp. 23-30.

MORIMOTO, N. 1988. Nomenclature of pyroxenes. *Mineralogical Magazine*, vol. 52, pp. 535-50.

MORRISON-LOW, A. D. 1992. William Nicol, FRSE c.1771-1851 Lecturer, Scientist and Collector. *Book of the Old Edinburgh Club*, new series, vol. 2, pp. 123-31.

MORRISON-LOW, A. D. and CHRISTIE, J. R. R. (eds) 1984. *'Martyr of Science': Sir David Brewster 1781-1868*. Proceedings of a Bicentenary Symposium, Royal Scottish Museum, 1981, together with a catalogue of scientific apparatus associated with Sir David Brewster and a bibliography of his published writings. Royal Scottish Museum, Edinburgh, 138pp.

MORRISON-LOW, A. D. and NUTTALL, R. H. 1984. A note on early fossil wood sections from the Allen Thomson collection. *Microscope*, vol. 32, pp. 23-28.

MROSE, M. E. and CHRISTIAN, R. P. 1969. The leadhillite-susannite relation (abstract only). *Canadian Mineralogist*, vol. 10, p. 141.

MUIR, A., HARDIE, H. G. M., MITCHELL, R. L. and PHEMISTER, J. 1956. The Limestones of Scotland. *Memoir Geological Survey, Special Reports on the mineral resources of Great Britain*, vol. 37, 150pp.

MURRAY, D. 1904. *Museums Their History and their Use*. James MacLehose and Sons, Glasgow, vol. 3, p. 227.

MURRAY, J. 1802. *A Comparative View of the Huttonian and Neptunian Systems of Geology: In Answer to the Illustrations of the Huttonian Theory of the Earth, by Professor Playfair*. Edinburgh, facsimile reprint, 1978, 256pp.

MURRAY, J. 1815. On the diffusion of heat at the surface of the Earth. *Transactions Royal Society Edinburgh*, vol. 7, pp. 411-34.

MURRAY, J. 1845, in *The New Statistical Account of Scotland*. Blackwood and Sons, Edinburgh, vol. 9, p. 153.

MYKURA, W. 1976. *Orkney and Shetland*. British Regional Geology, Institute of Geological Sciences, Edinburgh, 149pp.

NAGEL, J. 1994. Collections and the information age. *Mineralogical Record*, vol. 25, pp. 82-83.

NAWAZ, R. 1990. Brewsterite: re-investigation of morphology and elongation. *Mineralogical Magazine*, vol. 54, pp. 654-56.

NICHOLLS, G. D. 1950. The Glenelg-Ratagain igneous complex. *Quarterly Journal Geological Society London*, vol. 106, pp. 309-44.

NICHOLSON, K. 1983. Manganese mineralisation in Scotland. Unpublished Ph.D. thesis, University of Strathclyde.

NICHOLSON, K. 1988. Manganese cemented Quaternary gravels: birnessite, lithiophorite and quenselite from an example in Wigtownshire. *Scottish Journal of Geology*, vol. 24, pp. 194-200.

NICHOLSON, K. 1988a. Birnessite from Gourock, Renfrewshire, Scotland. *Mineralogical Magazine*, vol. 52, p. 415.

NICHOLSON, K. 1990. Stratiform manganese mineralisation near Inverness, Scotland: A Devonian sublacustrine hot-spring deposit? *Mineralium Deposita*, vol. 25, pp. 126-131.

NICOL, W. 1829. On a method of so far increasing the divergency of the two rays in calcareous-spar, that only one image may be seen at a time. *Edinburgh New Philosophical Journal*, vol. 6, pp. 83-84.

NICOL, W. 1839. Notice concerning an improvement in the construction of the single vision prism of calcareous spar. *Edinburgh New Philosophical Journal*, vol. 27, pp. 332-33.

ODERNHEIMER, F. 1841. On the mines and minerals of the Breadalbane Highlands. *Prize-Essays and Transactions of the Highland and Agricultural Society of Scotland*, new series, vol. 7, pp. 541-56.

OFFER, D. J. 1995. *Investigation of the Loch Roag sapphire deposit*. Unpublished M.Sc. thesis, University of Leicester, 195pp.

OKEN, L. 1813. *Oken's Lehrbuch der Naturgeschichte Iter Theil Mineralogie*. Leipzig.

PAAR, W. H., BRAITHWAITE, R. S. W., CHEN, T. T. and KELLER, P. 1984. A new mineral, scotlandite ($PbSO_3$) from Leadhills, Scotland; the first naturally occurring sulphite. *Mineralogical Magazine*, vol. 48, pp. 283-88.

PAAR, W. H., MEREITER, K., BRAITHWAITE, R. S. W., KELLER, P. and

DUNN, P. J. 1986. Chenite, $Pb_4Cu(SO_4)_2(OH)_6$, a new mineral, from Leadhills, Scotland. *Mineralogical Magazine*, vol. 50, pp. 129-35.

PALACHE, C., BERMAN, H. and FRONDEL, C. 1951. *The System of Mineralogy of James Dwight Dana and Edward Salisbury Dana*. 7th edn, New York, vol. 2, pp. 298-99.

PALACHE, C., BERMAN, H. and FRONDEL, C. 1966. *The System of Mineralogy of James Dwight Dana and Edward Salisbury Dana*. 7th edn, John Wiley and Sons, London, vol. 1, p. 762.

PANETH, F. A. 1960. The Discovery and Earliest Reproductions of the Widmanstätten Figures. *Geochimica et Cosmochimica Acta*, vol. 18, pp. 176-82.

PANHUYS-SIGLER, M. van, TREWIN, N. H. and STILL, J. 1996. Roscoelite associated with reduction spots in Devonian red beds, Gamrie Bay, Banffshire. *Scottish Journal of Geology*, vol. 32, pp. 127-32.

PARKER, R. T. G., CLIFFORD, J. A. and MELDRUM, A. H. 1989. The Cononish gold-silver deposit, Perthshire, Scotland. *Transactions of the Institution of Mining and Metallurgy*, vol. 98, pp. B51-52.

PARNELL, J. 1988. Mercury and silver-bismuth selenides at Alva, Scotland. *Mineralogical Magazine*, vol. 52, pp. 719-20.

PATTRICK, R. A. D. 1984. Sulphide mineralogy of the Tomnadashan copper deposit and the Corrie Buie lead veins, south Loch Tayside, Scotland. *Mineralogical Magazine*, vol. 48, pp. 85-91.

PATTRICK, R. A. D., BOYCE, A. and MacINTYRE, R. M. 1988. Gold-silver vein mineralization at Tyndrum, Scotland. *Mineralogy and Petrology*, vol. 38, pp. 61-76.

PAULY, H. 1963. 'Ikaite', a new mineral from Greenland. *Arctic*, vol. 16, pp. 263-64.

PEACH, B. N., HORNE, J., CLOUGH, C. T. and HINXMAN, L. W. 1907. The Geological Structure of the North-West Highlands of Scotland. *Memoirs of the Geological Survey of Great Britain*, 668pp.

PEACOCK, M. A. 1927. The geology of Iceland: the pioneer work of a Scottish geologist. *Transactions Geological Society Glasgow*, vol. 17, pp. 185-203 and 271-333.

PETERSEN, O. V. and SECHER, K. 1993. The minerals of Greenland. *Mineralogical Record*, vol. 24, pp. 4-65.

PHEMISTER, J. 1936 and 1948. *British Regional Geology Scotland: The Northern Highlands* (2nd edn, 1948), p. 17.

PHILLIPS, W. 1816, 1819, 1823. *An elementary introduction to the knowledge of Mineralogy*. 1st, 2nd, 3rd edns, London.

PHILLIPS, W. 1837. *An elementary introduction to Mineralogy*. 4th edn, London, 425pp.

PIPER, J. D. A. 1978. Palaeomagnetism and palaeogeography of the Southern Uplands block in Ordovician times. *Scottish Journal of Geology*, vol. 14, pp. 93-107.

PISANI, F. 1873. Analyse de la lanarkite de Leadhills (Écosse). *Compte Rendus*, vol. 76, pp. 114-16.

PORTEOUS, J. M. 1876. *God's Treasure House in Scotland*. Simpkin, Marshall and Co., London, pp. 96-98.

PRICHARD, H. M., POTTS, P. J., NEARY, C. R., LORD, R. A. and WARD, G. R.

1987. Development of techniques for the determination of the platinum-group elements in ultramafic rock complexes of potential economic significance: mineralogical studies. *Report commissioned by the European Economic Community Raw Materials Programme*, 163pp.

RACKHAM, R. A. 1983. Henry Witham (1779-1844) and Lartington Hall. *Society of History of Natural History Newsletter*, no. 19, pp. 8-9.

RAISTRICK, A. 1967. *The Hatchett Diary. A tour through the counties of England and Scotland in 1796 visiting their mines and manufactories.* D. Bradford Barton Ltd., Truro, 114pp.

RAPOPORT, P. A. and BURNHAM, C. W. 1973. Ferrobustamite: the crystal structure of two Ca, Fe bustamite-type pyroxenoids. *Zeitschrift für Kristallographie*, vol. 138, pp. 419-38.

READ, H. H. 1931. On corundum-spinel xenoliths in the gabbro of Haddo House, Aberdeenshire. *Geological Magazine*, vol. 68, pp. 446-53.

READ, H. H. 1931a. The geology of central Sutherland. *Memoir Geological Survey, Scotland*, 238pp.

RICHARDSON, J. B. 1974. *Industrial Archaeology*, 12, Metal Mining, Gold. Allen Lane, London, p. 25.

RICHEY, J. E., ANDERSON, E. M. and MacGREGOR, A. G. 1930. The Geology of North Ayrshire. *Memoirs of the Geological Survey, Scotland*, sheet 22, 417pp.

RICHEY, J. E., THOMAS, H. H., BAILEY, E. B., SIMPSON, J. B., EYLES, V. A. and LEE, G. W. 1930a. The Geology of Ardnamurchan, North-west Mull and Coll. *Memoirs of the Geological Survey, Scotland*, 393pp.

RICHMOND, W. E. and WOLFE, C. W. 1938. Crystallography of lanarkite. *American Mineralogist*, vol. 23, pp. 799-804.

ROLFE, W. D. I. 1985. Museum file 3: Hunterian Museum, University of Glasgow. *Geology Today*, vol. 1, Jul.–Aug., pp. 125-27.

ROMÉ DE L'ISLE 1783. *Cristallographie, ou description des formes propres a tout les corps du regne minéral, dans l'état de combinaison faline, pierreuse ou métallique, avec figures & tableaux synoptiques de tous les cristaux connus.* 2nd edn, Paris, vol. 1, 623pp., vol. 2, 659pp., vol 3, 611pp.

ROSE, A. 1847. *The Witness*, 30 June.

ROSE, H. 1827. On a chemical composition of zinkenite and jamesonite (and description and analysis of pyrochlore, a new mineral by F. Wöhler). *Edinburgh New Philosophical Journal*, vol. 2, pp. 341-43.

ROSE, M. T. S. (undated). *Alexander Rose Geologist and his grandson Robert Traill Rose Artist.* C. J. Cousland and Sons Ltd., Edinburgh, 67pp.

ROTHWELL, M. and MASON, J. 1992. Wulfenite in the British Isles. *UK Journal of Mines and Minerals*, no. 11, pp. 30-37.

RUSHTON, D. R. A. 1972. Arsenates of copper from Shetland. *Mineralogical Magazine*, vol. 38, pp. 626-27.

RUSSELL, A. 1944. Notes on some minerals either new or rare to Britain. *Mineralogical Magazine*, vol. 27, pp. 1-10.

RUSSELL, A. 1946. An account of the Struy lead mines, Inverness-shire, and of wulfenite, harmotome, and other minerals which occur there. *Mineralogical Magazine*, vol. 27, pp. 147-54.

RUSSELL, A. 1952. John Henry Heuland. *Mineralogical Magazine*, vol. 29, pp. 395-405.

RUSSELL, A. 1952a. Philip Rashleigh of Menabilly, Cornwall, and his mineral collection. *Journal Royal Institution of Cornwall*, vol. 1, pp. 96-118.

RUSSELL, A. 1954. John Hawkins, FGS, FRHS, FRS, 1761-1841. A distinguished Cornishman and early mining geologist. *Journal Royal Institution of Cornwall*, new series, vol. 2, pp. 98-106.

RUSSELL, J. D., FRASER, A. R. and LIVINGSTONE, A. 1984. The infrared absorption spectra of the three polymorphs of $Pb_4(SO_4)(CO_3)_2(OH)_2$ (leadhillite, susannite, and macphersonite). *Mineralogical Magazine*, vol. 48, pp. 295-97.

RUST, S. A. 1983. Notes on crystals of wulfenite from Leadhills, Scotland. *Mineral Realm*, vol. 3, no. 2/3, pp. 11-12.

RUST, S. A. 1994. Chenite from Llechwedd Helyg mine, Tir-y-Mynach, Dyfed, Wales. *UK Journal of Mines and Minerals*, no. 14, p. 9.

SAMSON, I. M. and BANKS, D. A. 1988. Epithermal base-metal vein mineralization in the Southern Uplands of Scotland: Nature and origin of the fluids. *Mineralium Deposita*, vol. 23, pp. 1-8.

SAUSSURE, H. B. 1779-1796. *Voyages dans les Alpes*. Neuchâtel, vols. 1 and 2, 1779-80; vols. 3, 4, 1796.

SAVAGE, K. D. 1995. Erionite from Edinbane, Isle of Skye, Scotland. *UK Journal of Mines and Minerals*, no. 15, p. 18.

SCHRAUF, A. 1877. Die Krystallographischen constanten des lanarkit. *Zeitschrift für Kristallographie*, vol. 1, pp. 31-38.

SCHUBNEL, H. J. 1987. *Giant crystals precious minerals*. Muséum National d'Histoire Naturelle, Paris, 62pp.

SCHUBNEL, H. J. 1992. *Les types d'espèces minérales des premiers cristallographes*. Muséum National d'Histoire Naturelle, Paris, 12pp.

SCOTT, B. 1967. Barytes mineralization at Gasswater mine, Ayrshire, Scotland. *Transactions Institution Mining Metallurgy*, vol. 76, pp. B40-51.

SCOTT, H. W. 1966. *Lectures on Geology, including hydrography, mineralogy, and meteorology with an introduction to biology by John Walker.* University of Chicago Press, Chicago and London, 280pp.

SCOTT, H. W. 1976. *Dictionary of Scientific Biography*. Charles Scribner's Sons, New York, pp. 131-33.

SHAW, M. H., GUNN, A. G., ROLLIN, K. E. and STYLES, M. T. 1994. Platinum-group element mineralisation in the Loch Ailsh alkaline igneous complex, north-west Scotland. *Mineral Reconnaissance Programme*, rpt. 131 WF/94/2, British Geological Survey, 88pp.

SHEARMAN, D. J. and SMITH, A. J. 1985. Ikaite, the parent mineral of jarrowite-type pseudomorphs. *Proceedings Geologists' Association*, vol. 96, pp. 305-14.

SIBBALD, R. 1697. *Auctarium Musaei Balfouriani, e Musaeo Sibbaldiano.* Edinburgh, 216pp.

SIMONS, W. V. 1866. *A catalogue of foreign minerals in the possession of the Mining Department, Melbourne, Victoria.* John Ferres, Melbourne, 74pp.

SIMPSON, A. D. C. 1982. Sir Robert Sibbald – The founder of the College. *Proceedings of the Royal College of Physicians of Edinburgh*, Tercentenary Congress, 1981. Edinburgh, pp. 59-91.

SIMPSON, J. B and MacGREGOR, A. G. 1932. The economic geology of the Ayrshire coalfields. Area IV; Dailly, Patna, Rankinston, Dalmellington, and New Cumnock. *Memoirs of the Geological Survey, Scotland*, p. 156.

SINCLAIR. J. 1796. *The Statistical Account of Scotland*. William Creech, Edinburgh, vol. 7, pp. 54-5, and vol. 17, p. 108.

SKINNER, B. J. and SKINNER, H. C. W. 1980. Is there a limit to the number of minerals? *Mineralogical Record*, vol. 11, pp. 333-35.

SMITH, C. J. 1986. *Historic South Edinburgh*. Charles Skilton Ltd., Edinburgh and London, vol. 3, 166pp.

SMITH, D. G. W. 1965. The chemistry and mineralogy of some emery-like rocks from Sithean Sluaigh, Strachur, Argyllshire. *American Mineralogist*, vol. 50, pp. 1982-2022.

SMITH, D. G. W. 1969. Pyrometamorphism of phyllites by a dolerite plug. *Journal of Petrology*, vol. 10, pp. 20-55.

SMITH, D. G. W. and McCONNELL, J. D. C. 1966. A comparative electron-diffraction study of sillimanite and some natural and artificial mullites. *Mineralogical Magazine*, vol. 35, pp. 810-14.

SMITH, G. F. H. 1907. Obituary of R. P. Greg. *Mineralogical Magazine*, vol. 14, pp. 268-71.

SMITH, G. F. H. 1919. Semseyite from Dumfriesshire. *Mineralogical Magazine*, vol. 18, pp. 354-59.

SMOUT, T. C. 1967. Lead-mining in Scotland, 1650-1850. In *Studies in Scottish Business History*, ed. P. L. Payne. Frank Cass and Co. Ltd., pp. 103-35.

SORBY, H. C. 1858. On the microscopical structure of crystals, indicating the origin of minerals and rocks. *Quarterly Journal Geological Society London*, vol. 14, pp. 453-500.

SOWERBY, J. 1804-17. *British Mineralogy: or Coloured Figures intended to elucidate The Mineralogy of Great Britain*. Richard Taylor and Co., London, vol. 1 (1804), 223pp., vol. 2 (1806), 199pp., vol. 3 (1809), 209pp., vol. 4 (1811), 184pp., vol. 5 (1817), 281pp.

SPEAKE, R. 1982. Kinfauns Castle. *Countrywide Holidays Association*, Manchester, p. 24.

SPENCER, A. M. 1971. Late Pre-Cambrian glaciation in Scotland. *Memoirs Geological Society London*, no. 6, 100pp.

SPENCER, L. J. 1931. Second supplementary list of British minerals. *British Association for the Advancement of Science Report*, p. 378.

STACE, H. E., PETTITT, C. W. A. and WATERSTON, C. D. 1987. *Natural Science Collections in Scotland* (Botany, Geology, Zoology). National Museums of Scotland, Edinburgh, 373pp.

STAPLES, L. W. 1964. Friedrich Mohs and the scale of hardness. *Journal of Geological Education*, vol. 12, pp. 98-101.

STARKEY, R. E. 1988. Phosgenite from Lossiemouth, Grampian Region: confirmation of the first Scottish occurrence. *Scottish Journal of Geology*, vol. 24, pp. 15-19.

STEACY, H. R. and ROSE, E. R. 1982. Bytownite – a legacy of early Ottawa, a Montreal Medical Doctor and a Royal Engineer. *Mineralogical Record*, vol. 13, pp. 101-5.

STEELE, I. M., PLUTH, J. J. and LIVINGSTONE, A. 1998. Crystal structure of macphersonite, $Pb_4SO_4(CO_3)_2(OH)_2$: comparison with leadhillite. *Mineralogical Magazine*, vol. 62, pp. 451-59.

STEELE, I. M., PLUTH, J. J. and LIVINGSTONE, A. 1999. Crystal structure of susannite, $Pb_4SO_4(CO_3)_2(OH)_2$: a trimorph with macphersonite and leadhillite. *European Journal of Mineralogy*, vol. 11, pp. 493-99.

STEELE, I. M., PLUTH, J. J. and LIVINGSTONE, A. 2000. Crystal structure of mattheddleite: a Pb, S, Si phase with the apatite structure. *Mineralogical Magazine*, vol. 64, pp. 915-21.

STEPHENSON, D. 1983. Hilderston Mine, West Lothian: Mining history and the nature of the vein mineralisation as deduced from old records. *Institute of Geological Sciences*, report NL83/4, 10pp.

STEWART, F. H. 1946. The gabbroic complex of Belhelvie in Aberdeenshire. *Quarterly Journal Geological Society London*, vol. 102, pp. 465-98.

STODDART, J. 1801. *Remarks on local scenery and manners in Scotland during the years 1799 and 1800*. William Miller, London, vol. 2, p. 166.

STORY-MASKELYNE and FLIGHT 1874. Mineralogical Notices (Caledonite and Lanarkite). *Journal London Chemical Society*, vol. 27, pp. 101-3.

STRACHEY 1911. *Memoirs of a Highland Lady*. John Murray, London, 427pp.

STRUNZ, H. and TENNYSON, C. 1956. Polymorphie in der gruppe der blätterzeolithe (heulandit-stilbit-epistilbit; brewsterit). *Neues Jahrbuch für Mineralogie Monatshefte*, vol. 11, pp. 1-9.

SULZER, R. 1791. Über den Strontianit, ein schottisches Foßil, das ebenfalls eine neue Grunderde zu enthalten scheint. *Bergmaernisches Journal*, Freiberg, vol. 1, part 3/5, pp. 433-35. (Aus einem Briefe des Herrn Rath Sulzer zu Ronneburg, mitgeteilt von J. F. Blumenbach.) (Concerning Strontianite, a Scottish 'Fossil' that also appears to contain a new element. [From a letter by Mr Rath Sulzer of Ronneburg, passed on by J. F. Blumenbach.])

SWEET, J. M. 1961. Tacharanite and other hydrated calcium silicates from Portree, Isle of Skye. *Mineralogical Magazine*, vol. 32, pp. 745-53.

SWEET, J. M. 1963. Robert Jameson in London, 1793. *Annals of Science*, vol. 19, no. 2, pp. 81-116.

SWEET, J. M. 1964. Matthew Guthrie (1743-1807): An eighteenth-century gemmologist. *Annals of Science*, vol. 20, pp. 245-302.

SWEET, J. M. 1967. The Wernerian Natural History Society in Edinburgh. Freiberger Forschungshefte, Gedenkschrift Abraham Gottlob Werner. Leipzig, pp. 205-18.

SWEET, J. M. 1970. William Bullock's collection and the University of Edinburgh, 1819. *Annals of Science*, vol. 26, pp. 24-32.

SWEET, J. M. 1970a. The collection of Louis Dufresne (1752-1832). *Annals of Science*, vol. 26, pp. 33-71.

SWEET, J. M. 1972. Instructions to collectors: John Walker (1793) and Robert Jameson (1817); with biographical notes on James Anderson (LL.D) and James Anderson (M.D.). *Annals of Science*, vol. 29, pp. 397-414.

SWEET, J. M. 1974. Robert Jameson and the Explorers: The search for the north-west passage, part 1. *Annals of Science*, vol. 31, no. 1, pp. 21-47.

SWEET, J. M. 1976. In The Wernerian theory of the Neptunian origin of rocks, by Robert Jameson. *Contributions to the History of Geology*, ed. G. W. White, vol. 9, 367pp.

TAYLOR, M. A. 1992. An entertainment for the enlightened; Alexander Weir's Edinburgh Museum of Natural Curiosities 1782-1802. *Archives of Natural History*, vol. 19, pp. 153-67.

TEALL, J. J. H. 1899. The natural history of cordierite and its associates. *Proceedings Geologists' Association*, vol. 16, pp. 61-74.

TEMPLE, A. K. 1954. The paragenetical mineralogy of the Leadhills and Wanlockhead lead-zinc deposits, Ph.D. thesis, University of Leeds.

TEMPLE, A. K. 1956. The Leadhills-Wanlockhead lead and zinc deposits. *Transactions Royal Society Edinburgh*, vol. 63, pp. 86-113.

THOMAS, H. H. 1922. On certain xenolithic Tertiary minor intrusions in the Island of Mull (Argyllshire). *Quarterly Journal Geological Society London*, vol. 78, pp. 229-60.

THOMSON, R. D. 1853. Memoir of the late Dr Thomas Thomson, F.R.S., M.W.S., etc., Professor of Chemistry in the College of Glasgow. *Edinburgh New Philosophical Journal*, vol. 54, pp. 86-98.

THOMSON, T. 1810. Experiments on allanite, a new mineral from Greenland. *Transactions Royal Society Edinburgh*, vol. 6, pp. 371-86.

THOMSON, T. 1810a. A chemical analysis of sodalite, a new mineral from Greenland. *Transactions Royal Society Edinburgh*, vol. 6, pp. 387-95.

THOMSON, T. 1836. *Outlines of Mineralogy, Geology, and Mineral Analysis.* Baldwin and Cradock, London, vol. 1, 726pp., vol. 2, 566pp.

THOMSON, T. 1836a. Minerals containing columbium. *Records of General Sciences*, vol. 4, p. 416.

THOMSON, T. 1840. On the minerals found in the neighbourhood of Glasgow. *London, Edinburgh and Dublin Philosophical Magazine*, vol. 17, 3rd series, pp. 401-18.

THOMSON, T. 1843. Notice of some new minerals. *London, Edinburgh and Dublin Philosophical Magazine*, vol. 22, 3rd series., pp. 188-94.

THOST, C. H. G. 1860. On the Rocks, Ores, and other Minerals on the Property of the Marquess of Breadalbane in the Highlands of Scotland. *Quarterly Journal Geological Society London*, vol. 16, pp. 421-28.

TILLEY, C. E. 1924. Contact metamorphism in the Comrie area of the Perthshire Highlands. *Quarterly Journal Geological Society London*, vol. 80, pp. 22-71.

TILLEY, C. E. 1948. Dolomite contact skarns of the Broadford area, Skye: a preliminary note. *Geological Magazine*, vol. 85, pp. 213-16.

TILLEY, C. E. 1948a. On iron-wollastonites in contact skarns: An example from Skye. *American Mineralogist*, vol. 33, pp. 736-38.

TILLEY, C. E. 1951. The zoned contact-skarns of the Broadford area, Skye: a study of boron-fluorine metasomatism in dolomites. *Mineralogical Magazine*, vol. 29, pp. 621-66.

TINDLE, A. G. and WEBB, P. C. 1989. Niobian wolframite from Glen Gairn in the Eastern Highlands of Scotland: a microprobe investigation. *Geochimica et Cosmochimica Acta*, vol. 53, pp. 1921-35.

TODD, J. G. 1989. Minerals of Trearne Quarry, Beith, Ayrshire, Scotland. *UK Journal of Mines and Minerals*, no. 6, pp. 18-20.

TODD, J. G. 1992. A late-Devensian marine fauna from the 'Clyde Beds', Linwood and Johnstone, Renfrewshire. *The Glasgow Naturalist*, vol. 22, no. 2, pp. 115-24.

TODD, J. G. 1992a. Boltwoodite stamps. *UK Journal of Mines and Minerals*, no. 11, p. 11.

TODD, J. G. and LAURENCE, D. W. A. 1989. Muirshiel Mine, Central Scotland. *UK Journal of Mines and Minerals*, no. 7, pp. 40-43.

TODD, J. G. and McMULLEN, M. J. 1991. Vein minerals of Mannoch Hill, Scotland, *UK Journal of Mines and Minerals*, no. 9, pp. 35-38.

TRAILL, T. S. 1806. Observations, chiefly mineralogical, on the Shetland Islands, made in the course of a tour through those islands in 1803. *A Journal of Natural Philosophy, Chemistry and the Arts*, vol. 15, pp. 353-67.

TRAILL, T. S. 1823 (read 1817). Account of a mineral from Orkney. *Transactions Royal Society Edinburgh*, vol. 9, pp. 81-92.

TRAQUAIR, R. H. 1877. The Ganoid Fishes of the British Carboniferous Formations. The Palaeontographical Society, London. (Published in 7 separate papers 1877-1914).

TRECHMANN, C. O. 1901. Ueber einen Fund von ausgezeichneten Pseudogaylussit (Thinolith-Jarrowit) Krystallen. *Zeitschrift für Krystallographie*, vol. 25, Band 3, Heft, pp. 283-85.

TRUCKELL, A. E. 1966. The Grierson collection, Thornhill, and its dispersal. *Transactions of the Dumfriesshire and Galloway Natural History and Antiquarian Society*, vol. 43, pp. 65-72.

TSCHERNICH, R. W. 1992. *Zeolites of the World.* Geoscience Press, Arizona, 563pp.

TURNBULL, J. 1997. Scottish Cobalt and Nicholas Crisp. *Transactions English Ceramic Circle*, vol. 16, pp. 144-51.

TURNER, R. 1927. *Descriptive catalogue of the Geological collection in the Chambers Institution, Peebles.* James Thin, Edinburgh, 206pp.

URE, D. 1793. *The History of Rutherglen and East Kilbride.* David Niven, Glasgow, 334pp.

WALKER, G. 1999. Snowball Earth. *New Scientist*, no. 2211, pp. 29-33.

WALKER, G. P. L. 1971. The distribution of amygdale minerals in Mull and Morven (western Scotland). *West Commerative Volume*, University of Saugar, India, pp. 181-94.

WATERSTON, C. D. 1953. An occurrence of harmotome in north-west Ross-shire. *Mineralogical Magazine*, vol. 30, pp. 136-38.

WATERSTON, C. D. 1959. Robert James Hay Cunningham (1815-1842). *Transactions Edinburgh Geological Society*, vol. 17, pp. 260-72.

WATERSTON, C. D. c.1960. *Hugh Miller, The Cromarty Stonemason.* The National Trust for Scotland, Edinburgh.

WATERSTON, C. D. 1965. William Thomson (1761-1806) a forgotten benefactor. *University of Edinburgh Journal*, pp. 122-34.

WATERSTON, C. D. 1972. Geology and the museum. *Scottish Journal Geology*, vol. 8, pp. 129-44.

WATERSTON, C. D. 1997. *Collections in Context.* National Museums of Scotland, Edinburgh, 212pp.

WATLING, R. J., HERBERT, H. K., DELEV, D. and ABELL, I. D. 1994. Gold fingerprinting by laser ablation inductively coupled plasma mass spectrometry. *Spectrochimica Acta*, vol. 49b, pp. 205-19.

WATSON, J. 1795. Parish of Dollar, in J. Sinclair, *The Statistical Account of Scotland*. William Creech, Edinburgh, vol. 15, p. 161.

WATSON, J. A. 1934. General History. *Transactions Edinburgh Geological Society*, vol. 13, 1931-38, pp. 231-41.

WAYMOUTH, C., THORNELY, P. C. and TAYLOR, W. H. 1938. An x-ray examination of mordenite (ptilolite). *Mineralogical Magazine*, vol. 25, pp. 212-16.

WEBB, P. C., TINDLE, A. G. and IXER, R. A. 1992. W-Sn-Mo-Bi-Ag mineralization associated with zinnwaldite-bearing granite from Glen Gairn, Scotland. *Transactions Institute Mining and Metallurgy*, vol. 101, pp. B59-72.

WILLIAMS, J. 1964. The Mineralogical Collection of the Dumfries Burgh Museum. *Transactions Dumfriesshire and Galloway Natural History and Antiquarian Society*, vol. 41, pp. 201-15.

WILLIAMS, J. 1965. Further notes on mineralogy in Dumfries and Galloway. *Transactions Dumfriesshire and Galloway Natural History and Antiquarian Society*, vol. 42, pp. 14-23.

WILLIAMS, J. 1970. New mineral localities for South-west Scotland. *Transactions Dumfriesshire and Galloway Natural History and Antiquarian Society*, vol. 47, pp. 191-92.

WILLIAMS, J. 1976. The minerals of south-west Scotland. *Journal Historical Metallurgy Society*, vol. 10, pp. 36-40.

WILLIAMSON, A. 1758. Extract from journal written by J. Williamson (brother) 28 March–5 April 1758 detailing clearing old mines at Alva. NLS Erskine-Murray papers, MS. 5098, f. 31.

WILSON, G. V. 1921. The lead, zinc, copper and nickel ores of Scotland. *Memoirs of the Geological Survey, Special Reports on the Mineral Resources of Great Britain*, vol. 17, Edinburgh, 160pp.

WILSON, G. V. 1924. Tertiary and post-Tertiary Geology of Mull, Loch Aline, and Oban. *Memoirs of the Geological Survey, Scotland*, Her Majesty's Stationery Office, 445pp.

WILSON, J. S. G. and CADELL, H. M. 1885. The Breadalbane Mines. *Proceedings Royal Physical Society Edinburgh*, vol. 8, pp. 189-207.

WILSON, M. J., RUSSELL, J. D., TAIT, J. M., CLARK, D. R. and FRASER, A. R. 1984. Macaulayite, a new mineral from north east Scotland. *Mineralogical Magazine*, vol. 48, pp. 127-29.

WILSON, W. E. 1994. The History of Mineral Collecting 1530-1799. *Mineralogical Record*, vol. 25, 243pp.

WINDLEY, B. F. 1977. *The Evolving Continents*. John Wiley and Sons, London, 385pp.

WINROW, A. 1986. The Mines of Strontian. *UK Journal of Mines and Minerals*, no. 1, pp. 40-43.

WITHERS, C. W. J. 1991. The Rev. Dr. John Walker and the practice of natural history in late eighteenth century Scotland. *Archives of Natural History*, vol. 18, pp. 201-20.

WOLFE, M. 1996. Mineral News. *UK Journal of Mines and Minerals*, no. 16, p. 4.

WOODWARD, J. 1728-29. *An attempt towards a natural history of fossils of England; in a catalogue of the English fossils in the collection of J. Woodward*, 2 vols, London.

YOUNG, B. R., HAWKES, J. R., MERRIMAN, R. J. and STYLES, M. T. 1978. Bazirite, $BaZrSi_3O_9$, a new mineral from Rockall Island, Inverness-shire, Scotland. *Mineralogical Magazine*, vol. 42, pp. 35-40.

---

ANON. 1842. *The Statistical Account of Ayrshire, by the Ministers of the Respective Parishes*. William Blackwood and Sons, Edinburgh, pp. 510-11.

ANON. 1857. Biographical notice of the late Thomas Thomson, M.D., F.R.S.S., L.&E., etc., Regius Professor of Chemistry in the University of Glasgow. *The Glasgow Medical Journal*, vol. 5, pp. 69-80, 121-53.

*The Gallovidian* 1910. Sons of the South. Patrick Dudgeon, F.R.S.E., Mineralogist, meteorologist, and antiquarian (1817-1895), vol. 12, no. 48, pp. 160-64.

*Edinburgh Evening Courant* The Edinburgh Room, Central Library, George IV Bridge, Edinburgh.

EDU. Edinburgh University Library, Special Collections, the University of Edinburgh, George Square, Edinburgh.

NLS. National Library of Scotland, George IV Bridge, Edinburgh.

*Notes and Queries*, 1904, series 10, 2, pp. 54, 155.

SRO. Scottish Records Office, General Register House, Princes Street, Edinburgh.

# INDEX